APPLICATIONS OF CONTINUOUS MATHEMATICS TO COMPUTER SCIENCE

THEORY AND DECISION LIBRARY

General Editors: W. Leinfellner (*Vienna*) and G. Eberlein (*Munich*)

Series A: Philosophy and Methodology of the Social Sciences

Series B: Mathematical and Statistical Methods

Series C: Game Theory, Mathematical Programming and Operations Research

SERIES B: MATHEMATICAL AND STATISTICAL METHODS
VOLUME 38

Scope: The series focuses on the application of methods and ideas of logic, mathematics and statistics to the social sciences. In particular, formal treatment of social phenomena, the analysis of decision making, information theory and problems of inference will be central themes of this part of the library. Besides theoretical results, empirical investigations and the testing of theoretical models of real world problems will be subjects of interest. In addition to emphasizing interdisciplinary communication, the series will seek to support the rapid dissemination of recent results.

APPLICATIONS OF CONTINUOUS MATHEMATICS TO COMPUTER SCIENCE

by

HUNG T. NGUYEN

New Mexico State University,
Las Cruces, New Mexico, U.S.A.

and

VLADIK KREINOVICH

University of Texas at El Paso,
El Paso, Texas, U.S.A.

KLUWER ACADEMIC PUBLISHERS
DORDRECHT / BOSTON / LONDON

A C.I.P. Catalogue record for this book is available from the Library of Congress

ISBN 978-90-481-4901-8

Published by Kluwer Academic Publishers,
P.O. Box 17, 3300 AA Dordrecht, The Netherlands.

Sold and distributed in the U.S.A. and Canada
by Kluwer Academic Publishers,
101 Philip Drive, Norwell, MA 02061, U.S.A.

In all other countries, sold and distributed
by Kluwer Academic Publishers,
P.O. Box 322, 3300 AH Dordrecht, The Netherlands.

Printed on acid-free paper

CONTENTS

PREFACE

This text grew out of a set of lecture notes for a course entitled "Applications of Continuous Mathematics in Computer Science" taught several times at the University of Texas at El Paso. This seems to be a unique type of special topic course, in which, instead of going deeply into a specific topic, students are exposed to a panorama of various topics in computer science.

Thus, this book, as a text for such a course, has three main purposes:

1. *To emphasize mathematics, and to make the researchers and students appreciate that mathematics can be really useful in computer science.*

    So, one of the purposes of this book is to give real-life examples of how continuous mathematics (algebra, differential equations, etc.) can be applied to actual important problems of Computer Science.

2. *To give an appropriate overview of Computer Science.*

    The second purpose of this book is to cover as many different areas of Computer Science as possible. Thus, unlike many others books, this book is not devoted to any *specific* area of CS, and therefore, it gives a unique opportunity to cover different areas just by picking the right examples.

3. *To expose researchers and students to some recent developments in mathematics for Computer Science.*

Computers are usually dealing with *discrete* objects, so *discrete* mathematics is used primarily in computer science. Students may even get an impression that discrete mathematics, with combinatorics, graphs, trees, etc., is all the mathematics they really need to solve computer science problems.

What we plan to show to students is that although most of the time, the mathematics used in computer science is *discrete* mathematics, in many cases, methods of *continuous* mathematics are also useful.

In this book, we give a gradual exposure of different applications of continuous mathematics to Computer Science.

We start with something that all computer science students are more or less familiar with: *analysis of algorithms and their complexity*. We take several known algorithms like merge sort and try to estimate their running time. This way we immediately run (after recalling what a logarithm is) into linear functional equations (they are similar to differential equations but somewhat easier) and solve these equations. Other divide-and-conquer algorithms lead to slightly more complicated equations, but these equations can also be solved by using precisely the same algebraic methods that students have already learned as undergraduates.

As every programmer knows, *designing* an algorithm is only part of the problem. After we have designed an algorithm, we must code it and check that the resulting program is indeed correct. We talk briefly about the existing methods of estimating the *program reliability*, point out what the problem is, and talk about how difficult it is even to formulate this problem in mathematical terms. And then, after we are ready to give up, we will show what the answer is. The formulation is somewhat tricky (we'll have to recall what a composition is, what an ordering relation is, and all that stuff), but the resulting mathematics will be the same good old functional equation. In a slightly different formulation we'll have an actual differential equation (a very simple one, easy to solve without using any tricks).

So, we learn how to estimate the quality of the algorithms and how to debug them. But where do these algorithms come from? In many practical problems for which we want to use a computer, we know exactly the system that we want to analyze, and we know exactly what we want to achieve. Such problems are naturally formalized as *optimization* problems. We show that method of continuous mathematics can help not only in solving *continuous* optimization problems, but they are also helpful in solving *discrete* optimization problems as well: namely, they help us to describe the method of *simulated annealing* and *genetic algorithms*.

In some practical problems, we do not have the complete knowledge of the analyzed system. For such problems, we must use *extrapolation* and *learning* techniques. We start with two reasonably simple computer-related problems that require extrapolation: *computer architecture* and the *growth of the Internet*. Then, we move to a slightly more difficult problem: the problem of *computer network congestion avoidance*. This problem requires that we learn some new mathematics: namely, something about transformation groups (mainly we'll talk about projective transformations; to some of the readers, they are familiar from technical drawing).

This new mathematics will prepare us for a universal learning tool: *neural networks*, that learn in a rather straightforward manner: by simulating the human brain.

This knowledge will allow us to analyze different algorithms of *intelligent control*. By this we mean methodologies that transform the experts' rules into a precise control. The readers will learn a few terms from the control folks' jargon. Again differential equations will be the main tool.

Differential equations work fine for simple models. In order to make models more realistic we'll have to use more complicated formalisms, including *fractals* (that are good in describing chaos). We do not promise to make the readers gurus in chaos, but at least they'll have the idea of what these buzzwords mean, and when to use them.

And finally, several appendices will be devoted to several additional areas.

This book is mainly designed as a one-semester course for seniors or beginning graduate students in Mathematics and Computer Science. Each lesson can be covered in, approximately, one week. The only pre-requisite for this course is the basic knowledge of computer science, at the level of the standard undergraduate curriculum, and basic mathematics that is usually taught as part of this curriculum.

The book can also be used as a reference source for researchers and as a text for a research seminar.

And, finally, our thanks:

- We would like to thank the students whose comments and notes helped us to write this text.

- Special thanks to Olga Kosheleva for taking notes and to Tran Cao Son for reading through the text to make sure that everything is understandable and correct.

- Vladik Kreinovich would like to acknowledge the partial support of NSF through grant No. EEC-9322370.

■ Hung T. Nguyen would like to express his appreciation to Professor H. Skala, Editor of the Series "Mathematical and Statistical Methods", for the constant support of his research and educational publications.

Hung T. Nguyen and Vladik Kreinovich

Las Cruces and El Paso, Spring 1997

1

ALGORITHM COMPLEXITY: TWO SIMPLE EXAMPLES

We start in this first lesson with the concept of algorithm complexity and use two simple examples to illustrate the applicability of continuous mathematics.

1.1. Algorithm complexity

Computer science, is, by definition, a science of *computing* performed by a computing device: a *computer*. Whatever a computer does, it implements a precise step-by-step sequence of unambiguous instructions, i.e., an *algorithm*. An algorithm is described as a text in a special formal language called a *programming language*; this text is called a *program*.

So, if we want to solve a real-life problem on a computer, we must:

- first, find an *algorithm* for solving this problem; and

- then, *code* this algorithm into a program.

Two types of computer science problems. With respect to the first step (finding an algorithm) we can roughly divide all problems into two classes:

- First, there exist *really complicated* problems, for which it is absolutely not clear how to solve them.

    For example, how to solve a system of equations coming from some version of quantum field theory?

1

For such problems, the main difficulty is to find *an* algorithm. For these problems, it is often useful to use methods from *discrete mathematics*.

Since the readers have some basic experience in programming, they must be familiar with the basic notions of *discrete mathematics* such as trees, graphs, tables, etc. By definition, discrete mathematics deals with *discrete* objects.

For example, if we want to find the shortest path, we can use the known algorithm (invented by Dijkstra) for computing the shortest path on a graph.

As we will see later, *continuous mathematics* is also sometimes useful for designing algorithms. By *continuous mathematics*, we mean mathematics that deals with *continuous* objects such as real numbers, real-valued functions, etc. Thus, continuous mathematics include calculus, differential equations, etc.

■ Computer professionals *typically* deal with *less complicated* problems, namely, with problems for which one or several algorithms are already known (or at least, if the problem is reasonably novel, such algorithms can be easily designed using the existing techniques).

  – If several algorithms are known, then we must choose the "best" one (best on some reasonable sense).
  – If only one algorithm is known, then we may need to check whether this algorithm is good enough, and, if it is not, try to propose another, better algorithm.

We want the best, but best in what sense? The above description seems correct but vague. To make it less vague, we must describe what "best" means and what "good" means.

Intuitively, "best" means the best for the purpose for which we are using the computer. So, to answer the question what "best" means we must recall what computers are used for.

In order to use a computer, we need an *algorithm*, i.e., a precise step-by-step sequence of unambiguous instructions. But if we have such an algorithm, we can simply implement it ourselves step-by-step. What's wrong with that? Well, nothing seems to be wrong with that until we remember that a typical computer performs 30, 60, 100, and more millions of operations per second. In other words, the main advantage of the computer is the enormous *speed* with which it performs the instructions. Even with this great speed, there is a constant need for faster and faster computers, because everyone who has ever used a computer knows that there are algorithms that simply run too slow. So, the main goal of a computer is to *speed up* computations. Thus, it is natural to say that an algorithm U is *better* than an algorithm V is U is *faster* than V, i.e., that U performs the required computations within a shorter *time period*.

Here is where algorithm complexity comes in. The actual running time of an algorithm depends not so much on the algorithm itself, but mainly on how fast the computer is. Crudely speaking, this running time is equal to the time required for a single elementary computer operation (e.g., assignment or addition), multiplied by the total number of these operations. Thus, if we fix a computer and compare the running time of two algorithms on this fixed computer, then an algorithm with fewer steps (elementary operations) is faster.

Thus, when we compare algorithms, it is sufficient to compare their numbers of elementary steps. This number of steps is called (somewhat misleading) a *computational complexity* of the algorithm[1].

So, when we have several algorithms for solving the same problem, we must choose a one with the smallest computational complexity (i.e., the smallest number of computation steps). To be able to make this choice, we must be able to *evaluate* computational complexity of different algorithms. We will show that in this, continuous mathematics can help.

1.2. First simple example of an algorithm: mergesort

Let us start with the problem that is indeed occurring very frequently: the problem of *sorting* a given list. One of the best known algorithms for sorting is the so-called *merge sort* (also known as *AVL*, by the initials of its inventors).

Main idea of mergesort. Mergesort is based on the following idea: A list consisting of a single element is automatically sorted. If we want to sort a list of $n > 1$ elements, we

- *divide* this list into two halves;

---

[1] We said "misleading" because an algorithm that adds 2 and 2 1,000,000 times is deemed complicated, while a truly difficult-to-describe algorithm (like mergesort that will be described in this lesson) that has fewer computational steps is called "simpler".

- *sort* each half (using the same algorithm); and then
- *merge* the two sorted halves into a single sorted list.

A simple example: a general description. Let us illustrate this idea on a simple example: Suppose that we are given $n = 8$ letters that form the word "computer":

C O M P U T E R

and we want to sort them in alphabetic order. According to the mergesort algorithm, we do the following (since $n = 8 > 1$):

- First, we *divide* this list into two halves: C O M P and U T E R.
- Then, we *sort* each half. For a while, let us skip the details of how these two lists are sorted, and just present the results of these sortings:

C M O P and E R T U.

- Finally, we must *merge* the sorted half-lists. This merger is done as follows: initially, we have two half-lists and 8 blank spaces of the merged list to fill:

C M O P E R T U   x x x x x x x x

Let us fill all eight places on the merged list one by one, from the smallest (first) to the largest (last).

  - Since both halves are already sorted, in order to find the smallest element of the original list, it is sufficient to compare the smallest elements of these two halves; in our example, C and E[2]. C is the smallest of the two, and, therefore, C goes to the first place of the resulting sorted list:

₵ M O P E R T U   C x x x x x x x

(As we place C on the resulting list, we cross it over from the list of elements that are not yet sorted.)

---

[2] If this sounds non-trivial, here is another example: suppose that we have two groups of soldiers who are standing on the parade grounds, each group ordered from the smallest on the right flank to the tallest on left flank. If we need to find the smallest soldier of the two groups, it is sufficient to compare the two right-flank guys from the two groups.

- The second element of the resulting list is the smallest of all the elements that have not yet been placed. These un-placed elements form the two sorted lists: M O P and E R T U. Both lists are sorted, so, the desired smallest-of-all is again the smallest of the two elements: M (which is the smallest of the first group) and E (which is the smallest of the second group). This time, E is the smallest, and therefore, E goes to the second place:

$$\cancel{C} \; M \; O \; P \quad \cancel{E} \; R \; T \; U \quad C \; E \; x \; x \; x \; x \; x \; x$$

- To fill the third place, we must find the smallest of M and R; as a result, we get

$$\cancel{C} \; \cancel{M} \; O \; P \quad \cancel{E} \; R \; T \; U \quad C \; E \; M \; x \; x \; x \; x \; x$$

Similarly, we get the following lists:

$$\cancel{C} \; \cancel{M} \; \cancel{O} \; P \quad \cancel{E} \; R \; T \; U \quad C \; E \; M \; O \; x \; x \; x \; x$$

and

$$\cancel{C} \; \cancel{M} \; \cancel{O} \; \cancel{P} \quad \cancel{E} \; R \; T \; U \quad C \; E \; M \; O \; P \; x \; x \; x$$

After this step, there are no elements left in the first half; this means that all the remaining elements of the second half must simply be added to the final list:

$$\cancel{C} \; \cancel{M} \; \cancel{O} \; \cancel{P} \quad \cancel{E} \; \cancel{R} \; \cancel{T} \; \cancel{U} \quad C \; E \; M \; O \; P \; R \; T \; U$$

The resulting sorted list is:

$$C \; E \; M \; O \; P \; R \; T \; U$$

Filling in the details. In the above explanation, we skipped the description of how exactly the halves are sorted. They are sorted by using the same mergesort algorithm:

■ The first half (C O M P) is sorted as follows:

- First, we *divide* the list into two halves: C O and M P.
- Then, we *sort* each half, getting C O and M P.

− Finally, we *merge* the two lists C O and M P, getting, consequently:

$$\cancel{C}\,O \quad M\,P \quad C\,x\,x\,x$$

$$\cancel{C}\,O \quad \cancel{M}\,P \quad C\,M\,x\,x$$

$$\cancel{C}\,\cancel{O} \quad \cancel{M}\,P \quad C\,M\,O\,x$$

and, finally,

$$C\,M\,O\,P$$

■ Similarly, the second half (U T E R) is sorted as follows:

− First, we *divide* the list into two halves: U T and E R.

− Then, we *sort* each half, getting T U and E R.

− Finally, we *merge* the two lists T U and E R, getting, consequently:

$$T\,U \quad \cancel{E}\,R \quad E\,x\,x\,x$$

$$T\,U \quad \cancel{E}\,\cancel{R} \quad E\,R\,x\,x$$

and finally,

$$E\,R\,T\,U$$

The resulting algorithm can be visualized in the following form:

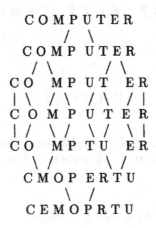

```
        COMPUTER
         /   \
      COMP   UTER
      / \     / \
    CO   MP UT   ER
    |\  /\ /\  / |
    C O M P U T E R
    |/  \/ \/  \ |
    CO   MP TU   ER
      \ /     \ /
      CMOP   ERTU
         \   /
        CEMOPRTU
```

1.3. Computational complexity of mergesort: an equation

It makes sense to only count comparisons. Mergesort consists of two types of elementary operations:

- we *compare* the two elements; and

- we *move* an element to a new location.

In a computer, moving a number or a letter is much faster than comparing two numbers or letters, so, to estimate the overall complexity of an algorithm, it is sufficient to only take into consideration the number of *comparisons*.

How many comparisons does mergesort require? To be precise, if we know the number n of elements in the list, how many comparisons do we need to sort it?

Worst case complexity vs. average complexity. The actual number of comparisons depends on the original list:

- for some lists (e.g., for lists that are already sorted), the mergesort algorithm requires fewer comparisons;

- for some other lists, we may need more comparisons.

For example, for the list C O M P U T E R, mergesort take the total of 14 comparisons. If we start with the sorted list, e.g., with the list C E M O P R T U, then the reader can check that we will need only 12.

In real life, we are interested in the *average* computations time, i.e., in the *average* number of computation steps:

$$t_{\mathcal{U}}^{\text{av}}(n) = \sum_x p(x) \cdot t_{\mathcal{U}}(x),$$

where:

- the summation is over all inputs x of length n;

- $p(x)$ is the frequency with which the input x occurs in real-life problems; and

- $t_\mathcal{U}(x)$ is the number of elementary computations steps that an algorithm $\mathcal{U}$ takes on the input x.

However, it is extremely difficult to compute this *average complexity*, for two reasons:

- First, to compute the average number of computation steps, we need to know the frequencies $p(x)$ with which different inputs x occur, and usually, we do not know these frequencies.

- Second, even if we know the current values of the real-life frequencies $p(x)$, these frequencies can change when we change the algorithm. Indeed:

  - Some inputs x can be rare because most users know that for these inputs, the existing algorithm works badly.
  - However, if we design a new algorithm that works well on these inputs, then they will occur much more frequently.

Since we do not know which inputs are more frequent and which inputs are less frequent, the only way to get a meaningful estimate of the algorithm complexity on inputs x of given length n ($\mathrm{len}(x) = n$) is to compute the *largest possible* number of computation steps for inputs of this length; this largest possible number is called a *worst-case* complexity:

$$\bar{t}_\mathcal{U}(n) = \max_{x:\mathrm{len}(x)=n} t_\mathcal{U}(x).$$

This worse-case estimate may sound pessimistic, but if we try to be more optimistic and take as an estimate the running time on the *second worst case*, then in the situation when most of the real-life cases are actually worst-case inputs, we will *underestimate* the running time of the algorithm.

In view of this argument, in the following text, we will mainly estimate the *worst-case complexity* of different algorithms.

We got an equation. The above description of mergesort:

- does *not* give an *explicit* description of this algorithm;

- instead, it describes the work on this algorithm on a given list *in terms* of its work on the two *halves* of the original list.

As a result, from this description of mergesort,

- it is difficult to immediately extract an *explicit* estimate for mergesort complexity, but
- we can extract a formula that relates the computational complexity of mergesort on a list of given list to its complexity for lists of half-size.

For *simplicity*, let us consider a situation when a list can be divided into two *equal* halves, i.e., when it has $2n$ elements for some integer n. For every integer m, let $t(m)$ denote the (worst-case) number of comparisons that the mergesort uses when it sorts a list of m elements.

According to the main idea of mergesort, in order to sort a list of $2n$ elements, we:

- *sort* two halves of this list; and then:
- *merge* the sorted halves into a sorted list.

So, the total time $t(2n)$ of sorting this list of $2n$ elements is a sum of the times that are necessary to sort the two halves and of the time required for merging.

- A half of a list of $2n$ elements has exactly n elements. Therefore, sorting a half-list requires $t(n)$ computational steps. We need to sort *two* halves, hence, we need time $2t(n)$ to sort both halves.
- How many comparisons do we need to merge two sorted lists of n elements each into a single sorted list of $2n$ elements? According to the mergesort algorithm:
  - we need exactly one comparison to find the *first* element of the sorted list;
  - we need exactly one comparison to find the *second* element of the sorted list;
  - etc., until all elements are found.

  How many comparisons total do we need for merging?

  - In *some* cases, we exhaust one half before the other; in these cases, we do not need any comparisons to fill the remaining list.

– In the *worst* case, however, we need one comparison to fill *each* of $2n$ elements of the final sorted list. (Except, of course, the very *last* element, that automatically goes into the only unfilled (last) place on the final sorted list.)

Thus, in the worst case, we need $2n-1$ comparisons to merge the two lists.

Thus, the total time $t(2n)$ can be represented as a sum of two terms that correspond to sorting and merging:

$$t(2n) = 2t(n) + 2n - 1. \tag{1.1a}$$

We also know that a list consisting of a single element ($n = 1$) is already sorted and thus, does not need any comparisons at all, i.e.:

$$t(1) = 0. \tag{1.1b}$$

Now, we do not have an explicit *expression* for $t(n)$; instead, we have an *equation* for $t(n)$. To find an explicit expression for $t(n)$, we need to *solve* this equation.

What kind of equation is it? This equation may be somewhat unusual for some readers. Indeed, we know:

■ *Algebraic* equations, like $x^3 + 3x + 1 = 0$, in which an unknown is a *number* x.

■ *Differential* equations, like

$$\frac{dy}{dx} + 3y = 0,$$

in which the unknown is a *function* $y(x)$, and the equation relates the values of this function and its derivative (or derivatives) for the same value x.

■ *Integral* equations, like $x(t) = \int x(s)e^{-(t-s)}ds$, in which a *function* $x(t)$ is unknown, and the equation relates the values of this unknown function and some integrals.

Our equation (1.1a) is:

■ on one hand, *similar* to differential and integral equations in the sense that in (1.1a), the unknown is a *function* ($t(n)$);

- on the other hand, *different* from differential and integral equations in the sense that in (1.1), we do not use differentiation or integration.

Such equations in which:

- the unknown is a function, and
- which relate the values of this function for different values of the argument (without using differentiation or integration)

are called *functional equations*.

(In this text, we will see lots of functional equations.)

Let's simplify the equation. In the equation (1.1a), we considered the simplest case of mergesort, when a list can be divided into two *equal* halves, i.e., when it has an *even* number of elements m: $m = 2n$ for some n.

According to mergesort, the sorting of the original list starts with sorting two half-lists of size n. If we want these sortings to be also simple, we must assume that each of the half-lists can also be divided into two equal halves, i.e., that n is also even: $n = 2n_1$ for some integer n_1. (In this case, $m = 2n = 2^2 n_1$).

Similarly, we can assume that n_1 is even, i.e., that $n_1 = 2n_2$ for some integer n_2, in which case $n = 4n_2 = 2^2 n_2$. We can repeat this argument again and again and get $n = 2^k n_k$ for some integer n_k, where n_k is getting smaller and smaller. The reduction ends when we get a half-list that has exactly one element ($n_k = 1$) and which is, therefore, automatically sorted. For such k, $n = 2^k n_k = 2^k$.

For such n, the equation (1.1a) takes the form

$$t(2^{k+1}) = 2t(2^k) + 2^{k+1} - 1. \tag{1.2}$$

Since this equation only relates the values $t(2^k)$ for different k, it makes sense to denote $t(2^k)$ by $f(k)$. For this new function $f(k)$, this equation (1.2) takes the following form:

$$f(k + 1) = 2f(k) + 2^{k+1} - 1. \tag{1.3a}$$

The equation (1.1b), in its turn, takes the form

$$f(0) = 0. \tag{1.3b}$$

Before we start solving these types of equations, let us give one more example.

1.4. Second simple example of an algorithm: binary search

Main idea. Binary search is a well-known method of searching for an element in a sorted list.

In binary search, we are given a sorted list of m elements $e_1, \ldots, e_m$, and an element e that we are looking for. The goal is to find the *place* where this element e occurs in the given list (if it occurs at all)[3].

There are, actually, several different variants of this algorithms, so, to be precise, let us describe what exactly variant we have in mind. In the analyzed variant, we take the middle element of the list and compare e with this middle element. Then:

- If the given element e coincides with the middle element, we have found a place for e.

- If e follows the middle element, then we have to look for e among all elements that are following this middle one.

- If e precedes the middle element, then we have to look for e among all elements that are preceding this middle one.

This "looking for" is done according to the same algorithm. The algorithm ends in two cases:

- either if we have found an element e,

- or if we have an empty list to look in, in which case we conclude that the given element e is not in the list.

To make this algorithm absolutely clear, let us run two examples.

Examples. As a list in which we will be looking, let us take the result of sorting the word SORTING, i.e., the 7-element list

---

[3] This makes sense, if we, e.g., we have a database of students in which all the data is ordered by their last names $e_1, \ldots, e_m$. Then, if we can find the place i of a given student e (i.e., i for which $e_i = e$), we will be able to easily extract all the information about this student e that is stored in this database.

GINORST

in which e_1 =G, e_2 =I, e_3 =N, e_4 =O, e_5 =R, e_6 =S, and e_7 =T.

Example 1. If we are looking for the letter S, we do the following:

- First, we compare S with the middle element of the list. The list starts with the first element, ends with the 7th, so the middle element is its element number $(1 + 7)/2 = 4$, i.e., O.

  Since S follows O in alphabetic order, we must look for S among elements that follow O, i.e., in the half-list

RST

- Next, we compare the given element S with the middle element of the new list. The new list starts with the 5-th element and ends with the 7-th. So, its midpoint is the element number $(5 + 7)/2 = 6$, which is S.

  Since the given element coincides with the midpoint, this means that we have found the desired letter in the list: it is the list's 6-th element.

Example 2. If we are looking for a letter H, the binary search algorithm will work as follows:

- First, we compare S with the middle element of the list. From the first example, we already know that the middle element is its element number 4, which is O.

  Since H precedes O in alphabetic order, we must look for H among elements that precede O, i.e., in the half-list

GIN

- Next, we compare the given element H with the middle element of the new list. The new list starts with the 1-st element and ends with the 3-rd. So, its midpoint is the element number $(1+3)/2=2$, which is I.

  Since H precedes I in alphabetic order, we must look for H among all elements that precede I, i.e., in the 1-element list G.

■   The new short list consists of only one element, so its middle element is exactly this same 1-st element G.

Since H follows G, we must look for H among all elements that follow G. But G was the only element of the list. So, we are left with an empty list to look in. Thus, we conclude that H is not an element of the original list.

1.5.   Computational complexity of binary search: an equation

We will only count comparisons. For binary search, we will also only count the number of comparisons.

We are interested in the worst-case complexity. The above two examples show that for binary search, the number of comparisons depends on what element we are looking for: we needed two comparisons when we looked for S, and three when we were looking for Q.

So, we want to know the *worst-case* complexity of binary search.

An equation. Let us denote by $t(m)$ the largest number of comparisons needed to find an element in a list of m elements.

In binary search, if an element e (for which we are looking) is not in the middle of the list, we look for it in one of the half-lists:

■   either in the half-list of all elements that *precede* the middle element,

■   or the half-list of all elements that follow the middle element.

For simplicity, let us consider situations in which both half-lists are of equal size.

If the size of each of these half-lists is n, then the size of the original list (two half-lists plus the middle element) is $2n+1$. To find an element e in the original list of this size $2n + 1$, we must:

- First, compare this element e with the midpoint of the original list. If we are lucky (i.e., if e is equal to the midpoint element), this is all we need to do. However, in other cases, we have to do the next step:

- We look for e in one of the half-lists of size n.

Thus, in the worst-case, the total time $t(2n+1)$ necessary to look for an element in a list of size $2n+1$, consists of the time for the comparison with the midpoint element (1 step) + the time require to look for this element in a half-size list ($t(n)$). So, we have the following functional equation:

$$t(2n + 1) = t(n) + 1. \qquad (1.4a)$$

If a list has no elements of all (i.e., if $n = 0$), then, whatever element e we are looking for, we do not need any comparisons at all to conclude that this element e is not on the given list. So, for $n = 0$, we need 0 comparisons:

$$t(0) = 0. \qquad (1.4b)$$

Let's simplify the equation. In the equation (1.4a), we considered the simplest case of binary search, when both half-lists are of equal size, i.e., when the original list contains an *odd* number of elements m: $m = 2n + 1$ for some n.

According to binary search algorithm, the search in the original list includes (in the worst case) searching in one of the half-lists of size n. If we want this search to be also simple, we must assume that each of the half-lists is also of odd size, i.e., that $n = 2n_1 + 1$ for some integer n_1. Similarly to the sorting case, we conclude that $n = 2^k - 1$ for some k (for details, see Appendix at the end of this lesson).

For such n, the equation (1.4a) takes the form

$$t(2^{k+1} - 1) = t(2^k - 1) + 1. \qquad (1.5)$$

Since this equation only relates the values $t(2^k - 1)$ for different k, it makes sense to denote $t(2^k - 1)$ by $f(k)$. For this new function $f(k)$, this equation (1.5) takes the following form:

$$f(k + 1) = f(k) + 1. \qquad (1.6a)$$

The equation (1.4b), in its turn, takes the form

$$f(0) = 0. \qquad (1.6b)$$

This equation is much simpler than the equation for mergesort, so, we will easily solve it (and find $f(k)$). While solving this equation, we will try to follow the ideas in the most general context, and thus, generate a general heuristic for solving these functional equations.

1.6. From simple equations to a general heuristic

Solving the simplest equation (that corresponds to binary search). The equation (1.6) is the easiest to solve:

- We know that $f(0) = 0$.

- From the equation (1.6a) for $k = 0$, we conclude that $f(1) = f(0) + 1$ and thus, $f(1) = 0 + 1 = 1$.

- From the equation (1.6a) for $k = 1$, we conclude that $f(2) = f(1) + 1$ and thus, $f(2) = 1 + 1 = 2$.

- From the equation (1.6a) for $k = 2$, we conclude that $f(3) = f(2) + 1$ and thus, $f(3) = 2 + 1 = 3$.

By induction, we can conclude that $f(k) = k$ for all integers k.

The resulting formula for computational complexity of binary search. Let us recall that for binary search, $f(k)$ meant $t(2^k - 1)$. Thus, we conclude that $t(2^k - 1) = k$.

To get a familiar formula for the binary search complexity, we must make one final step: express $t(n)$ in terms of n. For that, we must express k in terms of n. If $n = 2^k - 1$, then $2^k = n + 1$. To express k in terms of n, it is now sufficient to recall the definition of a logarithm $\log_a(b)$ as such a number that $a^{\log_a(b)} = b$. By this definition, the formula $2^k = n + 1$ means that $k = \log_2(n + 1)$, and thus,

$$t(n) = \log_2(n + 1).$$

This formula, in the form $t(n) \sim \log(n)$, is a well-known computational complexity estimate for binary search.

Analogy with continuous math: difference equations as approxima-tions to differential equations. So far, we only used discrete math: $f(k)$ is a function of discrete (integer-valued) variable k that took only discrete (integer-valued) values.

However, to solve this equation, we used its very special form. Based on this example only, it is difficult to design a more general method of solving functional equations.

To come up with such a method, we will use an analogy with similar equations in *continuous* math, i.e., with *differential* equations. This analogy will be helpful because continuous mathematics has been in existence for a long time, and many methods have been developed for solving its problems and equations, while discrete mathematics is a reasonably new area.

Let us describe this analogy. Equations (1.3) and (1.6) describe the value $f(k+1)$ of an unknown function f in the *next* value $k+1$ in terms of its value $f(k)$ for *previous* k. (Such functional equations are called *difference* equations.) But this is exactly how in numerical methods, we solve *differential* equations. Indeed, let us assume that have a differential equation

$$\frac{dy}{dx} = F(y(x), x),$$

with the known initial value $y(x_0) = y_0$. The main difficulty with solving this equation is that the derivative dy/dx is defined not as an explicit expression, but as a *limit*:

$$\frac{dy}{dx} = \lim_{h \to 0} \frac{\Delta y}{\Delta x},$$

where $\Delta y = y(x+h) - y(x)$ and $\Delta x = h$. Thus, to solve the differential equation numerically, we *approximate* the derivative dy/dx by the ratio $\Delta y/\Delta x$. The smaller h, the better this approximation. Thus, to get a good numerical solution, we must take a small "step" $h > 0$. After this approximation, the original differential equation takes the form

$$\frac{y(x+h) - y(x)}{h} \approx F(y(x), x),$$

which is equivalent to

$$y(x+h) \approx y(x) + h \cdot F(y(x), x).$$

We know the initial value of y: $y(x_0) = y_0$.

Using the above approximate equation, we can compute the estimate y_1 for the value $y(x_1)$ of the unknown function $y(x)$ at the "next" point $x_1 = x_0 + h$ as follows:

$$y_1 = y_0 + h \cdot F(y_0, x_0).$$

Similarly, we can compute the estimates $y_2, y_3, \ldots, y_k, \ldots$ for the values $y(x_k)$ at the points $x_2 = x_1 + h = x_0 + 2h$, $x_3 = x_2 + h = x_0 + 3h$, $\ldots$, $x_k = x_0 + kh$, $\ldots$ as follows:

$$y_{k+1} = y_k + h \cdot F(y_k, x_k).$$

Analogy illustrated on the binary search equation. One can easily see that our equation (1.6) has exactly this form, for $h = 1$ and $F(y, x) = 1$. In other words, the equation (1.6) is a discrete approximation to the differential equation $df/dx = 1$. Indeed, the expression $f(k + 1) - f(k)$, i.e.,

$$\frac{f(k + 1) - f(k)}{1}$$

is an approximation to the derivative

$$\frac{df}{dx} = \lim_{h \to 0} \frac{f(x + h) - f(x)}{h}.$$

The equation $dy/dx = 1$ (or, using the widely used denotation for the derivative as f', $f' = 1$, can solve in two different ways:

- First, we can solve this equation *explicitly*: since we know the derivative of the unknown function f (this derivative is equal to 1), the function f is equal to the integral of this derivative, i.e., $f(x) = \int 1 dx = x + C$.

  The integration constant C can be determined from the condition (1.6b) that $f(0) = 0$: for $x = 0$, we get $0 + C = 0$ and $C = 0$.

- We can also use a more *general* trick: *separation of variables*:

  - We start with the equation

  $$\frac{df}{dx} = 1.$$

  - To simplify this equation, we would like to *separate* its variables, i.e., move all terms containing x into one side, and all terms containing f into another side. For this equation, such a separation is easily achievable: it is sufficient to multiply both sides of this equation by dx. As a result, we get the equation

$$df = dx.$$

with separated variables.

– Integrating both sides of this equation, we conclude that

$$\int df = \int dx$$

i.e., that $f = x + C$.

Just like in direct integration, the constant C can be determined from the initial condition (1.6b).

In both cases, we get $f(k) = k$, which is *exactly* what we came up by using discrete methods.

How this analogy can help to solve general linear difference equations? An idea. We have just seen that a difference equation can be viewed as an approximation to a differential equation. It is therefore reasonable to try to use the experience (ideas and methods) of solving differential equations for solving difference equations.

For the simplest equation (1.6), the solution of the corresponding differential equation turned out to be *exactly* equal to the solution of the approximating difference equation. In general, the solution of a differential equation will be only an approximation, but hopefully, by using the *ideas* from differential equations, we will be able to come up with an *exact* solution.

We will discuss this in detail in Lesson 2.

Exercises

1.1 Use mergesort to sort (manually) your class list by first names (or, if the class list is not available, any other list).

1.2 Program the mergesort algorithm in such a way that it will not only sort the list, but also compute the number of comparisons. Run the resulting program on lists of different size, plot the results, and try to guess the dependence of $t(n)$ on the list size n.

Appendix: Simplification of the equation (1.4)

In the equation (1.4a), we considered the simplest case of binary search, when both half-lists are of equal size, i.e., when the original list contains an *odd* number of elements m: $m = 2n + 1$ for some n.

According to binary search algorithm, the search in the original list includes (in the worst case) searching in one of the half-lists of size n. If we want this search to be also simple, we must assume that each of the half-lists is also of odd size, i.e., that $n = 2n_1 + 1$ for some integer n_1. (In this case, $m = 2n + 1 = 4n_1 + 3 = 2^2 n_1 + (2^2 - 1)$.)

Similarly, we can assume that n_1 is odd, i.e., that $n_1 = 2n_2 + 1$ for some integer n_2, in which case $n = 2^2 n_2 + (2^2 - 1)$. We can repeat this argument again and again and get $n = 2^k n_k + (2^k - 1)$ for some integer n_k, where n_k is getting smaller and smaller. The reduction ends when we get a half-list that has no elements at all $(n_k = 0)$ and which is, therefore, automatically sorted. For such k, $n = 2^k n_k + (2^k - 1) = 2^k - 1$, and $m = 2n + 1 = 2(2^k - 1) + 1 = 2^{k+1} - 2 + 1 = 2^{k+1} - 1$.

2

SOLVING GENERAL LINEAR FUNCTIONAL EQUATIONS: AN APPLICATION TO ALGORITHM COMPLEXITY

In the study of algorithm complexity, we often need to solve linear functional equations. In this lesson, we show how continuous mathematics can help in solving such equations.

2.1. Introduction: why equations? what equations?

If we want to solve a problem, we first need to come up with an algorithm. However, having an algorithm is not enough: some algorithms require so much time to run that they are not practically useful. To estimate the running time of an algorithm, we must be able to estimate the algorithm *complexity*, i.e., the number of elementary steps that comprise this algorithm. In some cases, we have several algorithms for solving the same problem; in these cases, we would like to choose the *fastest*. To make this choice, we must also be able to estimate the complexity of different algorithms.

Some algorithms are straightforward. For such algorithms, we can simply count the elementary steps and thus, get the desired complexity. However, most algorithms are more complicated. Usually, these algorithms are *recursive* in the sense that instead of a straightforward solution, these algorithms allows us to *reduce* the original problem to one or several problems of smaller size (half-size, quarter-size, etc.). As a result, instead of an *explicit* formula for the algorithm complexity, we get an *equation* that expresses the desired complexity of problems of given size in terms of complexity of applying the same algorithm to simpler problems (of half-size, of quarter-size, etc.). So, to find an explicit expression for the algorithm complexity, we must solve this equation.

Usually, these equations are linear; they are called *linear functional equations*. In the previous lesson, we gave two examples of functional equations, (1.3) and (1.6), that stem from the analysis of algorithm complexity. We were able to easily solve the simpler equation (1.6).

In this lesson, we will show that continuous mathematics can help us solve arbitrarily complicated equations of this type, in particular, the more complicated equation (1.3).

If our only goal was to solve these equations, then we could as well produce a solution and require the readers to memorize it. Our goal, however, is not so much to teach how to solve *specific* functional equations, but rather to give an example of how mathematics can be applied. In view of this pedagogical goal, we want to *naturally arrive* at the method of solving different functional equations. To be able to do that, let us start with the simplest equations, solve them, and then increase the complexity of solved equations step by step.

Before we do that, let us first describe the class of equations that we will be solving.

Both equations (1.6) and (1.3) are of the type

$$f(k+1) = \text{const} \cdot f(k) + b(k),$$

where $b(k)$ is a given function: for the equation (1.6), it was $b(k) = 1$, and for the equation (1.3), it was $b(k) = 2^{k+1} - 1$.

In other words, in both equations, the value $f(k+1)$ of the unknown function f at a point $k+1$ depends linearly on the value $f(k)$ of this same function f at the previous point k. The general expression for a linear function is $y(x) = ax + b$; so, the general linear dependence of $f(k+1)$ on $f(k)$ is of the type $f(k+1) = a \cdot f(k) + b$.

- In the equation (1.6), both coefficients a and b of this linear dependence were constants (actually, they were both equal to 1).

- In the equation (1.3), the coefficient at $f(k)$ is still a constant ($a = 2$), that does not depend on k, while b is already a function of k.

So, to be able to solve equation (1.3), we must consider at least the cases when a is a constant, while b must be, in general, a function of k. It is not necessary to consider *arbitrary* functions $b(k)$, but this class of functions must be general enough to include a function $b(k) = 2^{k+1} - 1$.

The reason why we were able to get an equation that expresses $f(k+1)$ solely in terms of $f(k)$ is that in both algorithms from Lesson 1 (mergesort and binary search), the application of the algorithm to the list of $n = 2^{k+1}$ objects was reduced to one or several applications of this same algorithm to *half-lists* that contain $n/2 = 2^k$ objects. In more complicated cases, we would have a reduction to *quarter-lists* that contain $n/4 = 2^{k-1}$ objects, etc. To cover the functional equations that describe the algorithm complexity of these more complicated algorithms, we need to consider slightly *more general* equations, in which the value of the unknown function f at a given point depends (linearly) not only on one *previous* value of this same function f, but also on the *pre-previous, pre-pre-previous* values, etc. In other words, we would like to consider equations of the type

$$f(k+p) = a_1 \cdot f(k+p-1) + a_2 \cdot f(k+p-2) + \ldots + a_p f(k) + b(k). \quad (2.1)$$

2.2. The simplest functional equation

Which is the simplest functional equation. In order to be able to solve equations of type (2.1), with arbitrarily large p, and with complicated functions $b(k)$, let us first start with the *simplest* case of this equation:

- when $p = 1$, i.e., when $f(k+1)$ depends only on one previous value of the function f; and

- when the term $b(k)$ is the simplest possible, i.e., $b(k) = 0$.

In this case, the general equation (2.1) turns into the simple equation

$$f(k+1) = \alpha \cdot f(k) \quad (2.2)$$

(where, for simplicity, we omitted the subscript and denoted a_1 simply by α).

How to solve the simplest functional equation. This equation is easy to solve:

- If we know $f(0)$, then we can compute $f(1) = \alpha \cdot f(0)$.

- From $f(1)$, we can compute $f(2) = \alpha \cdot f(1) = \alpha^2 \cdot f(0)$.

- Similarly, from $f(2)$, we can compute $f(3) = \alpha \cdot f(2) = \alpha^3 \cdot f(0)$.

- etc.

In general, we get a solution $f(k) = \alpha^k \cdot f(0)$ for an arbitrary constant $f(0)$.

We can re-describe this situation as follows:

- We have a solution $f_0(k) = \alpha^k$.

- Since the equation is linear, for every constant C, the function $f(k) = C \cdot f_0(k)$ is also a solution.

If we know the value of $f(0)$ (as we did in both our equations (1.3) and (1.6)), then the constant C can be determined by comparing the value of the obtained general solution $f(k)$ for $k = 0$ with the known value $f(0)$.

Let us use this simple solution to go to slightly more complicated equations.

2.3. Next step: second simplest equations

What is the next step? The equation (2.2) was obtained under two simplifying assumptions:

- that $p = 1$ (i.e., that the current value of f depends only on a single previous value of this function), and

- that the free term $b(k)$ is equal to 0.

To get from the simplest equation (2.2) to the "next simplest" equations, we must relax one of these two assumptions. Which of these two assumptions should we relax? To find that out, let us informally compare the sizes of the classes that will be obtained when we relax each of these assumptions.

The class of the "simplest" equations described by the equation (2.2) depends on *one* parameter $\alpha\ (= a_1)$.

- If we allow $p > 1$ (but still do not allow any free terms), then we will have a class of equations that depends on p numerical parameters $a_1, \ldots, a_p$.

- On the other hand, if we allow arbitrary functions $b(k)$, then we will get a class of equations that depends on an arbitrary *function*.

Intuitively, there are, clearly, many more *functions* than *numbers*.

Indeed, there are, intuitively, more pairs of numbers (r_1, r_2) than numbers, more triples (r_1, r_2, r_3) than pairs, etc. To describe an arbitrary function $f(n)$ pf an integer argument n, we must describe all its values $f(1)$, $f(2)$, $f(3)$, ... There are, therefore, more functions than triples, pairs, etc., and hence, much more functions than numbers.

Thus, if we allow arbitrary functions, we get a much wider class than if we only allow numbers. Thus, to get the extension that will be the easiest to solve (i.e., to get the extension that is as narrow as possible), we should consider the first case: equation with arbitrary numerical parameters but no right-hand side.

In other words, the next step is to solve general equations of the type

$$f(k + p) = a_1 \cdot f(k + p - 1) + a_2 \cdot f(k + p - 2) + \ldots + a_p f(k). \qquad (2.3)$$

Example. A typical example of such an equation is an equation $f(k + 2) = f(k+1)+f(k)$ in which $p = 2$ and $a_1 = a_2 = 1$. If we start with $f(0) = f(1) = 1$, then this equation describes the so-called *Fibonacci numbers*, in which each number is a sum of two previous numbers:

- $f(0) = f(1) = 1$,
- $f(2) = f(1) + f(0) = 1 + 1 = 2$,
- $f(3) = f(2) + f(1) = 2 + 1 = 3$,
- $f(4) = f(3) + f(2) = 3 + 2 = 5$,
- $f(5) = f(4) + f(3) = 5 + 3 = 8$,
- etc.

Historical comment. This sequence was first described in the 13th century by the mathematician Leonardo of Pisa, better known as Fibonacci ("son of Bonaccio"). He described this sequence as a solution to the following problem:

"How many pairs of rabbits will be produced in a year, beginning with a single pair, if in every month each pair bears a new pair which becomes productive from the second month on?"

If we denote the number of pairs in k-th month by $f(k)$, then:

- In the beginning, we have only one pair ($f(0) = 1$).

- In the next month, we still have the same pair, because this pair is not yet productive ($f(1) = 1$).

- After two months, this pair becomes productive and gives birth to a second pair, so $f(2) = 2$.

- ...

- In k-th month, in addition to $f(k-1)$ pairs that were around in the previous month, every mature pair gives birth to a new pair of rabbits. Since rabbits mature in two months, there are as many mature rabbits in k-th month as there were rabbits two months before, i.e., $f(k-2)$. Each of these pairs gives birth to a new one, and, as a result, we get $f(k) = f(k-1) + f(k-2)$.

Rabbits were just a fun example, but it turned out that Fibonacci numbers really occur in different growth processes in nature, e.g., the number of leaves on a branch is often a Fibonacci numbers, etc. Thus, Fibonacci numbers are useful in *practical applications*.

Fibonacci numbers also turned out to be unexpectedly useful in *theoretical computer science*: namely, in 1970, Yuri Matiyasevich from St. Petersburg, Russia, used Fibonacci numbers to solve one of the most famous algorithmic problems of discrete mathematics, so-called Hilbert's Tenth Problem, one of the 23 problems that the 19-th century mathematicians formulated in 1900 for the 20-th century to solve. (Namely, Matiyasevich showed that no algorithm can check, for every polynomial $f(x_1, \ldots, x_n)$ with integer coefficients, whether the equation $f(x_1, \ldots, x_n) = 0$ has a solution with integer $x_1, \ldots, x_n$.)

Since the equation that defines Fibonacci numbers is very similar to the equations that we had for computational complexity, the reader will not be surprised to learn that Fibonacci sequence turned out to be useful in computational complexity problems as well (the relationship with one of our two examples is even closer, because the Fibonacci sequence turned out to be naturally appearing in some search problems). We will not go into details here, we just mention this connection to emphasize that rabbits or not rabbits, we are still solving equations related to algorithm complexity.

Now, back to the equation.

Analogy with continuous mathematics. Difference equations of type (2.3) are a natural analogue of *higher order* linear differential equations with constant coefficients. A general solution of such a differential equation is, usually, a linear combination of the *basic* solutions that are, normally, of the type $f(x) = \exp(\lambda x)$ (or, equivalently, of the type α^x, where $\alpha = \exp(\lambda)$). Let us, therefore, look first for solutions of the type $f(k) = \alpha^k$ for some α.

What α should we take? General idea. The main difference with the simple equation $f(k+1) = \alpha \cdot f(k)$ is that in this simplest equation, we knew what α to take, while for a more complicated equation (2.3), we need to find the appropriate α for which α^k is a solution.

To find this α, we can use the fact that $f(k) = \alpha^k$ is a solution of the simple equation $f(k+1) = \alpha \cdot f(k)$: namely, $f(k+1) = \alpha^{k+1} = \alpha \cdot \alpha^k = \alpha \cdot f(k)$; similarly, $f(k+2) = \alpha^{k+2} = \alpha^2 \cdot f(k)$, and in general, $f(k+l) = \alpha^l \cdot f(k)$. Thus, if we substitute $f(k) = \alpha^k$ into the equation (2.3), we get the following formula:
$$\alpha^p \cdot f(k) = a_1 \cdot \alpha^{k+p-1} \cdot f(k) + \ldots + a_p \cdot f(k).$$
Each term in both sides of this equation is proportional to $f(k)$.

Of course, for every α, the zero function $f(k) \equiv 0$ is a solution to the original equation. We are interested in *non-trivial* solutions, for whoch $f(k)$ is not always equal to 0. For such solutions, we can divide both sides by $f(k)$, we get the following equality:
$$\alpha^p = a_1 \cdot \alpha^{k+p-1} + a_2 \cdot \alpha^{k+p-2} + \ldots + a_p. \tag{2.4}$$
The only unknown in this equation is α. If we solve this equation and find α, then $f(k) = \alpha^k$ will be the solution to the equation (2.3)

Comment. This expression is not properly defined for $\alpha = k = 0$. For this case, we will assume that
$$f(0) = \alpha^0 = 0^0 = \lim_{\varepsilon \to 0} \varepsilon^0 = 1.$$

What α should we take? Example. Let us apply this idea to the equation that describes Fibonacci numbers. For this example, the equation (2.3) becomes a quadratic equation $\alpha^2 = \alpha + 1$. It is easy to solve this equation: just move all terms into the left-hand side $\alpha^2 - \alpha - 1 = 0$ and apply the standard formula for the solutions of the quadratic equation. As a result, we get two possible values of α:

$$\alpha_{1,2} = \frac{1 \pm \sqrt{1 - 4 \cdot 1 \cdot 1}}{2 \cdot 1} = \frac{1 \pm \sqrt{5}}{2}.$$

So, $\alpha_1 = (1 + \sqrt{5})/2$ and $\alpha_2 = (1 - \sqrt{5})/2$.

For both values, the corresponding functions

$$f_1(k) = \alpha_1^k = \left(\frac{1 + \sqrt{5}}{2}\right)^k$$

and

$$f_2(k) = \alpha_2^k = \left(\frac{1 - \sqrt{5}}{2}\right)^k$$

are the solution to the Fibonacci equation. This is good, but this is not yet what we want, because we want the function $f(k)$ not only to satisfy the *equation*, but also the *initial conditions* $f(1) = f(2) = 1$, and neither of the two functions $f_1(k)$ and $f_2(k)$ satisfies these initial conditions. So, how do we satisfy them? To answer this question, we will use the general property of equation (2.3) that is called *linearity*. To describe this property, let us go back to the general case.

Linearity. In short, an equation is called *linear* if an arbitrary linear combination of its solutions is also a solution. With respect to the equation (2.3), we can re-formulate this linearity property in more precise terms:

If the functions $f_1(k), \ldots, f_n(k)$ are solutions of the equation (2.3), then, for arbitrary constants $C_1, \ldots, C_n$, the corresponding linear combination $f(k) = C_1 \cdot f_1(k) + \ldots + C_n \cdot f_n(k)$ of these functions also satisfies the same equation (2.3).

It is easy to show that the equation (2.3) has this property: indeed, the fact that the functions $f_1(k), \ldots, f_n(k)$ satisfy this equation means that

$$f_1(k + p) = a_1 \cdot f_1(k + p - 1) + a_2 \cdot f_1(k + p - 2) + \ldots + a_p f_1(k);$$

$$\ldots$$

$$f_n(k + p) = a_1 \cdot f_n(k + p - 1) + a_2 \cdot f_n(k + p - 2) + \ldots + a_p f_n(k).$$

If we multiply the first equation by C_1, the second by C_2, ..., the n-th by C_n, and then add these equations, we get the desired formula

$$f(k + p) = a_1 \cdot f(k + p - 1) + a_2 \cdot f(k + p - 2) + \ldots + a_p f(k)$$

which shows that the linear combination $f(k)$ also satisfies the same equation.

Because of linearity, if we have found two solutions, we will get many more. By appropriately choosing the constants, we can try to find the solution that satisfies the given initial conditions. Let us show how this idea works in our example.

Example: application of linearity. In the Fibonacci example, we know two solutions $f_1(k) = \alpha_1^k$ and $f_2(k) = \alpha_2^k$. Thus, each linear combination $f(k) = C_1 \cdot f_1(k) + C_2 \cdot f_2(k) = C_1 \cdot \alpha_1^k + C_2 \cdot \alpha_2^k$ of these solutions is also a solution. Hence, to find the formula for Fibonacci numbers, we must find the values C_1 and C_2 for which $f(1) = 1$ and $f(2) = 1$. Substituting $k = 1$ and $k = 2$ into the expression for $f(k)$, we get the following two equations, from which we can determine C_1 and C_2:

$$f(0) = C_1 + C_2 = 1;$$

$$f(1) = C_1 \cdot \alpha_1 + C_2 \cdot \alpha_2 = 1.$$

We have a system of two linear equations with two unknowns C_1 and C_2. We know how to solve such systems. As a result, we get

$$C_1 = \frac{5 + \sqrt{5}}{10}; \quad C_2 = \frac{5 - \sqrt{5}}{10}.$$

(Exercise: check!). So, we get the following formula for the Fibonacci sequence:

$$f(k) = \frac{5 + \sqrt{5}}{10} \cdot (\frac{1 + \sqrt{5}}{2})^k + \frac{5 - \sqrt{5}}{10} \cdot (\frac{1 - \sqrt{5}}{2})^k.$$

This formula is helpful. What did we gain by deriving this formula?

- *Before* we knew this formula, the only known way to compute, say, $f(100)$ was to actually compute $f(1)$, $f(2)$, etc., all the way until $f(100)$. In general, to compute $f(k)$, we needed k additions.

- *After* we have derived this formula, we can compute $f(k)$ directly, without having to compute any previous Fibonacci number: indeed, α^k can be computed as $\exp(k \cdot \ln(\alpha))$. This is clearly faster that before.

So, continuous mathematics is useful. This is the first example where we can clearly state that yes, *continuous* mathematics (i.e., mathematics of objects that can take values from a *continuous* set), *is* useful for *discrete* problems.

- We started with a *discrete* problem, in which $f(1)$ and $f(2)$ are integers, and in which for every k, the value $f(k)$ is also an integer.

- It turned out that we were unable to find an easy formula to compute $f(k)$ for a given k if we only used these discrete objects (i.e., integers). However, we were able to find an efficient formula for $f(k)$ when we decided to consider, in addition to the *discrete* (integer) values, *continuous* (arbitrary real) values like $(1 \pm \sqrt{5})/2$.

This is just the first (and simple) example. Later on, we will see much more of them.

Now that we have successfully found a formula for the Fibonacci sequence, let us describe the corresponding general algorithm:

Algorithm: take one. Suppose that we have an equation (2.3) (with no free terms), and we want to find a solution of this equation that satisfies given initial conditions, i.e., which has the given values $f(0)$, $f(1)$, ..., $f(p-1)$. To find the corresponding solution $f(k)$, we do the following:

- First, we find *specific* solutions to the given equation, solutions of the type $f(k) = \alpha^k$. To find such solutions, we substitute the expression $f(k) = \alpha^k$ into the equation (2.3), and then find all possible values of α from the resulting equation (2.4).

- After we have found several solutions $f_1(k) = \alpha_1^k, \ldots, f_m(k) = \alpha_m^k$, we describe a *general* solution $f(k) = C_1 \cdot f_1(k) + \ldots + C_m \cdot f_m(k)$ and try to find the values $C_1, \ldots, C_m$ for which the resulting function satisfies the initial conditions, i.e., for which

$$C_1 + \ldots + C_m = f(0);$$

$$C_1 \cdot \alpha_1 + \ldots + C_m \cdot \alpha_m = f(1);$$

$$\ldots$$

$$C_1 \cdot \alpha_1^p + \ldots + C_m \cdot \alpha_m^p = f(p-1).$$

This can be done by solving the corresponding system of linear equations with unknowns $C_1, \ldots, C_m$.

Will this algorithm always lead to a solution? No, that's why we added "take one". In the following text, we will give the examples for which this natural (and rather simple) algorithm fails, and we will describe a more complicated algorithm that will always succeed. To get ready for this more complicated algorithm, let us somewhat simplify this one.

Slight reformulation. For this particular case, the above description seems to be OK. However, we will present a slight reformulation that does not change much here, but that will be very useful for more complicated equations.

The main idea behind this (slight) modification is that although we repeat the variable k p times in the equation (2.3), we really never use the actual value of k. All we need to know is that the value $f(k + 1)$ is the next value to $f(k)$, etc. Can we save omit k? To be able to do that, we need to represent the next value somehow. A natural idea is to introduce the following *shift* operation S: If a function f with the values $f(1), \ldots, f(k), \ldots$ is given, we define its shift Sf as a function with *shifted* values: $(Sf)(k) = f(k + 1)$. If we apply the shift S for the second time, then we get a function $S(Sf)$ (which we will denote by $S^2 f$) that is shifted by 2: $(S^2 f)(k) = f(k + 2)$; etc.

In these notations, $f(k + p) = (S^p f)(k)$, $f(k + p - 1) = (S^{p-1} f)(k)$, etc., and equation (2.3) takes the following form:

$$(S^p f)(k) = a_1 \cdot (S^{p-1} f)(k) + a_2 \cdot (S^{p-2} f) + \ldots + a_p \cdot f(k).$$

This is exactly the same equation as before, but this new form of this equation has the desired advantage: all terms refer to the same value k. So, we can omit this k, and describe this equation as simply an equality between the functions:

$$S^p f = a_1 \cdot S^{p-1} f + a_2 \cdot S^{p-2} f + \ldots + a_p \cdot f.$$

This is slightly simpler than before.

Can we simplify this equation even further? Definitely. If we look at this equation, we will see that all its terms contain f: either it is f itself, or the result of applying one or several shift operation to f. So, to simplify our notations, we can, formally, replace the right-hand side by a simplified expression $(a_1 \cdot S^{p-1} + a_2 \cdot S^{p-2} + \ldots + a_p)f$. If we do that, the above equation somewhat simplifies:

$$S^p f = (a_1 \cdot S^{p-1} + a_2 \cdot S^{p-2} + \ldots + a_p)f.$$

We can simplify this equation even more if we first move all the terms into the left-hand side. In this case, the original equation takes the form

$$P(S)f = 0,$$

where by $P(S)$, we denoted the result

$$P(S) = S^p - a_1 \cdot S^{p-1} - a_2 \cdot S^{p-2} - \ldots - a_p$$

of applying the polynomial

$$P(x) = x^p - a_1 \cdot x^{p-1} - a_2 \cdot x^{p-2} - \ldots - a_p$$

to the shift operator S.

If we represent the original equation (2.3) in this form, then, to find α, we only need to find the roots of the equation $P(\alpha) = 0$. In other words, we substitute α instead of the shift operator S. This substitution makes perfect sense, because for the sought solutions of the type $f(k) = \alpha^k$, shift is the same as multiplying by α: $(Sf)(k) = f(k+1) = \alpha \cdot f(k)$, i.e., $Sf = \alpha \cdot f$.

Comment. For this particular problem, the reformulation in terms of the shift operator S does not bring any clear advantage; we describe this reformulation because it will be useful in solving more complicated functional equations.

Exercise

2.1 Use the above algorithm to solve the functional equation $f(k+2) = 3f(k+1) + 2f(k)$ with the initial conditions $f(0) = 1$ and $f(1) = 2$. Check the resulting formula for $k = 0, 1, 2$, and 3.

2.4. A minor computational problem: complex roots

Complex roots are possible. The algorithm presented in the previous section requires us to solve a polynomial equation (namely, equation (2.4)). If the solutions of the corresponding equation (i.e., roots of the polynomial) are *real* numbers, we get the desired solution. (The corresponding formulas are easy to implement in any programming language.) It is well known, however, that some polynomials have *complex* roots.

For example, for the equation

$$f(k+2) = -f(k), \tag{2.5a}$$

with the initial condition

$$f(0) = 0, \quad f(1) = 1, \tag{2.5b}$$

both roots $\alpha_{1,2}$ of the corresponding equation $\alpha^2 = -1$ are complex numbers: $\alpha_1 = i$, $\alpha_2 = -i$.

We will use this simple equation to illustrate the complex-root case, because this equation is very easy to solve directly: we get

- $f(2) = -f(0) = 0$;

- $f(3) = -f(1) = -1$;

- $f(4) = -f(2) = 0$;

- $f(5) = -f(3) = 1$;

- $f(6) = -f(4) = 0$;

- etc.

The solution is a periodic sequence $(0, 1, 0, -1, 0, 1, 0, -1, \ldots)$ with a period $(0, 1, 0, -1)$ of length 4.

When operations with complex numbers are supported, complex roots present no problems. The formulas given in the previous section are valid for complex numbers as well. So, if a programming language that we use has built-in arithmetic (and other) operations with complex numbers, we can implement these formulas directly.

In particular, for the equation $(2.5a) - (2.5b)$, the above algorithm leads to the following solution:

- First, we must solve the corresponding equation (2.4). We have already done that, and we have found its two roots $\alpha_{1,2} = \pm i$. These roots lead to two *specific* solutions to the equation $(2.5a)$: $f_1(k) = i^k$ and $f_2(k) = (-i)^k$.

- Next, we find the solution $f(k) = C_1 \cdot i^k + C_2 \cdot (-i)^k$ that satisfies the given initial conditions by solving the following system of two linear equations with two unknowns:

$$C_1 + C_2 = 0;$$
$$C_1 \cdot i + C_2 \cdot (-i) = 1.$$

From the first equation, we conclude that $C_2 = -C_1$. Substituting this expression into the second equation, we get $2C_1 i = 1$, $C_1 = 1/(2i) = -i/2$, and $C_2 = -C_1 = i/2$.

Thus, the solution to the equation (2.5a) with the initial conditions (2.5b) is

$$f(k) = -\frac{i}{2} \cdot i^k + \frac{i}{2} \cdot (-i)^k. \tag{2.6}$$

In some programming languages, complex roots present a (minor) computational problem. Some programming languages have operations with *real* numbers but not with *complex* numbers. In such languages, in order to program (complex-valued) formulas of the type (2.6), we must first re-formulate them in terms of real numbers.

How to reformulate complex formulas in terms of real numbers. In order to reformulate complex-valued formulas in terms of real numbers, we can use the fact that every complex number $z = x + iy$ can be represented in the *polar* form $z = x + iy = \rho \cdot \exp(i\theta)$. The advantage of this representation is that if $z_1 = \rho_1 \exp(i\theta_1)$ and $z_2 = \rho_2 \exp(i\theta_2)$, then

$$z_1 \cdot z_2 = \rho_1 \cdot \rho_2 \cdot \exp(i(\theta_1 + \theta_2)).$$

In particular, if $z_1 = z_2 = \ldots = z_k$, we get $z^k = \rho^k \exp(ik\theta)$.

To find ρ and θ from x and y, we must use the "magic" formula (first discovered by de Moivre) according to which

$$\exp(i\theta) = \cos(\theta) + i \cdot \sin(\theta).$$

Using this formula, we can conclude that $x = \rho \cdot \cos(\theta)$ and $y = \rho \cdot \sin(\theta)$. From these two formulas, we get

$$\rho = \sqrt{x^2 + y^2},$$

and $y/x = \sin(\theta)/\cos(\theta) = \tan(\theta)$, so

$$\theta = \arctan\left(\frac{y}{x}\right).$$

So, to express the solution $f(k) = C_1 \cdot \alpha_1^k + \ldots + C_p \cdot \alpha_p^k$ with complex-values α_i and C_i in terms of real numbers, we must do the following:

- First, we represent each root α_i in the polar form, as $\alpha_i = \rho_i \cdot \exp(i\theta_i)$.

    Then, $f_i(k) = \alpha_i^k = \rho^k \cdot \exp(ik \cdot \theta_i)$.

- Next, we represent each coefficient C_i in the polar form, as $C_i = P_i \cdot \exp(i\Theta_i)$.

    Then, $C_i \cdot \alpha_i^k = P_i \cdot \rho_i^k \cdot \exp(i(\Theta_i + k\theta_i))$.

Adding these terms, we get $f(k) = f_{\text{Re}}(k) + if_{\text{Im}}(k)$, where

$$f_{\text{Re}}(k) = P_1 \cdot \rho_1^k \cdot \cos(k\theta_1 + \Theta_1) + \ldots + P_p \cdot \rho_p^k \cdot \cos(k\theta_p + \Theta_p);$$

$$f_{\text{Im}}(k) = P_1 \cdot \rho_1^k \cdot \sin(k\theta_1 + \Theta_1) + \ldots + P_p \cdot \rho_p^k \cdot \sin(k\theta_p + \Theta_p).$$

Comment. In this lesson, our main goal is to find the functions $f(k)$ that describe the algorithm complexity. By definition, an algorithm complexity is a real number (even an integer), so for these functions, $f(k)$ is a real number for all k. For such functions, it is, therefore, the desired function $f(k)$ coincides with its real part $f_{\text{Re}}(k)$; hence, it is sufficient to compute $f_{\text{Re}}(k)$ (because $F_{\text{Im}}(k) = 0$).

Example. Let us illustrate this method on the above simple example.

- For $\alpha_1 = i = 0 + 1 \cdot i$, we have $x_1 = 0$ and $y_1 = 1$. Thus, $\rho_1 = \sqrt{x_1^2 + y_1^2} = \sqrt{1} = 1$, and $\theta_1 = \arctan(1/0) = \arctan(+\infty) = \pi/2$. Hence, $\alpha_1 = \exp(i\pi/2)$. Similarly, $\alpha_2 = -i = \exp(-i\pi/2)$.

    Hence, $i^k = \exp(ik\pi/2)$ and $(-i)^k = \exp(-ik\pi/2)$.

- Similarly, $C_1 = -i/2 = (1/2) \cdot \exp(-i\pi/2)$ and $C_2 = i/2 = (1/2) \cdot \exp(i\pi/2)$.

    Hence, $(-i/2) \cdot i^k = (1/2) \cdot \exp(i(k\pi/2 - \pi/2))$ and $(i/2) \cdot (-i)^k = (1/2) \cdot \exp(i(-k\pi/2 + \pi/2))$.

Finally, $f(k) = f_{\text{Re}}(k) = (1/2) \cdot \cos(k\pi/2 - \pi/2) + (1/2) \cdot \cos(-k\pi/2 + \pi/2)$. This expression can be further simplified if we take into consideration that $\cos(x)$ is an even function, hence, $f(k) = \cos(k\pi/2 - \pi/2)$, and that $\cos(x - \pi/2) = \sin(x)$, hence $f(k) = \sin(k\pi/2)$. By substituting $k = 0, 1, 2, 3, 4, \ldots$, one can easily check that this is indeed the desired periodic solution.

Exercise

2.2 Solve the functional equation $f(k+2) = 2f(k+1) - 2f(k)$ with the initial conditions $f(0) = 1$ and $f(1) = 1$, and represent the complex-valued solution in terms of real numbers. Check the resulting formula for $k = 0, 1, 2$, and 3.

2.5. Multiple roots

The problem. We started the previous section 2.4 by mentioning that the above algorithm requires us to first find the roots of a polynomial, and that there may be a computational problem if some of these roots are complex numbers.

It turns out that another possibility – of a polynomial having *multiple* roots – also causes a problem. To illustrate this problem, let us give a simple example.

Example. For the equation

$$f(k+2) = 2f(k+1) - f(k) \tag{2.7a}$$

with the initial conditions

$$f(0) = 0, \quad f(1) = 1, \tag{2.7b}$$

the corresponding polynomial equation (2.4) is of the form $\alpha^2 = 2\alpha - 1$, which is equivalent to $(\alpha - 1)^2 = 0$. This polynomial equation has a double root $\alpha_1 = \alpha_2 = 1$, and therefore, we have only one specific solution $f_1(k) = \alpha_1^k = 1^k = 1$. Since we only have one solution, we can only have linear combinations of the type $f(k) = C_1 \cdot f_1(k) = C_1$. If we try to satisfy the first initial condition $f(0) = 0$, we get $C_1 = 0$, but the resulting solution $f(k) = 0$ does not satisfy the second initial condition $f(1) = 1$. So, the above algorithm cannot solve this linear functional equation. How can we modify this algorithm?

Idea. A polynomial with a *double* root $\alpha_1 = \alpha_2 = \alpha$ can be viewed as a *limit* case of polynomials with two close roots α and $\alpha + \varepsilon$ when $\varepsilon \to 0$. (Similarly, every equation with a *multiple* root can be described as a limit of non-degenerate equations in which all roots are different.) So, in order to find the solutions to the original *degenerate* equation (with multiple roots) we can:

- find the solution $f_\varepsilon(k)$ to the approximate *non-degenerate* equation (in which all roots are different), and then

- take the limit

$$f(k) = \lim_{\varepsilon \to 0} f_\varepsilon(k)$$

of these "approximate" solutions $f_\varepsilon(k)$ as the desired solution to the original (degenerate) equation.

This idea is *almost* an algorithm, but not yet: The main non-algorithmic step here is tending to a limit. Therefore, to turn this idea into an algorithm, we must find the algorithmic way of computing the desired limit.

From idea to algorithm. To find out how this idea can be converted into an algorithm, let us first consider the case of the *double* root α. The corresponding non-degenerate approximate equation has *two* different roots α and $\alpha + \varepsilon$ and therefore, two specific solutions α^k and $(\alpha + \varepsilon)^k$. Due to linearity, an arbitrary linear combination of these two solutions is also a solution; in particular, the difference $(\alpha + \varepsilon)^k - \alpha^k$ between the specific solutions of the equation is also a solution. If we tend $\varepsilon \to 0$, then the limit expression is a solution to the original functional equation.

This difference, by itself, tends to 0, and therefore, does not lead to any meaningful solution of the original equation. However, due to linearity, the result of multiplying this difference by any number is also a solution. We can, therefore, multiply it by, say, $1/\varepsilon$, to prevent it from having a 0 limit. As a result, we can conclude that the following limit function is a solution to the original equation:

$$f_{\text{new}}(k) = \lim_{\varepsilon \to 0} \frac{(\alpha + \varepsilon)^k - \alpha^k}{\varepsilon}.$$

The expression in the right-hand side is exactly the definition of the derivative:

$$f_{\text{new}}(k) = g'(\alpha) = \lim_{\varepsilon \to 0} \frac{g(\alpha + \varepsilon) - g(\alpha)}{\varepsilon}$$

for $g(\alpha) = \alpha^k$. The derivative of the function $g(\alpha) = \alpha^k$ is well known: $f_{\text{new}}(k) = g'(\alpha) = k\alpha^{k-1}$. So, we can conclude that for the case of a double root α, in addition to α^k, another function, $k \cdot \alpha^{k-1}$, is also a specific solution.

Since the equations are linear, multiplying a solution by a constant keeps it a solution. We can use this property to slightly simplify the new specific solution:

namely, instead of $f_{new}(k) = k \cdot \alpha^{k-1}$, we can consider $\tilde{f}_{new}(k) = \alpha \cdot f_{new}(k) = k \cdot \alpha^k$. So, *if a polynomial has a double root α, then, in addition to α^k, this equation has another specific solution $k \cdot \alpha^k$.*

Similarly, if we have a *triple* root, we can form a linear combination that is equal to the *second derivative* of $g(\alpha) = \alpha^k$, and thus, conclude that not only α^k, but also $k \cdot \alpha^k$ and $k^2 \cdot \alpha^k$ are solutions. If we have a root of multiplicity m, then $\alpha^k, k \cdot \alpha^k, \ldots, k^{m-1} \cdot \alpha^k$ are specific solutions.

Now, we are ready to describe the modified algorithm:

General algorithm for linear functional equations with zero right-hand side. Suppose that we have an equation (2.3) (with no free terms), and we want to find a solution of this equation that satisfies given initial conditions, i.e., which has the given values $f(0), f(1), \ldots, f(p-1)$. To find the corresponding solution $f(k)$, we do the following:

- First, we find *specific* solutions to the given equation. To find such solutions, we re-formulate the original equation in terms of the shift operator S, substitute α instead of S, and find all the roots of the resulting equation (2.4). Each root leads to *specific* solutions of the original functional equation:

  − For each *single* root α_i, we form a specific solution α_i^k.

  − For each *double* root α_i, we form *two* specific solutions α_i^k and $k \cdot \alpha_i^k$.

  − For each *triple* root α_i, we form *three* specific solutions α_i^k, $k \cdot \alpha^i$, and $k^2 \cdot \alpha_i^k$.

  − ...

  − For each root α_i of multiplicity m, we form m specific solutions α_i^k, $k \cdot \alpha^i, \ldots, k^{m-1} \cdot \alpha_i^k$.

  It is known that if we count each double root twice, each triple root three times, etc., then the total number of roots is equal to the degree of the original polynomial. Hence, we get exactly p specific solutions $f_1(k), \ldots, f_p(k)$.

- Now, we must find a solution $f(k) = C_1 \cdot f_1(k) + \ldots + C_p \cdot f_p(k)$ that the initial conditions, i.e., for which

$$C_1 \cdot f_1(0) + \ldots + C_p \cdot f_p(0) = f(0);$$
$$C_1 \cdot f_1(1) + \ldots + C_p \cdot f_p(1) = f(1);$$
$$\ldots$$
$$C_1 \cdot f_1(p-1) + \ldots + C_p \cdot f_p(p-1) = f(p-1).$$

The coefficients $C_1, \ldots, C_p$ can be found from this system of p linear equations with p unknowns.

Comment. One can prove that this system of linear equations is always non-degenerate, and that therefore, this algorithm will always find the desired solution.

Example. For the equation $(2.7a)-(2.7b)$, this algorithm leads to the following solution:

- First, we express the equation $(2.7a)$ in terms of the shift operator S, and substitute α instead of S. We have already done that, and we have shown that the resulting equation has a double root $\alpha = 1$. Thus, according to the algorithm, we must take *two* specific solutions: $f_1(k) = \alpha^k = 1^k = 1$, and $f_2(k) = k \cdot \alpha^k = k \cdot 1^k = k$.

- The resulting solution will now take the form $f(k) = C_1 \cdot f_1(k) + C_2 \cdot f_2(k) = C_1 + C_2 \cdot k$, where the coefficients C_1 and C_2 must be found from the initial conditions $f(0) = 0$ and $f(1) = 1$. Substituting the above expression for $f(k)$ into these initial conditions, we conclude that $C_1 = 0$ and $C_1 + C_2 = 1$. Hence, $C_1 = 0$, $C_2 = 1$, and $f(k) = C_1 + C_2 \cdot k = k$.

So, $f(k) = k$ is the desired solution. (The reader can easily check that it really satisfies both the initial conditions and the equation (2.7) for $k = 0, 1, \ldots$)

Exercise

2.3 Solve the functional equation $f(k+2) = 4f(k+1) - 4f(k)$ with the initial conditions $f(0) = 1$ and $f(1) = 1$. Check the resulting formula for $k = 0$, 1, 2, and 3.

2.6. Equations with non-zero right-hand sides

Equations with non-zero right-hand sides are necessary for algorithm complexity. We now know how to solve linear equations with *zero* right-hand sides. We also know, however, that for some algorithms, the equations that describe their complexity $f(k)$ do have a *non-zero* right-hand side: e.g., if we move all terms that contain f in equation (1.3) $f(k + 1) = 2f(k) + 2^{k+1} - 1$ into the left-hand side, we will get an equation $f(k + 1) - 2f(k) = 2^{k+1} - 1$ (or, in S–notations, $(S - 2)f = g$) which has a non-zero right-hand side $g(k) = 2^{k+1} - 1$.

How can we solve such equations?

Idea. The main idea of solving these equations is to reduce them to equations that we already know how to solve, i.e., to equations without any right-hand side. In other words, if we have an equation of the type $P(S)f = g$ with $g \neq 0$, we try to find an operator $X(S)$ that will annul the right-hand side, i.e., for which $X(S)g = 0$. If we find such $X(S)$, then, by applying this operator $X(S)$ to both sides of the original equation $P(S)f = g$, we get a new equation $X(S)P(S)f = 0$, which we already know how to solve.

The only remaining question is: how to find this operator $X(S)$? The answer is easy: we know, from the above-described algorithm, which equations annul which functions. This description can be easily reversed. For example:

- We know that the equation $(S - \alpha)f = 0$ with a single root α has a solution $C \cdot \alpha^k$. Thus, if we need to annul the expression $C \cdot \alpha^k$, we can take $X(S) = S - \alpha$.

- To annul a linear combination $C_1 \cdot \alpha^k + C_2 \cdot \alpha_2^k$, we can take an operator $X(S)$ that has two roots α_1 and α_2, e.g., $X(S) = (S - \alpha_1)(S - \alpha_2)$. Indeed, according to the above algorithm, the equation $X(S)g = 0$ has two specific solutions $f_1(k) = \alpha_1^k$ and $f_2(k) = \alpha_2^k$, and an arbitrary linear combination of these two specific solutions is also a solution (i.e., it is also annulled by the operator $X(S) = (S - \alpha_1)(S - \alpha_2)$).

- To annul a function $k \cdot \alpha^k$, we need an expression for which α is a double root, i.e., an expression of the type $X(S) = (S - \alpha)^2$.

- etc.

Let us illustrate this idea on several examples.

Example 1: equation (1.3) (complexity of mergesort). Let us first show how this idea can help us in solving the equation (1.3), i.e., in S-notations, the equation $(S - 2)f = 2^{k+1} - 1$. How can we annul the right-hand side $g(k) = 2^{k+1} - 1$ of this equation?

- The first term in $g(k)$ is of the type $2 \cdot 2^k$, i.e., of the type $C \cdot \alpha^k$ with $\alpha = 2$. So, to annul this term, we must have an expression $S - 2$.

- The second term in $g(k)$ is also of the same type: $1 = 1^k = \alpha^k$ for $\alpha = 1$. So, to annul this second terms, we need to take $X(S) = S - 1$.

To annul the entire linear combination $g(k)$, we must take $X(S) = (S - 2)(S - 1)$. If we apply this operator to both sides of the equation $(S - 2)f = g$, we get a new equation $(S - 2)(S - 1)(S - 2)f = 0$ with zero right-hand side.

We can now apply the algorithm from the previous section to this equation:

- If we substitute an unknown α instead of the shift operator S, we get an equation $(\alpha - 2)(\alpha - 1)(\alpha - 2) = 0$, which has one single root $\alpha = 1$ and one double root $\alpha = 2$. Thus, according to the algorithm, we get three specific solutions: $f_1(k) = 1^k = 1$ that correspond to the single root, and $f_2(k) = 2^k$ and $f_3(k) = k \cdot 2^k$ that correspond to the double root.

- Hence, the desired solution has the form $f(k) = C_1 + C_2 \cdot 2^k + C_3 \cdot k \cdot 2^k$. According to the above algorithm, to find the three coefficients C_i, we must know the first three values of $f(k)$. From the initial condition (1.3b), we only know the value $f(0) = 0$. So, we must apply the equation (1.3a) and find the values $f(1) = 2f(0) + 2^{0+1} - 1 = 1$ and $f(2) = 2f(1) + 2^{1+1} - 1 = 2 \cdot 1 + 4 - 1 = 5$. Substituting into the above expression for $f(k)$ (with the coefficients C_i) the values $k = 0$, 1, and 2, we get the system of three linear equations with three unknowns C_1, C_2, and C_3:

$$C_1 + C_2 = 0;$$

$$C_1 + 2C_2 + 2C_3 = 1;$$

$$C_1 + 4C_2 + 8C_3 = 5.$$

From these equations, we get the values $C_1 = C_3 = 1$, $C_2 = -1$.

Hence, $f(k) = 1 - 2^k + k \cdot 2^k$.

Let us now recall that this equation describes the algorithm complexity of *mergesort*, and that k stands for $k = \log_2(n)$. For such k, $2^k = n$, and the resulting algorithm complexity of mergesort is $t(n) = 1 - n + n \cdot \log_2(n)$ (i.e., $\approx n \cdot \log_2(n)$).

Comment. Before we use this idea for estimating algorithm complexity of other algorithms, let us apply this same idea to a couple of problems for which we already know the solutions.

Example 2: the sum of a geometric progression. Let us show that this method can help us deduce the formula for the sum $f(k) = 1 + q^1 + q^2 + \ldots + q^k$ of the geometric progression. Indeed, the next sum

$$f(k + 1) = 1 + q + \ldots + q^k + q^{k+1}$$

is obtained from the previous sum $f(k)$ by adding a new term q^{k+1}. In other words, we get a linear functional equation $f(k + 1) = f(k) + q^{k+1}$. In S-form, this functional equation takes the form $(S - 1)f = q^{k+1}$. The right-hand side $g(k) = q^{k+1}$ of this equation is of the type $C \cdot \alpha^k$, with $C = q$ and $\alpha = q$. Thus, to annul the right-hand side, we can use the operator $X(S) = S - q$. If we apply this operator to both sides of the equation $(S - 1)f = q^{k+1}$, we get the equation $(S - q)(S - 1)f = 0$ with a zero right-hand side.

For this new equation, the algorithm from the previous section leads to the following solution: First, we form an equation $(\alpha - q)(\alpha - 1) = 0$. This equation has two non-degenerate roots $\alpha = q$ and $\alpha = 1$; therefore, we have two specific solutions $f_1(k) = q^k$ and $f_2(k) = 1^k = 1$. The desired solution is, therefore, a linear combination $f(k) = C_1 \cdot q^k + C_2$, where the coefficients C_1 and C_2 can be obtained from the condition that $f(0) = 1$ and $f(1) = 1 + q$. As a result, we get $C_1 = q/(q - 1)$, $C_2 = -1/(q - 1)$, and the known formula

$$f(k) = \frac{q}{q - 1} \cdot q^k - \frac{1}{q - 1} = \frac{q^{k+1} - 1}{q - 1}.$$

Example 3: the sum of an arithmetic progression. Let us now compute the sum $f(k) = 1 + 2 + \ldots + k$. It is well known that $f(k) = k(k + 1)/2$. There is a known legend related to this formula: when a famous 19 century mathematician Karl Friedrich Gauss was in elementary school, he used to solve all the mathematical problems in no time, and to ask the teacher for more and more. In desperation, the teacher asked him to add all integers from 1 to 100, hoping that this addition would give him a few minutes of rest. Alas, in a few seconds, Karl gave the correct answer: 5,050. "How could you add these numbers so fast?" asked the teacher. "I did not", replied Karl, I simply noticed that if you add 1 and 100, you get 101; if you add 2 and 99, you also get 101, etc. So, we have 50 pairs, each of which adds to 101, and the result is $50 \cdot 101 = 5050$".

For the sum of k numbers, we get $k/2$ pairs with a sum of $k + 1$, so, the total sum is $f(k) = k(k + 1)/2$. Let us show that this same formula can be derived by using our general method.

For this sum, $f(k + 1) = f(k) + k + 1$. In S-form, this functional equation takes the form $(S - 1)f = k + 1$. The right-hand side of this equation is a linear combination of two terms, of types α^k and $k \cdot \alpha^k$ (with $\alpha = 1$). To annul both terms, we, thus, need to use $X(S) = (S - \alpha)^2 = (S - 1)^2$. Applying this $X(S)$ to both sides of the equation $(S - 1)f = k + 1$, we get a new equation $(S - 1)^3 f = 0$ with a zero right-hand side.

If we substitute α instead of S into this functional equation, we get a polynomial equation $(\alpha - 1)^3 = 0$ with a triple root $\alpha = 1$. Thus, the specific solutions are $f_1(k) = \alpha^k = 1^k = 1$, $f_2(k) = k \cdot \alpha^k = k$, and $f_3(k) = k^2$. The desired solution is of the form $f(k) = C_1 + C_2 \cdot k + C_3 \cdot k^2$, where the the coefficients C_i can be determined from the conditions that $f(0) = 0$, $f(1) = 1$, and $f(2) = 1 + 2 = 3$. From these conditions, we conclude that $C_1 = 0$ and $C_2 = C_3 = 1$, i.e., we get the known formula $f(k) = k(k + 1)/2$.

Example 4: mergesort on a parallel computer. The equation (1.3) describes the complexity of mergesort on a normal (sequential) computer: the time $t(2n) = f(k + 1)$ that is necessary to sort a list of size $2^{k+1} = 2 \cdot 2^k = 2n$ consists of:

- the time $2f(k)$ that is necessary to sort the two half-lists, of size $n = 2^k$ each, plus

- the time $2n - 1 = 2^{k+1} - 1$ for merging the two sorted half-lists into a single sorted list.

If we have an unlimited number of processors that can work in parallel, then we can use two different machines to sort the two halves. Since these two machines work in parallel, we need exactly the same time to sort both halves as we need to sort one half-list, i.e., $t(n) = f(k)$. Hence, the time $t(2n) = f(k + 1)$ that is necessary to sort a list of size $2n = 2^{k+1}$ on a *parallel* computer can be represented as the sum of two terms:

- the time $f(k)$ that is necessary to sort both two half-lists, of size $n = 2^k$ each, plus

- the time $2n - 1 = 2^{k+1} - 1$ for merging the sorted half-lists into a single sorted list.

So, we get the equation

$$f(k + 1) = f(k) + 2^{k+1} - 1. \qquad (2.8a)$$

A list of size $n = 2^0 = 1$ is automatically sorted, so we need 0 steps for $k = 0$:

$$f(0) = 0. \qquad (2.8b)$$

Exercises

2.4 Solve the functional equation $(2.8a) - (2.8b)$ and find the number of computational steps necessary for parallel mergesort.

2.5 Find the formula for the sum $1^2 + 2^2 + \ldots + k^2$.

3

PROGRAM TESTING: A PROBLEM

Now that we know how to solve simple functional equations, we will start the discussion of another computer science area where these equations are useful: the problem of software reliability. In this lesson, we will briefly describe the problem, and describe its current (semi-heuristic) solutions.

In the next lesson, we will re-formulate the problem of software reliability as a mathematically precise optimization problem, and solve the simplest case of the corresponding optimization problem.

3.1. Software testing: a (very) brief introduction

Software testing is important. In the first two lessons, we considered the situation when we have a working (i.e., practically error-free) program (or several working programs), and the only thing we need is to check whether the given programs are fast enough, and if they are, which of these programs is the fastest. We have shown that to answer these questions, it is often helpful to use continuous mathematics.

In this lesson, we will see that continuous mathematics is also helpful at the previous stage: of making a given program practically error-free.

Software testing: how? How can we make sure that the program is practically error-free?

- For a sufficiently *simple* program, it is possible to eliminate *all* the errors. Moreover, there exist methods (called "proving program correctness") that can guarantee that all the errors has indeed been removed and the resulting program is "flawless".

- However, the experience of many programmers shows that it is actually *impossible* to extract *all* the faults from complicated software, such as:

  − system-type software, that is commonly involved in resource contention,

  − programs with a sophisticated user interface, etc.

If after some debugging such a program works fine, it simply means that we have not yet reached the point where it will start erring. So we are unable to debug such a program completely; moreover we are sure that sooner or later the program will err. Therefore it is necessary to estimate the time interval during which the remaining faults will not influence the program.

At present about three dozen statistical models (called *software reliability models*) are used to get such estimates. The most widely accepted models are basic (exponential) and the logarithmic Poisson models proposed by Musa, Iannino, and Okumoto. In many situations these models proved to be a good fit. However, these statistical models lack convincing theoretical explanation. They are semi-heuristic, and often look like curve-fitting.

What we are planning to do. In this and in the following lesson, we formulate the problem of choosing the best interval software reliability model as a mathematical optimization problem, and solve this problem. As a result, we explain Musa's experiments.

Brief historical comment on program verification. Formal methods of program verification were initiated by John McCarthy (best known as the father of Artificial Intelligence) in his papers "Towards a mathematical theory of computation" (*Proceedings IFIP Congress 62*, North Holland, Amsterdam, 1963, pp. 21–28) and "A basis for the mathematical theory of computation" (In: P. Braffort, D. Hirschenberg (eds.) *Computer programming and formal systems*, North-Holland, Amsterdam, 1963, pp. 33–70). These methods became widely applicable starting from the breakthrough 1969 paper of C. A. R. Hoare "An axiomatic basis for computer programming", (*Communications of the ACM*, 1969, Vol. 12, pp. 576–580). For more recent surveys on program correctness see, e.g.,

- J. de Bakker. *Mathematical theory of program correctness.* Englewood Cliffs, NJ, 1980.

- E. M. Clarke and R. P. Kurshan (eds.). *Computer-aided verification'90. Proceedings of a DIMACS Workshop*, American Mathematical Society, Providence, RI, 1991.

- P. Cousot, "Methods and logics for proving programs", In: J. van Leeuwen (ed.) *Handbook of Theoretical Computer Science*, Vol. B, Elsevier, Amsterdam, 1990, pp. 843–982.

3.2. Software testing: a problem

It is possible to eliminate all the errors from a small simple program. For a sufficiently *simple* program it is possible to eliminate all the errors.

Moreover, there exist methods (called "proving program correctness") that can guarantee that all the errors has indeed been removed and the resulting program is "flawless". These methods were initiated by the pioneering 1962–63 works of John McCarthy, and became widely applicable starting from the breakthrough 1969 paper of C. A. R. Hoare.

For example, if you have a program with a loop, you may have a loop invariant and you may be able to prove that this invariant is really preserved from iteration to iteration and thus, that the program does compute what it is supposed to compute. For example, in a program that computes the sum $\sum_{i=1}^{n} a_i$ by starting with $sum \leftarrow 0$ and adding a_1 on the first iteration, a_2 on the second iteration, ..., and a_n on the n-th iteration, the i-th iteration is $sum \leftarrow sum + a_i$. After i-th iteration, we get $sum = a_1 + \ldots + a_i$. One can easily prove that if this equality is satisfied before an iteration, it will also be true after the iteration. Thus, after n iteration, we indeed get the desired sum.

Alas, these methods do not work for more complicated programs.

For more complicated programs, it is difficult to decide when to stop testing. For *complicated* programs it is much more *difficult* to decide when to stop testing, since it is extremely difficult (and often practically impossible) to prove formally the correctness of such a program.

A reasonable idea is to stop testing when several tests in a row do not discover any more bugs. However, to rely on such a criterion would be sometimes misleading. It is often practically impossible to perform a fully exhaustive test, and therefore failures often occur during the customer's usage, long after the product has been tested and released.

Moreover, the experience of many programmers shows that it is actually *impossible* to extract all the faults from system-type software, that is commonly involved in resource contention, or from the programs with a sophisticated user interface. So if such a program works fine it simply means that we have not yet reached the point where it will start erring.

For example, Robert L. Glass, the editor-in-chief of the *Systems and Software* journal, in his bestseller book *Software conflict. Essays on the art and science of software engineering* (Yourdon Press, Englewood Cliffs, NJ, 1991) summarizes (on pp. 33–36) his *"over 35 years of experience in the industrial and academic worlds of software engineering"*, and concludes that *"not all software errors are found EVER"* and so *"we cannot remove all errors from software"*. And he is not alone in this opinion: this viewpoint is widely spread in the literature and is also shared by the users who still find bugs in the old compilers, operating systems, and in other software.

The only thing that we can do is to estimate the time before a next failure. If we take seriously this wide-spread opinion that it is impossible to eliminate all the errors, then when do we stop testing?

Whenever we stop there will be some bugs remaining, so we can never guarantee that the released product will *always* work fine. The only thing that we can try to guarantee is that it will work fine *for some* given *period of time*. For example:

- If we are launching an automated mission to another planet, then we would like to be able to guarantee that the software work correctly during the entire mission (i.e., for several years).

- On the other extreme, if we are preparing a new operating system for beta testing, we do not want to wait until all bugs are already eliminated: finding these bugs is, after all, one of the main purpose of beta-testing. The only thing that we need to guarantee is that the system more or less works, with some reasonable limit on possible faults (like no more than one fault a day or no more than one fault a week) that would prevent the testers from being too irritated.

So, in order to decide when to stop testing, we must:

- fix some desired time interval T_0 and

- stop testing when the program is guaranteed to work fine during this time interval.

In order to do this we must be able to estimate the time interval during which the remaining errors will not influence the program.

3.3. Existing methods of software testing

Software testing: the basic idea. How can we predict the time during which the program will work fine? This can be done in the same way as any other prediction:

- we try to find the *equations* that describe the changes;

- when we extract the possible equations from the existing data, we *check* the corresponding hypotheses on the existing data;

  - if this hypothesis is disproved by the other data, we go back to step 1 and try to find a better hypothesis;

  - if a hypothesis is confirmed by the other data, we use this hypothesis to make predictions.

(This is what high school science textbooks call the *scientific method*.)

For software testing, we want to predict the time of the next program fault. To do that, it is natural to use the information about the times of the previous program faults. The software debugging process is usually well documented, and therefore, we know the moments of time $t(1), t(2), ..., t(m)$ at which the previous faults occurred and when, as a result of these faults, the first, second, ..., m-th bug were discovered. So, all we need to do is to:

- find a reasonable formula $f(n)$ for the dependency of $t(n)$ on n that is consistent with this data (i.e., for which $f(n) \approx t(n)$), and

- use this formula to predict the expected time $f(m+1) - f(m)$ between the last detected (m-th) fault and the next fault.

This means that in its present state, the program is expected to work fine for the time $f(m+1) - f(m)$.

- If this expected period is shorter than the period of time T_0 during which we want the program to work fine (i.e., if $f(m+1) - f(m) < T_0$), this means that the program is not yet ready for release, and needs to be tested further.

- If, on the other hand, the expected period of good performance is equal to (or greater than) the required time T_0 ($f(m+1) - f(m) \geq T_0$), we can release the program.

What we have described is just the basic idea, not the full picture. We are trying to describe the *basic idea*, not its actual implementation. Let us give two examples of important aspects of program testing that we did not take into consideration:

- We only talked about the *average* predicted time $f(n) \approx t(n)$. In addition to that, we can also try to predict the *probabilities* of different deviations $f(n) - t(n)$ from this "average" model. If we know these probabilities, then we can be more precise in guaranteeing that the *actual* time $t(m+1) - t(m)$ before the next fault exceed T with a certain probability p_0.

   Many existing models of software behavior actually take these probabilities into consideration.

- The above-described approach gives the reliability estimates that do not take into consideration the environment in which the analyzed software will be used.

   There exist more accurate estimates that take this environment into consideration.

What do we need to do to implement this idea? To implement the above-described basic idea, we need to choose a formula $f(n)$ that fits the experimental data. Usually, to fit the experimental data, we:

- choose an expression $f(n, C_1, \ldots, C_p)$ with several parameters $C_1, \ldots, C_p$, and

- find the values of these parameters for which for all $n = 1, \ldots, m$, the values $f(n, C_1, \ldots, C_p)$ predicted by the chosen formula, are the closest to $t(n)$, i.e., for which $f(n, C_1, \ldots, C_p) \approx t(n)$ for $n = 1, \ldots, m$.

Then the value $f(n, C_1, \ldots, C_p)$ for these values C_i will be used for prediction.

Let us describe these two problems one by one.

In general, we can use different models for prediction. In general, we can try different formulas for prediction:

- For *example*, we may try to assume that the dependence of t on n is *linear* (i.e., the simplest possible), and take, as $f(n, C_1, C_2)$, the general expression for a linear function: $f(n, C_1, C_2) = C_1 + C_2 \cdot n$ with unknown parameters C_1 and C_2.

- If none of the linear functions provides a good fit for the experimental data, we can try *quadratic* functions $f(n; C_1, C_2, C_3) = C_1 + C_2 \cdot n + C_3 \cdot n^2$, etc.

- Instead of using the polynomial functions, we can try some more complicated dependencies.

In program testing, standard prediction models do not work well. In many physical situations, *linear* functions work well; however, in program testing, a linear function, usually, does not give a good description of the faults' timing.

Indeed, what would that mean that $t(n)$, i.e., the total amount of faults uncovered by time n, grows linearly with n, as $t(n) = C_1 + C_2 \cdot n$? It means that we debug and debug and we still get the same amount of faults per time unit as before. No matter how many bugs we correct, the predicted time $t(m + 1) - t(m)$ between the next correction will remain the same and never grow to the required value T_0:

$$t(m + 1) - t(m) = (C_1 + C_2 \cdot (m + 1)) - (C_1 + C_2 \cdot m) = C_2.$$

The program will never be ready for a release, and we can as well throw it away.

Quadratic functions are not that automatically bad, but the traditional motivation for quadratic formulas in science and engineering is that linear formulas usually work *well*, but *not perfectly* well, so we need a small adjustments to the linear formulas. It is natural to take a quadratic terms as an adjustment to the linear formula. Since for program testing, linear formulas do not work well at all, there is no specific reason in using quadratic formulas either.

So, for program testing, instead of more traditional linear, quadratic, and other similar formulas, we need to use some specific formulas $f(n, C_1, \ldots, C_p)$. Such formulas are used to predict how reliable the software is, and are, therefore, called *software reliability models*.

Main existing models of software reliability: briefly. Currently, many different software reliability models are in use. Most of these models are described in detail e.g., in a monograph *"Software reliability: measurement, prediction, application"* by J. D. Musa, A. Iannino, K. Okumoto (Mc-Graw Hill Co., 1987).

This book also contains the results of the experimental comparison of different models. The authors used different criteria for their choice:

- computational simplicity,

- predictive quality,

- etc.

It turns out (see Section 13.5 of Musa's book) that with respect to all these natural criteria two models are superior:

- the *basic* Poisson execution time model (proposed first in J. D. Musa in 1975) and

- the *logarithmic* Poisson execution time model (proposed first by J. D. Musa and K. Okumoto in 1984).

Both models describe failures as a random process. In particular, for the expected total number of failures $n(t)$ that have occurred before time t, these models give the following formulas:

- in the *basic* model, $n(t) = n_0(1 - A \cdot \exp(-kt))$ (for some constants n_0, A, and k);

- in the *logarithmic* model, $n(t) = a\ln(1 + bt)$ for some constants a and b (the very name of this model, *logarithmic* model, actually comes from this formula).

Comment. The parameters n_0 and k of the basic model have a clear intuitive meaning:

- n_0 is the total number of bugs in the program.

  Indeed, when time t tends to ∞, the total number of uncovered bugs tends to n_0.

- k describes our ability to discover the bugs.

  Indeed, if we increase k, then $n(t)$ approaches n_0 faster.

Comment. After the experimental comparison of various models, whose results are given in the above-mentioned 1987 monograph, several others researchers undertook similar comparisons. Some of them came out with different results, namely that in their experiments some other methods turned out to be better than both the basic and the logarithmic models. (Actually on practically every conference several papers with competing experiments appear). However, to the best of our knowledge, the amount of experimental data and varieties of data covered by the 1987 book, is still greater than by any of the competing experiments. Therefore, we used Musa's results as the most general ones.

The main difference between the two models. There is an important difference between these two models:

- In the *basic* model, as $t \to \infty$, the total amount of uncovered bugs tends to a constant (to n_0). This means that in this model, a program has finitely many bugs, and, in principle, it is quite *possible to debug it completely* and make it error-free.

  This is typical for a relatively *small* program.

- In the *logarithmic* model, as $t \to \infty$, the total amount of uncovered bugs tends to infinity. This means that no matter how many bugs we have uncovered, *new faults are inevitable.* The average intervals between the two faults increases, but still, the program is never error-free.

This behavior is typical for *complicated* software, especially for software that is interacting with other programs, such as operating systems, networking software, etc.

Models reformulated. Both basic and logarithmic model are described in terms of the dependency $n(t)$. This is a convenient way to describe a random process, but for our purposes, we need the *inverse* dependence: of t on n. To get the dependence of t on n, we can "invert" the above formulas and get the dependency of the average time $\bar{t}(n)$ of n-th failure on n:

- In the *basic* model it is $\bar{t}(n) = (-1/k) \cdot \ln((n_0 - n)/A)$.

- In the *logarithmic* model it is $\bar{t}(n) = (1/b) \cdot (\exp(n/a) - 1)$.

Two possible uses of these models. Let us first assume that we have already found the values of the parameters that describe the actual program debugging. How can we then use these models?

- First, we can check whether the software is ready for the release.

- Second, if the program is not yet ready for release, we can estimate how much more time is needed for the further debugging.

Let us consider these two tasks.

Checking whether a program is ready for the release.

- If we are dealing with the *basic* model, in which it is possible to delete all the bugs and make the program error-free, then, ideally, we should do it. In this sense, the program is ready for release if the total number m of already un-covered bugs m is equal to the expected amount of bugs n_0 in this program (*reminder*: we are assuming that we have already estimated the parameters of the model). If $n < n_0$, then further debugging is necessary.

   In real life, we are often pressed for time; in this case, it may be necessary to release a program with a few bugs possibly remaining. We can consider the program ready for release if $f(m+1) - f(m) \geq T_0$, i.e., if

$$\left(-\frac{1}{k}\right) \cdot \ln\left(\frac{n_0 - (m+1)}{A}\right) - \left(-\frac{1}{k}\right) \cdot \ln\left(\frac{n_0 - m}{A}\right) \geq T_0.$$

- For a *logarithmic* model, as we have already mentioned, it is simply impossible to get the program completely error-free. So, what we can hope for is that the time interval $t(m+1) - t(m)$ between the last uncovered (m-th) fault and the next ($(m+1)$-st) fault will be at least as large as the required time T_0. In this case, we release the program if

$$\frac{1}{b} \cdot (\exp((m+1)/a) - \exp(m/a)) \geq T_0.$$

Example. Suppose that the experimental data on the fault times is consistent with the logarithmic model with $a = 1$ and $b = 2$ (in other words, $n(t) \approx 1 \cdot \ln(1 + 2t)$), and we want the program to run error-free for a year $T_0 = 365$). Then, we can make the following conclusions:

- After the first *day* of testing ($t = 1$), we have uncovered $\approx \ln(3) \approx 1$ bug. The expected time before the next bug is

$$(1/2) \cdot (\exp(2) - \exp(1)) \approx 2 \ll 365.$$

   This is a very reasonable conclusion: if we have uncovered a bug in the very first day of testing, it is hardly probable that after that, the program will work well for the whole year.

- After the first *week* of testing ($t = 7$), we have uncovered $\approx \ln(31) \approx 3$ bugs. The expected time before the next bug is

$$(1/2) \cdot (\exp(4) - \exp(3)) \approx 20 \ll 365.$$

   This conclusion is also very much consistent with common sense: if we have been un-covering, on average, about a bug every two days, there is a low chance that the program will work well for a year.

- After the first *month* of testing ($t = 30$), we have uncovered $\approx \ln(61) \approx 4$ bugs. The expected time before the next bug is

$$(1/2) \cdot (\exp(5) - \exp(4)) \approx 60 < 365.$$

- After the first *quarter* of testing ($t = 90$), we have uncovered $\approx \ln(181) \approx 5$ bugs. The expected time before the next bug is

$$(1/2) \cdot (\exp(6) - \exp(5)) \approx 130 < 365.$$

- After the first *year* of testing ($t = 365$), we have uncovered $\approx \ln(731) \approx 7$ bugs. The expected time before the next bug is

$$(1/2) \cdot (\exp(8) - \exp(7)) \approx 500 \geq 365.$$

This means that the software is finally ready for release.

Number of days of testing	Number of bugs uncovered	Expected time before the next bug
1	1	2
7	3	≈ 20
30	4	≈ 60
90	5	≈ 130
365	7	≈ 500

Comment. Please keep in mind that we are using the *simplified* formulas. For real-life applications, more complicated formulas are needed.

Predicting the remaining debugging time.

- In the *basic* model, we want to predict the time t when

$$n(t) = n_0 \cdot (1 - A\exp(-kt)) \geq n_0 - 1/2.$$

This would mean that the *predicted* number of uncovered bugs is greater than or equal to $n_0 - 1/2$, and therefore, that the *actual* number of uncovered bugs will be most probably equal to their total number n_0.

Let us use the above inequality to find the desired time t. The unknown t is only in the left-hand side, and the only problem with it is that t is not explicitly given in the left-hand side. So, to extract an inequality for t, we must apply some equivalent transformations to both sides of this inequality so that we will simplify the left-hand side until it is simply equal to t. The first possible simplification is that we can get rid of n_0 in the left-hand side. To do that, we divide both sides of the above inequality by n_0, and get

$$1 - A\exp(-kt)) \geq 1 - 1/(2n_0).$$

To get rid of 1 in the left-hand side, we subtract both sides of this inequality from 1, and conclude that

$$A\exp(-kt) \leq 1/(2n_0).$$

We can further simplify this inequality by eliminating A; for that, we must divide both sides by A:

$$\exp(-kt) \leq 1/(2n_0 \cdot A).$$

As a further simplification, we apply logarithm to both sides to eliminate exp in the left-hand side:

$$-kt \leq \ln(1/(2n_0 \cdot A)).$$

The logarithm of a product is known to be equal to the sum of logarithms, so,

$$-kt \leq -(\ln(2) + \ln(n_0) + \ln(A)).$$

To get an inequality with t as the left-hand side, we must make a final transformation: divide both sides by the negative number $-k$. As a result, we get the following inequality:

$$t \geq \frac{1}{k} \cdot (\ln(2) + \ln(n_0) + \ln(A)).$$

Thus, if we are currently at the moment of time t_0, then the remaining debugging time T_{rem} can be determined from the requirement that the above inequality be true for $t = t_0 + T_{\text{rem}}$. The smallest T_{rem} for which this inequality is true is:

$$T_{\text{rem}} = \frac{1}{k} \cdot (\ln(2) + \ln(n_0) + \ln(A)) - t_0.$$

■ In the *logarithmic* model, we want to find t for which

$$t(n+1) - t(n) \geq T_0.$$

To simplify the desired inequality, let us express

$$t(n+1) = (1/b) \cdot (\exp((n+1)/a) - 1)$$

in terms of $t(n) = (1/b) \cdot (\exp(n/a) - 1)$. If for some n, we know $t(n) = (1/b) \cdot (\exp(n/a) - 1)$, then we can determine $\exp(n/a)$ as $b \cdot t(n) + 1$. Then, $\exp((n+1)/a) = \exp(n/a) \cdot \exp(1/a)$. If we now the expression for $e^{n/a}$ in terms of $t(n)$, we can conclude that

$$t(n+1) = \frac{1}{b} \cdot \left(\exp(1/a) \cdot (b \cdot t(n) + 1) - 1 \right) =$$

$$\exp(1/a) \cdot t(n) + \frac{\exp(1/a) - 1}{b}.$$

Thus, the condition $t(n+1) - t(n) \geq T_0$ is first satisfied at a time $t = t(n)$ for which

$$\exp(1/a) \cdot t + \frac{\exp(1/a) - 1}{b} - t = T_0,$$

i.e., for

$$t = \frac{T_0}{\exp(1/a) - 1} - \frac{1}{b}.$$

At any given moment of time t_0, the remaining debugging time T_{rem} can be now estimated as $t - t_0$, i.e., as

$$T_{\text{rem}} = \frac{T_0}{\exp(1/a) - 1} - \frac{1}{b} - t_0.$$

Example. If we are at time $t_0 = 30$ (after a month of debugging), $a = 1$, $b = 2$, and we want the program to work for a year $T_0 = 365$), then we need at least

$$T_{\text{rem}} = \frac{360}{\exp(1) - 1} - 1/2 - 30 \approx 170$$

days for testing. In other words, at the current rate, we need at least a half-year more of testing if we want a year of fault-free work.

Comments.

- If this half-year period is too long, we may want, e.g., to add other people to the testing team and thus, increase the rate of testing.

- This is one more example of how continuous mathematics is helpful for computer science: in order to predict the values of an integer-valued function $t(m)$ of an integer argument m, we use expressions in which the parameters can take arbitrary real values.

Exercise

3.1 Suppose that we are debugging a complicated system software (and that, therefore, we are dealing with a logarithmic model). Suppose that based on the debugging history, we have estimated the parameters of this model as $a = 1.6$ and $b = 5.0$. Estimate the remaining debugging time if we have already been debugging for two months $t_0 = 60$), and we want the program to work for two years.

3.4. How do we determine the parameters of the model?

In the above formulas, we assumed that we have already determined the values of the model's parameters. The question remains: how do we determine them?

General idea. From the debugging protocol, we know the times $t(1), \ldots, t(m)$ when we un-covered, correspondingly, the 1-st, the 2-nd, ..., and the m-th faults. If we use a software reliability model $t(n) \approx f(n, C_1, \ldots, C_p)$ with unknown parameters $C_1, \ldots, C_p$, then we must determine the values $C_1, \ldots, C_p$ of these parameters from the condition that

$$t(1) \approx f(1, C_1, \ldots, C_p);$$

$$\ldots$$

$$t(m) \approx f(m, C_1, \ldots, C_p).$$

Example. For example, for a logarithmic model, we must determine the coefficients a and b from the following conditions:

$$t(1) \approx (1/b) \cdot (\exp(1/a) - 1);$$

$$\ldots$$

$$t(m) \approx (1/b) \cdot (\exp(m/a) - 1).$$

Geometric re-formulation of the problem. This problem can be naturally reformulated in geometric terms if we plot the values (n, t) on a 2-D graph. In terms of this graph, the problem means that we must find a curve of the type $t = f(n, C_1, \ldots, C_p)$ that goes through (or near) all m experimental points $(1, t(1)), \ldots, (m, t(m))$.

How can we do it? To answer this question, let us formulate the general problem of determining the parameters from the experimental data.

A general problem of determining the parameters: formulation. The above problem is a special case of the general problem of determining parameters from the experimental data:

- we know that the dependency of some quantity y on the quantities $x_1, \ldots, x_n$ is described by a formula $y = f(x_1, \ldots, x_n, C_1, \ldots, C_p)$ for some (*a priori* unknown) unknown values of $C_1, \ldots, C_p$;

- in m different situations ($1 \le i \le M$), we know the values $x_1^{(i)}, \ldots, x_n^{(i)}$ of the parameters $x_1, \ldots, x_n$, and the corresponding (approximate) values $y^{(i)}$ of y;

- we must estimate the values of the parameters C_j.

The simplest case is when the dependence of the function $f(x_1, \ldots, x_n, C_1, \ldots, C_p)$ on the parameters C_j is *linear*, i.e., when

$$f(x_1, \ldots, x_n, C_1, \ldots, C_p) =$$

$$f_0(x_1, \ldots, x_n) + C_1 \cdot f_1(x_1, \ldots, x_n) + \ldots + C_p \cdot f_p(x_1, \ldots, x_n).$$

This is true in most practical examples. In software reliability models, $n = 1$, $x_1^{(i)} = i$, $y^{(i)} = t(i)$, and the dependence of y on C_j is *non-linear*.

A general problem of determining the parameters: main idea. For each set of values $C_1, \ldots, C_p$, the quality of the corresponding approximation is describe by m differences

$$e_i = y^{(i)} - f(x_1^{(i)}, \ldots, x_n^{(i)}, C_1, \ldots, C_p).$$

The main idea of determining the values C_j can be formulated as follows:

The closer each of these differences to 0, the better the approximation.

How can we express this idea in formal terms?

- This idea means, in particular, that if for some set of values $C_1, \ldots, C_p$, *each* difference e_i *is closer to 0* than the corresponding difference e_i' obtained by using some other set of parameters $C_1', \ldots, C_p'$, then the set $C_1, \ldots, C_p$ is clearly better than the set $C_1', \ldots, C_p'$.

- But what can we conclude in a realistic situation when:

  – for some i, the values e_i are better, while

  – for some other i, the values e_i' are better?

Which of the sets of values is then better?

To compare such sets of values, we need to have a *single* numerical criterion $J(e_1, \ldots, e_m)$ that incorporates all the differences $e_1, \ldots, e_m$. If we have such a criterion, then we need to find the values $C_1, \ldots, C_m$ for which this criterion takes the smallest possible value: $J \to \min$.

In search of a criterion: motivations for the least squares method. There are different ways to combine the differences e_i into a single criterion $J(e_1, \ldots, e_m)$. It is reasonable to choose a function J for which the resulting optimization problem is the *simplest possible*. Let us show, step-by-step, that this idea leads to a very specific form of the function J.

1. When the function J is differentiable, the optimization problem $J \to \min$ is the easiest to solve: In this case, the desired values are the values for which all p partial derivatives $\partial J / \partial C_j$ are equal to 0. Therefore, we will assume that the function J is indeed *differentiable*.

2. For a differentiable function J, to find the values $C_1, \ldots, C_p$, we must solve a system of p equations

$$\frac{\partial J}{\partial C_j} = 0, \quad j = 1, \ldots, p$$

with p unknown $C_1, \ldots, C_p$. Out of all possible systems of equations, systems of *linear* equations are the easiest to solve. Thus, it is desirable to choose the function J in such a way that all its derivatives are linear functions of C_j (at least for the case when the function f itself is a linear function of C_j).

   For the derivatives to be linear, the function J itself has to be *quadratic*. Thus, we will consider quadratic functions.

3. The general form of a quadratic function is:

$$J(e_1, \ldots, e_m) = J_0 + J_1 \cdot e_1 + \ldots + J_m \cdot e_m + \sum_{i=1}^{m} \sum_{j=1}^{m} J_{ij} \cdot e_i \cdot e_j.$$

   The function J should achieve its smallest value when $e_1 = \ldots = e_m = 0$. Thus, in the point $e_1 = \ldots = e_m = 0$, all partial derivatives $\partial J / \partial e_i = J_i$ should be equal to 0. Hence,

$$J(e_1, \ldots, e_m) = J_0 + \sum_{i=1}^{m} \sum_{j=1}^{m} J_{ij} \cdot e_i \cdot e_j.$$

4. Since the function J is only used for minimization, we can as well replace it with a function $J'(e_1, \ldots, e_m) = J(e_1, \ldots, e_m) - J_0$ that attains minimum for exactly the same values. Thus, we can assume, without loss of generality, that $J_0 = 0$ and

$$J(e_1, \ldots, e_m) = \sum_{i=1}^{m} \sum_{j=1}^{m} J_{ij} \cdot e_i \cdot e_j.$$

5. If we simply change the sign of one of the differences, i.e., replace e_i by $e'_i = -e_i$, then all the values e'_i are still as close to 0 as before. Therefore, the overall quality $J(e_1, \ldots, e_n)$ of the approximation should not change:

$$J(e_1, \ldots, e_{i-1}, -e_i, e_{i+1}, \ldots, e_n) = J(e_1, \ldots, e_{i-1}, e_i, e_{i+1}, \ldots, e_n).$$

If we substitute the expression $\sum \sum J_{ij} \cdot e_i \cdot e_j$ into both sides of this equality, then, from the equality of the two quadratic forms, we will conclude that their coefficients must coincide. Thus, for every $j \neq i$, $-J_{ij} = J_{ij}$, and therefore, $J_{ij} = 0$. Hence, the above quadratic form takes the form

$$J(e_1, \ldots, e_m) = \sum_{i=1}^{m} J_{ii} \cdot e_i^2.$$

6. There is no *a priori* reason why we should take different differences with different weights; therefore, the weights J_{ii} should all be equal to the same number (in particular, to J_{11}). Hence,

$$J(e_1, \ldots, e_m) = J_{11} \cdot \sum_{i=1}^{m} e_i^2.$$

7. Again, the only reason why we use the function J is that we want to minimize it. If we multiply a function by a positive constant, it will still attain its minimum at the same point as before. Thus, minimizing J is the same as minimizing J/J_{11}. Thus, without loss of generality, we can assume that

$$J(e_1, \ldots, e_m) = \sum_{i=1}^{m} e_i^2.$$

In other words, it is reasonable to choose the values C_j for which *the sum of the squares of differences e_i is the smallest possible*. This approach to finding the values C_j is called the *least squares method*.

Historical comment. The least squares method was *originally* proposed by the famous 19th century mathematician Karl Friedrich Gauss for situations in which the errors e_i are normally distributed *random variables*. At present, this method is also widely used in practice in cases when (like for program debugging) we do not know the probabilities of different values of e_i.

Pedagogical example: Least squares method for a simple linear model. Let us describe how the least squares method works for a simple linear model, in which $n = 1$, and $y = C_1 + C_2 \cdot x$. In this case, $e_i = y^{(i)} - C_1 - C_2 \cdot x^{(i)}$, and the least squares method for determining the values C_1 and C_2 takes the following form:

$$\sum_{i=1}^{m} (y^{(i)} - C_1 - C_2 \cdot x^{(i)})^2 \to \min_{C_1, C_2}.$$

The minimum can be found from the condition that both partial derivatives of the minimized function must be equal to 0:

$$\frac{\partial}{\partial C_1} \sum_{i=1}^{m} (y^{(i)} - C_1 - C_2 \cdot x^{(i)})^2 = \sum_{i=1}^{m} 2(y^{(i)} - C_1 - C_2 \cdot x^{(i)})(-1) = 0;$$

$$\frac{\partial}{\partial C_2} \sum_{i=1}^{m} (y^{(i)} - C_1 - C_2 \cdot x^{(i)})^2 = \sum_{i=1}^{m} 2(y^{(i)} - C_1 - C_2 \cdot x^{(i)})(-x^{(i)}) = 0.$$

From the first equation, we conclude that

$$C_1 \cdot m + C_2 \cdot \sum_{i=1}^{m} x^{(i)} = \sum_{i=1}^{m} y^{(i)}.$$

From the second equation, we conclude that

$$C_1 \cdot \sum_{i=1}^{m} x^{(i)} + C_2 \cdot \sum_{i=1}^{m} (x^{(i)})^2 = \sum_{i=1}^{m} x^{(i)} \cdot y^{(i)}.$$

Thus, we get a system of two linear equation with two unknowns C_1 and C_2, from which we can easily determine both C_1 and C_2.

Least squares method for software reliability models. If we use the least squares method, then:

■ For the *logarithmic* model, we must find the parameters a and b from the condition that

$$\sum_{i=1}^{m} (t(i) - (1/b) \cdot (\exp(i/a) - 1))^2 \to \min_{a,b}.$$

■   For the *basic* model, we must find the parameters k, n_0, and A from the
    condition that

$$\sum_{i=1}^{m}(t(i) - (-1/k) \cdot \ln((n_0 - n)/A))^2 \to \min_{k,n_0,A}.$$

Since the models are non-linear, the functions that we need to minimize are
not quadratic. Optimization of such functions is a rather complicated computational problem. To minimize these functions, we cannot simply use simple
analytical methods; we must use (more complicated) optimization software
tools.

Exercise

3.2 Use the least squares method to find the parameters C_1 and C_2 of the best
     linear dependence $y = C_1 + C_2 \cdot x$, assuming that the measurement results
     lead to the following four points:

$$x^{(1)} = 0.0, \quad y^{(1)} = 3.1;$$

$$x^{(2)} = 1.0, \quad y^{(2)} = 4.9;$$

$$x^{(3)} = 0.0, \quad y^{(3)} = 7.1;$$

$$x^{(4)} = 0.0, \quad y^{(4)} = 8.9.$$

4

OPTIMAL PROGRAM TESTING

In this lesson, we re-formulate the problem of program testing as a mathematically precise optimization problem, and solve the simplest case of the corresponding optimization problem.

4.1. How to choose a software reliability model?

In principle, different software reliability models can be used. It is known that often, the use of different models leads to different quality of the resulting predictions. Therefore, it is important to choose the *right* software reliability model.

Currently, this choice is mainly made largely *ad hoc*, at best, by testing a few possible models and choosing the one that performs the best on a few benchmarks. Since only a few models are analyzed, we are not sure that we did not miss the real good model. (And since only a few benchmarks are used for comparison, we are not sure that the chosen family is indeed the best one.) It is, therefore, desirable to find the *optimal* software reliability model.

In this lesson, we will:

- first, *formulate the problem* of choosing the "best" software reliability model (for approximating the expected time of n−th failure) as an optimization problem, and

- then, *solve this problem*.

65

As a result, we will explain why the existing models work fine.

Comment. We want to warn the readers that the main goal of this chapter is to describe new ideas, and not the solution that would always work. In spite of the visible progress, the problem of testing software has not yet achieved a complete solution.

For example, in these two lessons, we consider an "optimistic case", when every bug is eventually discovered, and as a result, the program runs better and better, and fewer and fewer failures occur. In this case, the main problem is when to stop testing software. However, every programmer working in a real world knows that in many practical cases, the situation is not that rosy. E.g., in some cases, in spite of all the debugging efforts, failures continue to occur with a non-decreasing rate. In this case, the problem is *not when* to stop, but *how to change* debugging so that the program will start working better (and is this program worth working at at all).

4.2. Auxiliary (minor) problem: how to count the errors

There are several aspects of this problem.

■ First, do we count the bugs discovered before we initially executed a program on a computer?

  – Some programmers prefer to scrutinize the code first and find lots of errors *before* touching a terminal.

  – Other programmers type the raw code in and start debugging on the computer.

Since the usual statistical procedures use only the bugs that were discovered during the computer run, these procedures will *underestimate* the total number of bugs for the programmers who do some preliminary code analysis and thus end up with the *biased* predictions of the program's reliability.

This problem is not easy to deal with, because it is difficult to force a programmer to write down all the errors he revealed in his mind, (before touching the computer). Moreover, even if we succeed in forcing him to do it, this additional activity will seriously slow him down and will certainly badly interfere with his ability of creative thinking.

So what we need is not an *administrative* solution to this problem, but some kind of a *mathematical* solution that will somehow take care of these possible additional revealed bugs without actually counting them (i.e., in some indirect way).

- Second, how do we count bugs:

  - do we take one failure for one bug or
  - do we count the number of lines that demanded a change?

Or do we somehow else take into consideration the fact that, citing Glass's book again, *"errors are not equal"*. If we count just failures, then we can count them automatically, but we lose some essential part of the information. This problem is discussed in detail in Section 9.1.4 of Musa's 1987 monograph. However, we do not have the feeling that this problem has been completely and satisfactorily solved.

4.3. Towards a formal formulation of the optimization problem

Basic idea. Our goal is to choose the family of functions $f(n, C_1, \ldots, C_p)$ that is the best software reliability model. As soon as a family is chosen, the answer to the question "when to stop testing?" is as follows:

- first, based on the known values $t(1), \ldots, t(m)$, we find the values of the parameters $C_1, \ldots, C_p$ for which $t(n) \approx f(n, C_1, \ldots, C_p)$ for all n;

- for these $C_1, \ldots, C_p$, we produce $f(m + 1, C_1, \ldots, C_p) - t(m)$ as the guaranteed time during which the software will function properly.

  - If this time exceeds the desired value T_0, then we stop testing, else, we continue testing.
  - In case we continue testing, we can estimate the necessary additional testing time as follows:
    * estimate the required number of uncovered bugs N as the the first integer N for which

    $$f(N + 1, C_1, \ldots, C_p) - f(N, C_1, \ldots, C_p) \geq T_0;$$

* estimate the expected time of discovering N-th bug as $f(N, C_1, \ldots, C_p) - t(m)$;
* estimate the necessary additional testing time as the difference $f(N, C_1, \ldots, C_p) - t(m)$ between the moment of time $f(N, C_1, \ldots, C_p)$ when the testing will be completed, and the current moment of time $t(m)$.

We must choose a family of functions. Before we start discussing what is the best choice (and what do we mean by "the best") let's first make the following remark (that will later on prove to be helpful).

■   Suppose that a function $f_0(n)$ describes the results $t(n)$ of some actual debugging process performed by an actual programmer (i.e., $t(n) \approx f_0(n)$ for all n). As Robert Glass remarks in his 1991 book (that we cited in Lesson 3), *"not all software error finders are equal"*. The differences range in magnitude up to 30 : 1. So for another programmer, who is c times faster, $c \cdot f(n)$ will be a good fit.

■   In addition to different debugging rates, programmers also start debugging at different times. If a programmer started debugging d days later, then for this programmer, a good fit will be not $c \cdot f_0(n)$, (where c is his relative rate of debugging), but $c \cdot f_0(n) + d$.

Therefore, it is reasonable to demand that if a function $f_0(n)$ is a good fit (i.e., belongs to the desired family of functions $f(n, C_1, \ldots, C_p)$), then all the functions $d + c \cdot f_0(n)$ (for $c > 0$) must also belong to this family.

This family of functions must, therefore, depend on at least two parameters ($p \geq 2$). The smallest possible number of parameters is two, when the family consists of the functions $d + c \cdot f_0(n)$ for different c, d and some fixed function $f_0(n)$.

In the following test, we will consider such simplest 2-parameter families.

What is a criterion for choosing a family of functions? What does it mean to choose a *best* family of functions? It means that we have some *criterion* that enables us to choose between the two families.

This criterion can describe different properties of the software reliability model, such as:

- its computational simplicity;

- its prediction ability;

- etc.

Most frequently, optimality criteria are *numerical*, when to every alternative F, we assign some value $J(F)$ expressing its performance, and choose a family for which this value is maximal (i.e., when $J(F) \geq J(G)$ for every other alternative G). However, it is not necessary to restrict ourselves to such numeric criteria only.

For example, if we have several different families F that have the same prediction ability $P(F)$, we can choose between them the one that has the minimal computational complexity $C(F)$. In this case, the actual criterion that we use to compare two families is not numeric, but more complicated:

A family F_1 is better than the family F_2 if and only if

  – either $P(F_1) > P(F_2)$,

  – or $P(F_1) = P(F_2)$ and $C(F_1) < C(F_2)$.

A criterion can be even more complicated.

The only thing that a criterion *must* do is to allow us, for every pair of families (F_1, F_2), to make one of the following conclusions:

- the first family is better with respect to this criterion (we'll denote it by $F_1 \succ F_2$, or $F_2 \prec F_1$);

- with respect to the given criterion, the second family is better ($F_2 \succ F_1$);

- with respect to this criterion, these families have the same quality (we'll denote it by $F_1 \sim F_2$);

- this criterion does not allow us to compare the two families.

Of course, it is necessary to demand that these choices be consistent.

For example, if $F_1 \succ F_2$ and $F_2 \succ F_3$ then $F_1 \succ F_3$.

The criterion must be final, i.e., it must pick the unique family as the best one. A natural demand is that this criterion must choose a *unique* optimal family (i.e., a family that is better with respect to this criterion than any other family).

The reason for this demand is very simple:

■ If a criterion *does not choose* any family at all, then it is of no use.

■ If *several* different families are the best according to this criterion, then we still have a problem to choose among those best. Therefore we need some additional criterion for that choice, like in the above example:

> If several families $F_1, F_2, \ldots$ turn out to have the same prediction ability ($P(F_1) = P(F_2) = \ldots$), we can choose among them a family with minimal computational complexity ($C(F_i) \to \min$).

So what we actually do in this case is abandon that criterion for which there were several "best" families, and consider a new "composite" criterion instead: F_1 is better than F_2 according to this new criterion if:

− either it was better according to the old criterion,

− or they had the same quality according to the old criterion and F_1 is better than F_2 according to the additional criterion.

In other words:

> **If**
>
>> a criterion does not allow us to choose a unique best family,
>
> **then**
>
>> it means that this criterion is not final, we'll have to modify it until we come to a final criterion that will have that property.

This criterion must not depend on the (unknown) number of errors that we revealed before typing the program in. The next natural demand on the criterion is connected with the above-mentioned uncertainty in counting the number of errors.

Suppose that we have two programmers who have the same abilities to debug (so they reveal the same errors in the same order), but:

- the first programmer discovers the first 20 faults in his mind and only then starts testing the program on the computer, while

- the second programmer starts executing his program from the very beginning.

Then:

- what is the error number 3 for the *first* programmer,

- for the second programmer it will be the error number 23 (because the second programmer also counts the 20 errors that the first programmer discovered before he started computer testing).

On the other hand, an error number 4 for the second programmer was discovered by the first programmer before he went to the computer and started computer testing. If the first programmer assigns number 1 to the first error that he discovered during computer testing, then it is natural to assign numbers $0, -1, -2, ..., -19$ to the errors that he discovered before typing the program into the computer. In particular:

- an error number 4 for the second programmer is

- error number -16 for the first one.

Comment. This example shows that:

- although at first glance, the expression $t(n)$ makes sense not only for *positive n*,

- in all the cases when some errors were detected *before* the testing (and it is a very frequent case) it is quite reasonable to consider *negative* values of n as well.

How can we describe this independence formally? Let's denote by s the total number of errors that the first programmer discovered before he started testing on a computer. Then:

- an error number n for a first programmer is

■   error number $n + s$ for the second one.

So if we denote the moment of time, when i-th programmer discovers his error number n, by $t_i(n)$, we can conclude that $t_1(n) = t_2(n + s)$.

It is natural to demand that the fact that a family is better (in some reasonable sense) that some other family should not depend on the number of bugs that were discovered before we started computer testing.

In more formal terms:

If

> a family $\{f(n), g(n), \ldots\}$ is better than some other family $\{h(n), k(n), \ldots\}$:

$$\{f(n), g(n), \ldots\} \succ \{h(n), k(n), \ldots\},$$

then

> the "shifted" family $\{f(n+s), g(n+s), \ldots\}$ is better than the "shifted" family $\{h(n + s), k(n + s), \ldots\}$:

$$\{f(n + s), g(n + s), \ldots\} \succ \{h(n + s), k(n + s), \ldots\},$$

It is also reasonable to demand that the criterion should not depend on how we define a bug. Another reasonable demand is associated with another uncertainty in counting bugs: depending on how detailed we are, we can:

■   either count failures,

■   or count the lines that demanded correction, etc.

Crudely speaking, if we count *lines* of code that need correction instead of counting *failures*, then we count C bugs where we initially counted just one (where by C, we denoted an average number of changed lines per failure). So:

■   what was $t(n)$ for one programmer

- turns out to be $t(Cn)$ for another.

Comment. This transformation means that we have changed a *scale* for measuring bugs just like we change scales from kilograms to pounds: what was 1 kg is now ≈ 2.2 lb; likewise what was 1 bug in the old scale is C bugs in the new one.

It may also seem reasonable to demand that the fact that one of the families is better should be still true if we change the way we count bugs.

This consideration causes an *additional problem*:

- Previously, the number of bugs was always a positive *integer*.

- Now, we must consider "fractions" of bugs, because, e.g., a failure in average means that several lines should be changed, and therefore a line of changed code correspond in average to only part of a failure. In this case what was 1 bug for one programmer is *part of the bug* for another one. So we come to the notion of a *fractional bug* (see, e.g., Musa's 1987 book).

From the mathematical viewpoint it means that in this case, the argument n of the desired function $t(n)$ is not necessarily an integer, as before, but it can be an *arbitrary* real number.

Now we are ready to introduce formal definitions.

4.4. Definitions and the first result

Definition 4.1.

- Let $f_0(n)$ be a function from integers to real numbers. By a *software reliability model (or a model, for short)* that corresponds to this function $f_0(n)$, we mean a family of all functions of the type $d + c \cdot f_0(n)$, where:
  - d is an arbitrary real number and
  - c is an arbitrary positive number
- Two models are considered *equal* if the corresponding families coincide (i.e., the corresponding families consist of the same functions).

Denotation. *Let's denote the set of all possible models by* Φ.

Comment. In order to formalize the notion of an optimality criterion, we must describe that for some pairs of models (F, G), F is better than G, and for some other pairs, F is not better than G. To describe this "relation" *better*, we must, thus, describe the *set* of all possible pairs (F, G) for which F is better than G. In mathematics, if a set X is given:

- the set of all pairs (x_1, x_2) of elements $X_1 \in X$, $x_2 \in X$, is usually denoted by $X \times X$.

- An arbitrary subset R of a set of pairs $X \times X$ is called a *relation* on the set X. If $(x_1, x_2) \in R$, it is said that x_1 and x_2 are in relation R; this fact is denoted by $x_1 R x_2$.

Definition 4.2. *A pair of relations* $(\prec, \sim)$ *on a set* Φ *is called* consistent *if it satisfies the following conditions, for every* $F, G, H \in \Phi$:

(1) *if* $F \prec G$ *and* $G \prec H$ *then* $F \prec H$;

(2) $F \sim F$;

(3) *if* $F \sim G$ *then* $G \sim F$;

(4) *if* $F \sim G$ *and* $G \sim H$ *then* $F \sim H$;

(5) *if* $F \prec G$ *and* $G \sim H$ *then* $F \prec H$;

(6) *if* $F \sim G$ *and* $G < H$ *then* $F < H$;

(7) *if* $F \prec G$ *then it is not true that* $G \prec F$, *and it is not true that* $F \sim G$.

Comment. The intended meaning of these relations is as follows:

- $F \prec G$ means that with respect to a given criterion, G is better than F;

- $F \sim G$ means that with respect to a given criterion, F and G are of the same quality.

Under this interpretation, conditions (1)–(7) have simple intuitive meaning:

(1) if G is better than F, and H is better than G, then H is better than F;

(2) every alternative F is of the same quality as itself;

(3) if G is of the same quality as F, then F is of the same quality as G;

(4) if F is of the same quality as G, and G is of the same quality as H, then F is of the same quality as H;

(5) if G is better than F, and H is of the same quality as G, then H is also better than F;

(6) if H is better than G, and F is of the same quality as G, then H is better than F;

(7) if G is better than F, then F cannot be better than G and F cannot be of the same quality as G.

Definition 4.3. *Assume a set Φ is given. Its elements will be called alternatives.*

■ *By an optimality criterion we mean a consistent pair $(\prec, \sim)$ of relations on the set Φ of all alternatives.*

  – *If $F \succ G$ we say that F is better than G;*

  – *if $F \sim G$ we say that the alternatives F and G are equivalent with respect to this criterion.*

■ *We say that an alternative F is optimal (or best) with respect to a criterion $(\prec, \sim)$ if for every other alternative G either $F \succ G$ or $F \sim G$.*

■ *We say that a criterion is final if there exists an optimal alternative, and this optimal alternative is unique.*

Comment. In this chapter, we will consider optimality criteria on the set Φ of all families.

Definition 4.4. *Let s be an integer.*

■ *By a s-shift of a function $f(n)$ we mean a function $g(n) = f(n + s)$.*

■ *By a s-shift of a family of functions F we mean the family consisting of s-shifts of all functions from F.*

Reminder. We need a shift to describe the possibility that some bugs have been uncovered before going to the computer.

Denotation. *s-shift of a model F will be denoted by $S_s(F)$.*

Definition 4.5. *We say that an optimality criterion on Φ is shift-invariant if for every two families F and G and for every integer s, the following two conditions are true:*

i) *if F is better than G in the sense of this criterion (i.e., $F \succ G$), then $S_s(F) \succ S_s(G)$;*

ii) *if F is equivalent to G in the sense of this criterion (i.e., $F \sim G$), then $S_s(F) \sim S_s(G)$.*

Comment. As we have already remarked, the demands that the optimality criterion is final and shift invariant are quite reasonable. At first glance they may seem rather trivial and therefore weak, because these demands do not specify the exact optimality criterion. However, these demands are strong enough, as the following theorem shows:

Theorem 4.1. *If a model F is optimal in the sense of some optimality criterion that is final and shift-invariant, then it:*

■   *either contains functions $d + c \cdot \exp(kn)$ (that corresponds to the logarithmic model),*

■   *or if consists of functions of the type $d + c \cdot n$.*

Comments.

■   In addition to the logarithmic model we get an additional case, when $t(n) = a + c \cdot n$. This case corresponds to a disastrous situation when the bugs are found again and again with a non-decreasing rate. Such situations happen sometimes; the usual reaction is to throw away this software as non-repairable and write everything anew.

■   Instead of describing which model is optimal with respect to a *certain* criterion, we have actually done much more: we have actually described *all* models that can possibly be the best under *different* reasonable criteria.

Which of these families is the best for each particular criterion, still has to be decided.

- In every *specific* application, if we have a good understanding of the program and of its debugging process, then other models may turn out to be better than the models described in Theorem 4.1.
- However, if we want to choose the best model for *general* use, so that this model will be applicable to any debugging situation, to any starting point for measuring bugs, then it is natural to require shift invariance and thus, for such a general application, one of the models described in Theorem 4.1 will be indeed the best.

4.5. Proof of the first result

This proof is based on the following auxiliary result of independent interest:

Proposition 4.1. *If an optimality criterion is final and shift-invariant then the optimal model F_{opt} is also shift-invariant, i.e., $S_s(F_{opt}) = F_{opt}$ for every integer s.*

Comment. This proposition shows that if we use an optimal approximation family then it does not matter how you count bugs:

- if you start counting them from m-th bug you still get the same model;
- therefore if you use this family for extrapolation, you get the same extrapolation results!

This irrelevance to the choice of the starting point for bugs explains why it is *so difficult* to choose such a point.

On the other hand, what this result says is that there is no need to worry about that: no matter how we count, the reliability estimates will still be the same.

Let's prove the Proposition 4.1.

Proof of Proposition 4.1. Since the optimality criterion is final, there exists a unique model F_{opt} that is optimal with respect to this criterion, i.e., for every other F:

- either $F_{opt} \succ F$

- or $F_{opt} \sim F$.

To prove that $F_{opt} = S_s(F_{opt})$, we will first show that the shifted family $S_s(F_{opt})$ is also optimal, i.e., that for every family F:

- either $S_s(F_{opt}) \succ F$

- or $S_s(F_{opt}) \sim F$.

If we prove this optimality, then the desired equality will follow from the fact that our optimality criterion is final and therefore, there is only one optimal model (so, since the models F_{opt} and $S_s(F_{opt})$ are both optimal, they must be the same model).

Let us show that $S_s(F_{opt})$ is indeed optimal. How can we, e.g., prove that $S_s(F_{opt}) \succ F$? Since the optimality criterion is shift-invariant, the desired relation is equivalent to $F_{opt} \succ S_{-s}(F)$. Similarly, the relation $S_s(F_{opt}) \sim F$ is equivalent to $F_{opt} \sim S_{-s}(F)$.

These two equivalences allow us to complete the proof of the proposition. Indeed, since F_{opt} is optimal, we have one of the two possibilities:

- either $F_{opt} \succ S_{-s}(F)$

- or $F_{opt} \sim S_{-s}(F)$.

In the first case, we have $S_s(F_{opt}) \succ F$; in the second case, we have $S_s(F_{opt}) \sim F$. Thus, whatever model F we take, we always have either $S_s(F_{opt}) \succ F$, or $S_s(F_{opt}) \sim F$. Hence, $S_s(F_{opt})$ is indeed optimal and thence, $S_s(F_{opt}) = F_{opt}$. Q.E.D.

Proof of Theorem 4.1. Since the criterion is final, there exists an optimal model $F_{opt} = \{d + c \cdot f_0(n)\}$ for some function $f_0(n)$. The corresponding function $f_0(n)$ belongs to the family F_{opt} (for $d = 0$ and $c = 1$).

Due to Proposition 4.1, the optimal family is shift-invariant, i.e., $F_{opt} = S_s(F_{opt})$. In particular, for $s = 1$, we conclude that $F_{opt} = S_1(F_{opt})$. Therefore, the function $S_1(f_0(n)) = f_0(n+1)$ (that belongs to $S_1(F_{opt})$), must also belong to F_{opt}. By definition, the family F_{opt} consists of all functions of the type $d + c \cdot f_0(n)$, therefore there exists c and d, for which the functions $f_0(n+1)$ and $d + c \cdot f_0(n)$ coincide (i.e., their values are equal for all n). Hence,

$$f_0(n+1) = d + c \cdot f_0(n)$$

for all n. We get a functional equation of the type that we know how to solve. So let us solve it.

In terms of a shift operator, the equation takes the form $(S - c)f_0 = d$. According to the general algorithm of solving linear functional equations (described in Lesson 2), we first need to eliminate the right-hand side. This right-hand is a constant, i.e., it is of the form $d \cdot 1^k$. Hence, to eliminate this right-hand side, we must apply the operation $S - 1$ to both sides of the original functional equation. As a result, we get the equation $(S-1)(S-c)f = 0$. The corresponding polynomial equation is $(\alpha - c)(\alpha - 1) = 0$. The solution of this functional equation depends on whether the corresponding polynomial equation has two different roots or a single double root:

- If $c \neq 1$, then this equation has two different roots $\alpha = c$ and $\alpha = 1$. The general solution is, therefore, $f(n) = c_1 \cdot c^n + c_2 \cdot 1^n = c_1 \cdot c^n + c_2$. If a model contains a single function of this type, it, therefore, contains a function $c^n = \exp(kn)$ (where $k = \ln(c)$), and therefore, coincides with the logarithmic model.

- If $c = 1$, then this equation has a double root $\alpha = 1$. According to the general algorithm, the general solution of the corresponding functional equation is $f(n) = c_1 \cdot 1^n + c_2 \cdot n \cdot 1^n = c_1 + c_2 \cdot n$. Thus, if $c = 1$, we get a linear model.

Q.E.D.

4.6. Second result: using scale-invariance

In the two previous sections, we have formalized and used *shift-invariance*. Let us now express the demand of *scale-invariance* in mathematical terms. Before we write down the definitions we have to make two remarks related to the problem of how to count bugs.

■ Scale invariance corresponds to the question "what is a bug":

  − from one point of view this particular error is a bug,

  − from another point of view the same error is a part of a bug.

As a result, the number of bugs n is not necessarily an integer, it can be a real number.

■ The second remark corresponds to the fact that different people can start counting bugs in different moments of time:

  − a *bolder* programmer can start debugging at a point which

  − a *more cautious* person will still consider a part of the design process, when a product is not yet ready for real testing and debugging.

In the *shift-invariant* case we were lucky enough to have Proposition 4.1, that allowed us to disregard this difference. But in the *scale-invariant* case this proposition may be not true any longer (and it really turns out to be not true), so we better take this possible difference into consideration in our definitions.

With these remarks in mind we come to the following definitions.

Definition 4.1'.

■ *Let $f_0(n)$ be a monotonic function from real numbers to real numbers. By a software reliability model (or a model, for short) that corresponds to the function $f_0(n)$, we mean a set of all functions of the type $d + c \cdot f_0(n)$, where:*

  − *d is an arbitrary real number and*

  − *c is an arbitrary positive number.*

■ *Two models are considered equal if the corresponding families coincide (i.e., if they consist of the same functions).*

Definition 4.4'. *Assume some value s is fixed.*

■ *By an s-shift (or simply a shift, for short) of a function $f(n)$ we mean a function $f(n + s)$.*

- By an *s-shift* of a model F we mean a model that consists of s-shifts of all the functions from F.

Comment. This value s corresponds to the difference between the two starting points for counting bugs:

- some fixed starting point that we assumed when we deduced the specific type of functions from the family, and

- a starting point that was actually used in the debugging documentation to which we want to apply these functions.

Denotation. The set of all the families (in the sense of Definition 4.1′) will be denoted by Φ'.

Definition 4.6. *Let C be a positive real number.*

- By a *C-rescaling* of a function $f(n)$ we mean a function $g(n) = f(Cn)$.

- By a *C-rescaling* of a model F we mean a model consisting of C-rescalings of all functions from F.

Denotation. C-rescaling of a family F will be denoted by $R_C(F)$.

Definition 4.7. *We say that an optimality criterion on Φ' is scale-invariant if for every two models F and G and for every real number $C > 0$ the following two conditions are true:*

i) *if F is better than G in the sense of this criterion (i.e., $F \succ G$), then $R_C(F) \succ R_C(G)$;*

ii) *if F is equivalent to G in the sense of this criterion (i.e., $F \sim G$), then $R_C(F) \sim R_C(G)$.*

Models that are optimal with respect to some scale-invariant criterion are described by the following theorem:

Theorem 4.2. *If a model F is optimal in the sense of some optimality criterion that is final and scale-invariant, then each s-shift of this model:*

- *either consists of the functions of the type $f(n) = a + b \cdot \ln(k - n)$ for some real numbers a, b, and k (these functions correspond to the basic model),*

- *or consists of the functions of the type $f(n) = a + b \cdot (n + k)^c$ for some real numbers a, b, k, and c.*

Comments.

- Models with $t \sim n^\alpha$ (*power models*) have really been proposed and experimentally confirmed (see, e.g., Chapter 11 of Musa's 1987 monograph.

  - In every *specific* application, if we have a good understanding of the program and of its debugging process, then other models may turn out to be better than the models described in Theorem 4.2.
  - However, if we want to choose the best model for *general* use, so that this model will be applicable to any debugging situation, to any way of counting bugs, then it is natural to require scale invariance and thus, for such a general application, one of the models described in Theorem 4.2 will be indeed the best.

To prove Theorem 4.2, we will first prove the following proposition that is also of independent interest:

Proposition 4.2. *If an optimality criterion is final and scale-invariant then the optimal model F_{opt} is also scale-invariant, i.e., $R_C(F_{opt}) = F_{opt}$.*

Comments.

- This Proposition shows that if you use an optimal approximation family then it does not matter what scale you choose for counting bugs: you still get the same approximation family; therefore if you use this family for extrapolation, you get the same extrapolation results.

- Proposition 4.2 is proved just like Proposition 4.1, the only difference is that we consider $R_{C^{-1}}$ instead of S_{-s}.

Let us now prove Theorem 4.2. Like in the proof of Theorem 4.1, from the scale-invariance of the optimal model $F_{opt} = \{d + c \cdot f_0(n)\}$, we conclude that for every $\lambda > 0$ there exist real numbers d and c, depending on λ, such that

$$f_0(\lambda \cdot n) = d(\lambda) + c(\lambda) \cdot f_0(n) \qquad (4.1)$$

for all $n \geq 0$ and for all $\lambda \geq 0$.

This new functional equation is different from the ones we had before, for two reasons:

- First, previously, we had equations with only *one* unknown function $(f(n))$, and now we have *three* unknown functions: $f_0(n)$, $d(\lambda)$, and $c(\lambda)$.

- Second, previously, the unknown function depended on an *integer* variable n, while now, we have unknown functions of a *real* variable.

How can we solve such a functional equation? There are two ways to do it:

- First, we can look into some of the "encyclopedia" books that contain all known solutions of functional equations. The best of these books is, probably, J. Aczel, *Lectures on functional equations and their applications* (Academic Press, N.Y.–London, 1966). In particular, the equation (4.1) in which we are currently interested has been already solved about a century ago (see Section 3.1 of Aczel's book).

- Looking for known solutions works for most simple equations, but sometimes we may run into new equations that have not been explicitly solved before. To be prepared for such situations, we would like to explain general methods of solving these equations.

As we will see, functional equations are the easiest to solve if we know that the unknown functions are differentiable; then, we can usually differentiate and get *differential* equations that are, usually, easier to solve. Therefore, the easiest way to solve a functional equation is to:

- first, prove that the solution is differentiable;

- second, differentiate and get a differential equation;

- and finally, solve the corresponding differential equation.

The first part is the most difficult one; for this part, we refer the readers to Aczel's book.

In this text, we will, for simplicity, assume that the functions $f(n)$, $d(\lambda)$, and $c(\lambda)$ are differentiable. The main idea of differentiating is that, instead of arbitrary scalings $n \to \lambda \cdot n$, we consider "infinitesimal" scalings $n \to \lambda \cdot n$ with $\lambda \approx 1$. To describe such "infinitesimal" scalings, we must consider the values $\lambda \approx 1$. In other words, to get a differential equation, we must do two things:

■   differentiate both sides of the equation (4.1) with respect to λ, and

■   substitute $\lambda = 1$.

As we differentiate the equation, we get the formula

$$n \cdot \frac{df_0}{dn} = d + c \cdot f_0,$$ (4.2)

where we denoted

$$c = \frac{dc(\lambda)}{d\lambda}\Big|_{\lambda=1}; \quad d = \frac{dd(\lambda)}{d\lambda}\Big|_{\lambda=1}.$$

To facilitate the solution of the system (4.2), we would like to *separate* the variables, i.e., re-arrange the terms in such a way that:

■   all the terms that contain n and dn move to one side, and

■   all the terms that contain f_0 and df_0 move to the other side.

This can be done rather easily if we:

■   divide both sides by $d + c \cdot f_0$, thus moving all the terms that contain f_0 and df_0 to the left-hand side, and

■   divide both sides by n/dn, thus moving all the terms that contain n and dn to the right-hand side.

As a result, we get the following equation:

$$\frac{df_0}{d + c \cdot f_0} = \frac{dn}{n}.$$ (4.3)

To solve the equation, it is now sufficient to integrate both parts of the resulting equation (4.3), i.e., to explicitly integrate the equation

$$\int \frac{df_0}{d + c \cdot f_0} = \int \frac{dn}{n},$$ (4.4)

and to express $f_0(n)$ from the resulting equation.

■ The right-hand side is the easiest to integrate, its integration leads to

$$\int \frac{dn}{n} = \ln(n) + C$$

for some constant C.

■ The integration of the left-hand side is slightly more complicated, because the formulas are different depending on whether $c = 0$ or $c \neq 0$:

– If $c = 0$, then

$$\int \frac{df_0}{d + c \cdot f_0} = \int \frac{df_0}{d} = \frac{f_0}{d}.$$

In this case, the equation (4.4) takes the form

$$\frac{f_0(n)}{d} = \ln(n) + C,$$

and $f_0(n) = d \cdot \ln(n) + \text{const}$. This case corresponds to the basic model.

– If $c \neq 0$, then we can compute the integral in the left-hand side of (4.4) by introducing a new variable $z = d + c \cdot f_0$. For this new variable, $f_0 = (z - d)/c$, $df_0 = dz/c$, and

$$\int \frac{df_0}{d + c \cdot f_0} = \int \frac{dz}{cz} = \frac{1}{c} \cdot \ln(z) = \frac{1}{c} \cdot \ln(d + c \cdot f_0).$$

The resulting equation (4.4) is

$$\ln(d + c \cdot f_0) = c \cdot \ln(n) + \text{const}.$$

Taking exp of both sides, we conclude that

$$d + c \cdot f_0 = \text{const} \cdot \exp(c \cdot \ln(n)) = \text{const} \cdot n^c.$$

This equation leads to the power model, with $f_0(n) = a + b(n + k)^c$.

Q.E.D.

Exercise

4.1 Describe the proof of Proposition 4.2 in detail.

4.7. Basic conclusions

Theorems 4.1 and 4.2 solve the following problems:

- *We have solved the problem of what software reliability model to choose: whatever reasonable criterion we use, we shall get either the basic or the logarithmic model.*

- *We have thus explained why the basic model and the logarithmic model are experimentally the best.*

These theorems also address the auxiliary problem: that it is difficult to choose one of the possible reasonable ways to count bugs. Namely, the corresponding Propositions explain that the optimal family does not change if we use a different way to count bugs, and therefore the extrapolation results do not change. So the answer to this problem is as follows:

- *One can count bugs in any reasonable way, the approximation function and hence the extrapolation results will not depend on that choice.*

5

OPTIMAL CHOICE OF A PENALTY FUNCTION: SIMPLEST CASE OF ALGORITHM DESIGN

In the previous lessons, we learned how continuous mathematics is useful in choosing and debugging the existing algorithms and programs. With this lesson, we will start learning how continuous mathematics can be useful in designing new algorithms. We will consider the simplest case of design problems, in which both the constraints and the objective are well defined and described by differentiable functions. For such problem, we describe the penalty function method, and then formulate and solve the problem of choosing the best penalty function.

5.1. Formulation of the problem: what is a penalty function, why do we need them, and how can we choose a penalty function

There is a constant need to design new algorithms. In the previous lessons, we assumed that we *already have* an algorithm (or even several algorithms) for solving a given real-life problem, and the only remaining computer problems were:

- to select the best of the given algorithms, and
- to make sure that the resulting programs do not contain any errors.

We have shown that continuous mathematics can be really helpful in solving both computer problems.

In many situations, however, we do not yet have an algorithm for solving a problem. So, before we can start choosing an algorithm or debugging the resulting program, we still need to *design* an algorithm for solving the real-life problem. In this lesson (and in a few following lessons), we will show that continuous mathematics can help in designing algorithms as well:

- In this lesson, we will start with the simplest problems.

- Then, in the following lessons, we will show how similar ideas can be used in more complicated problems as well.

Simple practical problems are naturally described as problems of constrained optimization. In general, to describe a real-life problem, we must describe:

- what is known, and
- what we want.

In some problems, we only have a subjective description.

> For example, if we are designing the train that guarantees the smoothest ride, or if we are designing a pretty bridge, the term "smoothest" or "pretty" describe subjective impressions, and it is not easy to describe them in formal terms that a computer can understand.

The formalization of a subjective description is not an impossible task: there are methods of formalizing such informal problems, and in the following lessons, we will describe these methods (and show that continuous mathematics helps to formalize these subjective impressions). All we want to say is that the necessity to formalize the subjective impressions makes a problem *more complicated*.

Therefore, among all possible problems, *the simplest* are the ones in which both what we know and what we want is described *in precise terms*.

The information about what the user wants usually consists of two parts:

- First, we have constraints that need to be satisfied.

    For example, if we are planning a space trip to the Moon, then the resulting trajectory must bring the spaceship to the Moon.

- There are, usually, *many* possible solutions that satisfy these constraints. So, to make a problem well formulated, we must describe an *optimality criterion* that will enable us to *choose* one of these possible solutions.

    We have already discussed, in Lesson 4, that optimality criteria can be quite complicated:

    * the *simplest* class of optimality criteria consists of *numerical* criteria, when we have a function $J(x)$ that describes a quality of each alternative x, and x is better than y if $J(x) > J(y)$;
    * there   can   be *more complicated multi-criteria* formulationsoptimization,mutlictiteria, in which, e.g., two functions $J_1(x)$ and $J_2(x)$ are given, and x is better than y if either $J_1(x) > J_1(y)$, or $J_1(x) = J_1(y)$ and $J_2(x) > J_2(y)$;
    * even more complicated criteria are possible.

    Since in this lesson, we are interested in the problems from the *simplest possible* class, we will consider the problems in which the optimalizty criterion is from the simplest possible class, i.e., the problems in which we are looking for an alternative x for which $J(x) \to \max$ for some numerical function $J(x)$.

    For example, for controlling the spaceship, $J(x)$ may be the time saved, or the fuel saved, or the probability of a flawless mission.

As a result of our analysis, we conclude that the simplest practical problems can be formulated as follows:

From a certain given set X, find x for which $J(x) \to \max$ among all the alternatives $x \in X$ that satisfy certain (known) constraints.

Such problems of optimization under constraints are called, in mathematics, problems of *constrained optimization*.

Comment. In many situations, we are interested in *minimizing* a certain function $f(x)$, rather than in *maximizing* it. From the mathematical point of view,

however, we can restrict ourselves to maximization problems only, because minimizing $f(x)$ is the same as maximizing a new function $J(x) = -f(x)$.

The simplest problems of unconstrained optimization. The simplest
case of an optimization problem is when we have no constraints at all. In this
case (as we have already mentioned in Lesson 3), the easiest case of optimization
is when we have a *differentiable* objective function $J(x)$, i.e., when:

- a possible solution x is usually characterized by one or several numerical parameters $x_1, \ldots, x_n$ (so that we can identify x with the tuple $(x_1, \ldots, x_n)$);
  and

- the objective function $J(x) = J(x_1, \ldots, x_n)$ is a differentiable function of
  n variables.

In this case, we can simply find the desired optimal solution x by looking for
a point $x = (x_1, \ldots, x_n)$ in which all n partial derivatives of the function $J(x)$
are equal to 0:
$$\frac{\partial J}{\partial x_i} = 0.$$

Constrained optimization is also necessary. In many situations, the objective function itself does not contain all the requirements on x.

For example, when $J(x)$ is the flight time of a space mission to the Mars,
we must not only guarantee that this time is the smallest possible, but we
must also make sure that the mission really lands on the Mars, and that
on its way, it does not, e.g., hit the Moon.

In such situations, we have additional requirements on the values of x that need
to be satisfied. These requirements are usually of two types:

- *equalities*, of the type $g(x) = 0$ for some function $g(x)$; and

- *inequalities*, of the type $g(x) \geq 0$.

In the above example:

- the requirement that we reach the desired destination is an example
  of an equality, while

      – the requirement that we must avoid a certain area is an example of
an inequality.

When we have constraints, the optimization problem becomes a more complicated *constraint* optimization:

$$J(x) \to \max \text{ under the constraint that } g(x) = 0 \text{ or that } g(x) \geq 0.$$

How can we solve such problems?

**Constrained optimizations under constraints of equality type: La-
grange multipliers.** A natural idea of solving *constrained* optimization problems is to reduce them to a simpler *un-constrained* optimization.

This idea has first implemented, for constraints of the equality type, by the
19th century French mathematician Lagrange. His method is now called the
Lagrange multipliers method, and it is based on the following result: Every
solution to the constrained optimization problem $J(x) \to \max$ under the constraint that $g(x) = 0$ is at the same time the solution to an unconstrained
optimization problem $J(x) + \lambda \cdot g(x) \to \max$ for some constant λ (this constant
is called *Lagrange multiplier*). Therefore, in order to solve the constrained optimization problem, we must find, among all the solutions to the unconstrained
optimization problem $J(x) + \lambda \cdot g(x) \to \max$ (that correspond to different values
of λ), a solution for which $g(x) = 0$.

We know how to express the maximum of a smooth function of n variables in
terms of n equations. Therefore, Lagrange multipliers methods actually means
that we reduce the original constrained optimization problem to the following
system of $n + 1$ equations with $n + 1$ unknowns $x_1, \ldots, x_n, \lambda$:

$$\frac{\partial J}{\partial x_i} + \lambda \frac{\partial g}{\partial x_i} = 0, \quad i = 1, \ldots, n;$$

$$g(x) = 0.$$

Example. As an example of Lagrange multiplier method, let us consider the
following simple geometric problem: on a 2D plane, find a point on the 0-
centered unit circle that is the closest to the point $(1, 1)$.

- In this problem, possible solutions are points on the plane; each point can be uniquely characterized by its two Cartesian coordinates; so, we have $n = 2$.

- The objective function is the distance between the desired point x with coordinates x_1 and x_2 and the point $(1, 1)$. This distance is described by a formula $\sqrt{(x_1 - 1)^2 + (x_2 - 1)^2}$. Is this distance the desired function $J(x)$? No, for two reasons:

  - First, we want a *differentiable* objective function, while the distance function is, alas, not everywhere differentiable (its derivative at the point $(1, 1)$ is infinite).

  - Second, we want to formulate a *maximization* problem, while in the original problem, the distance has to be *minimized*.

  It is easy to make the objective function differentiable: for that, it is sufficient to take the *square* of the distance function, i.e., $f(x) = (x_1 - 1)^2 + (x_2 - 1)^2$. We can do that because the distance is the smallest possible if and only if the square of the distance is the smallest possible.

  To replace a minimization problem by a maximization one is even easier: we will take $J(x) = -f(x)$, i.e.,

  $$J(x) = -((x_1 - 1)^2 + (x_2 - 1)^2).$$

- We want to maximize the function $J(x)$ under the constraint that the point $x = (x_1, x_2)$ lies on the unit circle. The unit circle is described by the equation $x_1^2 + x_2^2 = 1$, i.e., equivalently, by the equation

  $$g(x) = x_1^2 + x_2^2 - 1 = 0.$$

For this problem,

$$J(x) + \lambda \cdot g(x) = -((x_1 - 1)^2 + (x_2 - 1)^2) + \lambda \cdot (x_1^2 + x_2^2 - 1 = 0).$$

Now that we have $J(x)$ and $g(x)$, we can explicitly differentiate the expression for $J(x) + \lambda \cdot g(x)$, and describe the corresponding system of $n + 1 = 3$ equations with three unknowns x_1, x_2, and λ:

$$-2(x_1 - 1) + 2\lambda \cdot x_1 = 0; \qquad (5.1a)$$

$$-2(x_2 - 1) + 2\lambda \cdot x_2 = 0; \qquad (5.1b)$$

$$x_1^2 + x_2^2 - 1 = 0. \qquad (5.1c)$$

From the first equation, we can express x_1 in terms of λ. Namely:

- If we move all terms that do not contain x_1 to the right-hand side, we get an equation $x_1 \cdot (2\lambda - 2) = 2$.

- Since the right-hand side of this equation is different from 0, the coefficient at x_1 in the left-hand side cannot be equal to 0. Hence, we can divide both sides of this equation by this coefficient, and conclude that $x_1 = 2/(2\lambda - 2) = 1/(\lambda - 1)$.

Similarly, from the second equation (5.1b), we conclude that $x_2 = 1/(\lambda - 1)$, and therefore, that $x_2 = x_1$. Substituting $x_2 = x_1$ into the third equation, we conclude that $2 \cdot x_1^2 = 1$, $x_1^2 = 1/2$, and $x_1 = \pm\sqrt{1/2} = \pm\sqrt{2}/2$. Thus, we have two candidates for the solution:

- $x_1 = x_2 = \sqrt{2}/2$.
- $x_1 = x_2 = -\sqrt{2}/2$.

To make a final choice, it is sufficient to compare the values of the objective function $J(x)$ for these two solutions. The first one has a larger value of $J(x)$ and therefore, it is the desired solution.

One can easily check that this is indeed the closest point to $(1, 1)$.

Constrained optimizations under constraints of inequality type: Lagrange multiplier approach. In principle, the Lagrange multiplier approach can be also used if we have constraints of inequality type.

Indeed, if we, e.g., have a single constraint of this type $g(x) \geq 0$, then we can have two possible situations:

- It could happen that the constrained maximum is achieved exactly when $g(x) = 0$. In this case, according to Lagrange multiplier method, for some real number λ, this *constrained* maximum coincides with the *unconstrained* maximum of a function $J(x) + \lambda \cdot g(x)$.

- However, it can also happen that the maximum is attained when $g(x) > 0$. In this case, this maximum is, at the same time, an unconstrained (local) maximum of the problem $J(x) \to \max$.

So, to solve the constrained optimization problem $J(x) \to \max$ with a single inequality-type constraint $g(x) \geq 0$, we can do the following:

- First, solve the unconstrained optimization problem $J(x) \to$ max.

  - If the unconstrained maximum is attained when $g(x) \geq 0$, this unconstrained maximum is, at the same time, the maximum to our constrained optimization problem as well.

  - If the global unconstrained maximum is attained when $g(x) < 0$, then:

    * We check whether any local unconstrained maxima satisfy this constraint, and if any of them does, we remember the one with the largest values of $J(x)$.

    * After that, we use the Lagrange multiplier method to solve the *constrained* optimization problem $J(x) \to$ max under the condition $g(x) = 0$.

    * Finally, we compare the values $J(x)$ for the local unconstrained maxima and for the maxima on the boundary $g(x) = 0$, and, as the solution to the original problem, choose the value x for which the value $J(x)$ is the largest.

Example. As an example of Lagrange multiplier method, let us consider a simple geometric problem similar to the one considered above: on a 2D plane, find a point on the 0-centered unit *disk* that is the closest to the point $(1, 1)$.

In this problem, the (differentiable) objective function is $J(x) = -((x_1 - 1)^2 + (x_2 - 1)^2)$, and the constraint takes the form $x_1^2 + x_2^2 \leq 1$, or, equivalently, $g(x) \geq 0$, where $g(x) = 1 - x_1^2 - x_2^2$.

Hence, according to our method, we must solve the following two problems:

- The unconstrained optimization problem $J(x) \to$ max. Equating the partial derivatives of the function $J(x)$ with 0, we get the solution $x_1 = 1$, $x_2 = 1$. This point is *outside* the unit disk, and therefore, does not have to be considered.

- The constrained optimization problem $J(x) \to$ max under the constraint $g(x) = 0$. We already know the solution to this problem: it is $x = (x_1, x_2) = (\sqrt{2}/2, \sqrt{2}/2)$.

Since the first problem does not have any solution, the solution to the second problem is the desired solution to the original problem.

Lagrange multiplier approach to constrained optimizations under several constraints of inequality type. If we have *several* constraints of inequality type $g_1(x) \geq 0, \ldots, d_m(x) \geq 0$, then we can apply a similar approach. Indeed, at whatever point $x_{\max}$ the constrained maximum is attained, for some of the constraints, we have $g_i(x_{\max}) = 0$ and for some others, we have $g_i(x_{\max}) > 0$. If we denote by $I = \{i_1, \ldots, i_k\} \subseteq \{1, \ldots, m\}$ the set of all indices i for which $g_i(x_{\max}) = 0$, then the desired constrained maximum $x_{\max}$ coincides with the (maybe local) solution to one of the following constrained optimization problems: $J(x) \to \max$ under the constraints that $g_{i_1}(x) = 0, \ldots, g_{i_k}(x) \geq 0$.

There are 2^m possible sets I, and we do not know which set we are dealing with. Therefore, to make sure that we do not miss the desired constrained maximum, we must consider all 2^m possible sets. The resulting algorithm is as follows:

- For each of 2^m sets $I = \{i_1, \ldots, i_k\} \subseteq \{1, 2, \ldots, m\}$, we use the Lagrange multiplier method to solve the following auxiliary constrained optimization problem: $J(x) \to \max$ under the constraints $g_{i_1}(x) = 0, \ldots, g_{i_k}(x) = 0$, and keep all (local and global) maxima for which $g_i(x) \geq 0$ for all $i = 1, 2, \ldots, m$.

- Then, we compare the solutions to these 2^m problems, and find the solution for which the value of the objective function $J(x)$ is the largest possible.

Lagrange multiplier approach to constrained optimizations under constraints of inequality type: main drawback. The main drawback of this method is that for large m, this method is not feasible.

For reasonably small m, this methods is quite feasible. However, in many real-life problems, there are lots of constraints of inequality type. If we have a problem with $m = 300$ inequality constraints, then, to apply the above method, we need to solve $2^{300} = 10^{90}$ different optimization problems. This is not feasible at all. To solve such problem, we need a new idea.

Constrained optimizations under constraints of inequality type: penalty functions. The number of constraints was not an issue with constraints of equality type: no matter how many equality-type constraints $g_1(x) = 0, \ldots, g_m(x) = 0$ we have, we only have to solve a *single* unconstrained optimization problem $J(x) + \lambda_1 \cdot g_1(x) + \ldots + \lambda_m \cdot g_m(x)$. It is natural to try a similar idea in case the restriction on x is of the *inequality* type $g(x) \geq 0$, so that for several restrictions, we would simply add several terms to a single objective function.

To describe the possible additions to the objective function, let us consider the case of a single constraint $g(x) \geq 0$. In this case:

- It could happen that the constrained maximum is achieved exactly when $g(x) = 0$. In this case, according to Lagrange multiplier method, for some real number λ, this *constrained* maximum coincides with the *unconstrained* maximum of a function $J(x) + \lambda \cdot g(x)$.

- However, it can also happen that the maximum is attained when $g(x) > 0$.

To cover this case, we would like to add an *extra* term (depending on $g(x)$) to the resulting objective function, and to consider optimization problems of the type $J(x) + \lambda \cdot g(x) + \mu \cdot P(g(x)) \to \max$ for some function $P(y)$. This function $P(y)$ is called a *penalty function*.

Example. The idea of a penalty function is not just a mathematical trick, it is actually used in economics, and it is from economics that the name "penalty" came. Let us give an economic example.

In the 19th century, manufacturing plants usually operated according to the necessity to maximize profits. In other words, the choice of the manufacturing process x was dictated by the constraint that $J(x) \to \max$, where $J(x)$ is the profit that correspond to the process x. With this profit-oriented economy, some plants were actually heavily *polluting* the environment.

To avoid this pollution, most countries have adopted environmental protection laws, according to which the pollution level $p(x)$ of each plant cannot exceed a certain limit p_0: $p(x) \leq p_0$. This inequality-type constraint can be expressed in the form $g(x) \geq 0$ if we take $g(x) = p_0 - p(x)$.

- Ideally, we would like the plants to maximize its profits not unconditionally, but under these environmental constraints, i.e., to solve a constrained optimization problem $J(x) \to \max$ under the constraint $g(x) \geq 0$.

- However, we have the following problem: the decrease in pollution usually comes at a certain cost, so the manufacturers try to save money on that. It would have been impractical and unrealistically costly to police each and every plant.

So what governments do instead is impose heavy *penalties* on polluting manu-facturers. The larger the pollution level, the higher the penalty. This penalty can be described thus as function $F(p(x))$ of the pollution level. As a result, the actual profit of the plant is no longer the original profit $J(x)$, but the profit from which these penalties have been subtracted, i.e., $J(x) - F(p(x))$. For an appropriately chosen penalty function, plants will stop polluting.

The expression $F(p(x))$ is not exactly the penalty function that we are looking for, because we want a penalty function to depend on $g(x)$; however, from the definition of $g(x) = p_0 - p(x)$, we can conclude that $p(x) = p_0 - p(x)$ and therefore, that $F(p(x)) = F(p_0 - g(x)) = -P(g(x))$, where we denoted $P(y) = -F(p_0 - y)$. In these terms, we replace the original *constrained* optimization problem

$$J(x) \rightarrow \max \text{ under the constraint } g(x) \geq 0$$

by an *unconstrained* optimization problem

$$J(x) + P(g(x)) \rightarrow \max.$$

This is exactly what penalty method is about.

This example shows the three things:

- first, the example of countries that achieved the desired decrease in first, pollution shows that in principle, *the penalty functions method can work*;

- second, the example of less successful countries shows that for an inade-quate choice of a penalty function, *we may not get good results*;

- third, the complexity of the existing penalty laws shows that it is *not* that *simple to design a* successful *penalty function*.

These observations lead us to the following problem:

How to choose a penalty function? Which function $P(y)$ should we choose? From common sense arguments, we can conclude only one thing: that the penalty function $P(y)$ must be *non-linear*: Indeed, adding a linear expression $P(g(x)) = c \cdot g(x)$ is exactly what Lagrange multiplier method does, and this method only covers constraints of equality type.

Different non-linear penalty functions have been proposed, and these functions lead to different success rates in solving constrained optimization problems. So, a natural question is: *which is the best penalty function?*

In the following text, we will formulate this problem in precise mathematical terms and solve the resulting mathematical problem.

Exercise

5.1 Apply the Lagrange multiplier method to solve the following problem: find the closest point to $x = (1, 2)$ on a hyperbola $x_1^2 - x_2^2 = 1$.

5.2 Apply the Lagrange multiplier method to solve the following problem: find the closest point to $x = (2, 1)$ in the area $x_1^2 - x_2^2 \leq 1$.

5.2. How to choose the best penalty function: Mathematical formulation of the problem

A penalty function must be differentiable. The main objective of using a penalty function was to reduce the original *constrained* optimization problem $J(x) \to$ max under the condition $g(x) \geq 0$ to an easier-to-solve *unconstrained* optimization problem $\tilde{J}(x) \to$ max with a differentiable objective function $\tilde{J}(x) = J(x) + \lambda \cdot g(x) + \mu \cdot P(g(x))$. For this new objective function to be differentiable, we must only consider differentiable function $P(y)$.

Actually, since we are interested in guaranteeing that $g(x) \geq 0$, it is sufficient to require that the function $P(y)$ be differentiable only for $y \geq 0$.

Moreover, if $y = g(x) = 0$, then a simple Lagrange multiplier method works, and we do not need any additional penalty function at all. So, we may restrict the differentiability condition even further: to the cases when $y > 0$. (To be more precise, we would like to be able to have *some* derivative for $y = 0$, but we should not mind having an infinite value for $y = 0$.)

Without loss of generality, we can assume that $P(0) = 0$. From the fact that for $y = g(x) = 0$, we do not need any penalty function at all, we can also conclude that $P(0) = 0$.

- The differentiability condition actually limits our choices of penalty functions.

- In contrast to that, the condition $P(0) = 0$ is not restrictive at all: indeed, whatever penalty function $P(y)$ we choose, we can always take a new function $P'(y) = P(y) - P(0)$ for which $P'(0) = P(0) - P(0) = 0$. When we add a constant to $P(y)$, this adds a constant to the resulting objective function $J'(y)$, and adding a constant to all the values does not change where the maximum is.

Thus, we can, without loss of generality, assume that $P(0) = 0$.

We need to choose a family of penalty functions, not a single function. Choosing a penalty function $P(y)$ mean that we will try to reduce a constrained optimization problem $J(x) \to \max$ under the constraint $g(x) \geq 0$ to the unconstrained optimization problem $J(x) + \lambda \cdot g(x) + \mu \cdot P(g(x)) \to \max$ for some constants λ and μ.

This expression can be re-written as $J(x) + A(g(x)) \to \max$, where we denoted $A(y) = \lambda \cdot y + \mu \cdot P(y)$. So, when we choose a function $P(y)$, we actually choose the entire *family* of functions $A(y) = \lambda \cdot y + \mu \cdot P(y)$. The efficiency of the choice depends on this family only, so, if two different functions lead to the same family, they are, basically, the same choice. For example, we may choose $P(y) = y^2$ and $\tilde{P}(y) = y^2 - y$, the resulting classes of functions are the same.

So, strictly speaking, the problem is not to choose *a* penalty function $P(y)$, but rather to choose a *family* of penalty functions. This argument leads to the following definition:

Definition 5.1.

- *Let $P(y)$ be a continuous function from non-negative real numbers to real numbers that is differentiable for $y > 0$. By a family of penalty functions (or a family, for short) that corresponds to this function $P(y)$, we mean a family of all functions of the type $\tilde{P}(y) = \lambda \cdot y + \mu \cdot P(y)$, where:*

  - *λ is an arbitrary real number and*
  - *μ is an arbitrary positive number.*

- *Two families are considered equal if they coincide, i.e., consist of the same functions.*

Denotation. *Let's denote the set of all possible families by Φ.*

In this lesson, we will consider optimality criteria on the set Φ of all families of penalty functions. Similar to the problem of choosing the optimal software reliability model, we have natural *symmetries* here. Namely, the inequality-type constraint is usually of the form $g(x) \geq 0$, where $g(x)$ is the value of some physical quantity y that depends on the parameters x of the solution. For example, for pollution, $g(x)$ is the amount of pollution.

The numerical value of each quantity depends on the choice of a unit in which we measure this quantity. For example, we may measure the amount of pollution in the atmosphere in SI units of kilograms per cubic meters, or in CGS units of grams per cubic centimeter, or in American units of pounds per square foot. In all these units, the amount of pollution will be miniscule (much smaller than 1), so, to make the resulting numbers more convenient and easier to understand, we can use some special units instead.

It is reasonable to require that the relative quality of different penalty functions should depend on which of many possible units we choose. How can we express this requirements in precise terms?

When we change a unit for measuring a certain quantity to another unit that is C times smaller, then each numerical value x is replaced by a new value Cx that is C times larger. For example, when we change a unit of length from meter to centimeter, a $C = 100$ times smaller unit, then 1.5 meters becomes $1.5 \cdot 100 = 150$ centimeters.

In particular, if we change a unit for measuring the quantity $y = g(x)$ to a new unit that is C times smaller, then the new numerical value $\tilde{y}$ of y will be C times larger, i.e., we will have $\tilde{y} = \tilde{g}(x) = C \cdot y = C \cdot g(x)$. If we apply the same penalty function $P(y)$ to the value of y expressed in the new units, we will get the value $P(\tilde{y}) = P(C \cdot y)$. We will get exactly the same expression if we used the *old* units and the *new* penalty function $\tilde{P}(y) = P(C \cdot y)$. So, from the viewpoint of the constrained optimization problem, replacing a unit is equivalent to replacing the original penalty function $P(y)$ by a new penalty function $\tilde{P}(y) = P(C \cdot y)$. Thus, our requirement can be reformulated as follows:

- if a penalty function $P(y)$ is better than the penalty function $Q(y)$, then a new penalty function $\tilde{P}(y) = P(C \cdot y)$ is better than the new penalty function $\tilde{Q}(y) = Q(C \cdot y)$;

- if a penalty function $P(y)$ is of the same quality as the penalty function $Q(y)$, then a new penalty function $\tilde{P}(y) = P(C \cdot y)$ is of the same quality as the new penalty function $\tilde{Q}(y) = Q(C \cdot y)$.

We already had formulas in Lesson 4. Namely, according to Definition 4.6, the transition from the original function $P(y)$ to the new function $\tilde{P}(y) = P(C \cdot y)$ was called *C-rescaling*. The above two requirements on an optimality criterion were described, in Definition 4.7, under the name of *scale-invariance*. In these terms, all we require is that the optimality criterion is scale-invariant.

It turns out that this requirement describes a reasonably small class of possibly optimal penalty functions:

Theorem 5.1. *If a family F is optimal in the sense of some optimality criterion that is final and scale-invariant, then this family corresponds to $P(y) = y \cdot \ln(y)$ or to $P(y) = y^\alpha$ for some real value α.*

Comments.

- This result is in good accordance with the experience of numerical methods, that show that $y\ln(y)$ is indeed the most widely used and the most successful penalty function, with y^2 coming second.

- Similar to the result from Lesson 4, this result does not necessarily mean that in every possible situation, one of these penalty functions will be better than any other possible choice.

  - In every *specific* application, if we have a good understanding of the function $g(x)$, of the criterion $J(x)$, etc., then other penalty functions may turn out to be better than the functions described here.

  - However, if we want to choose the best method for *general* use, so that this method will be applicable to any situation, to any units for measuring $g(x)$, then it is natural to require scale invariance and thus, for such a general application, one of these functions will be indeed the best.

- In the remaining section, we will show how this theorem can be proven. In the process of proving it, we will reduce the problem to the problem of solving a simple differential equation. In the next lesson, we will learn (or recall if you already knew that) how to solve such differential equations. This knowledge will be useful in the following lessons as well.

5.3. Proof of the main result: reduction to a differential equation

The outline of the proof. In proving this result, we will follow the sequence of steps similar to the ones used in the proof of Theorem 4.2:

- first, we will find a functional equation describing the desired optimal penalty function;

- then, we will show that the functions used in this equations are indeed differentiable;

- after that, we will deduce the differential equation for this desired function $P(y)$;

- finally, we will solve this differential equation and get the desired expression for the optimal penalty function $P(y)$.

First step: deducing a functional equation. According to Proposition 4.2, the optimal family $\{\lambda \cdot y + \mu \cdot P(y)\}$ must be scale-invariant. Therefore, for every $C > 0$, there must exist values $\lambda(C)$ and $\mu(C)$ such that

$$P(C \cdot y) = \lambda(C) \cdot y + \mu(C) \cdot P(y). \qquad (5.2)$$

This is the desired functional equation.

Second step: proving differentiability. The functional equation (5.2) contains three functions: $P(y)$, $\lambda(C)$, and $\mu(C)$. We want to prove that all three functions are differentiable. We already know, from the very definition of a penalty function, that $P(y)$ must be differentiable for all $y > 0$. So, it remains to prove that the functions $\lambda(C)$ and $\mu(C)$ are differentiable.

To prove this, we will express both functions $\lambda(C)$ and $\mu(C)$ in terms of the function $P(y)$. Then, from the fact that $P(y)$ is differentiable, we will be able to conclude that the functions $\lambda(C)$ and $\mu(C)$ are differentiable as well.

Let us fix a value $C > 0$. How can we express the two $\lambda(C)$ and $\mu(C)$ in terms of the function $P(y)$? The only relationship that we have between $\lambda(C)$, $\mu(C)$, and the function $P(y)$ is the equation (5.2): for every $y > 0$, we get the relation between $\lambda(C)$, $\mu(C)$, and certain values of the function $P(y)$. To uniquely describe two values $(\lambda(C)$ and $\mu(C))$, we need at least two equations. So, let us take the two equations that are obtained from (5.2) by taking two different values y_1 and y_2. As a result, we get the following system of two linear equations with the two unknowns:

$$P(C \cdot y_1) = \lambda(C) \cdot y_1 + \mu(C) \cdot P(y_1); \tag{5.3a}$$

$$P(C \cdot y_2) = \lambda(C) \cdot y_2 + \mu(C) \cdot P(y_2). \tag{5.3b}$$

In linear algebra, there is a formula (called *Cramer's rule*) that explicitly expresses the solution of a system of linear equations in terms of the coefficients of this system: Namely, the j-th component x_j of the solution of a system of equations $\sum a_{ij} x_j = b_i$ is equal to the fraction D_j/D, where D is the determinant of the matrix consisting of the coefficients a_{ij}:

$$\begin{pmatrix} a_{11} & a_{12} & \cdots & a_{1n} \\ a_{21} & a_{22} & \cdots & a_{2n} \\ \cdots & & & \\ a_{n1} & a_{n2} & \cdots & a_{nn} \end{pmatrix},$$

and D_j is the determinant of the matrix in which the elements of j–th column is replaced by the values b_i:

$$\begin{pmatrix} a_{11} & \cdots & a_{1,j-1} & b_1 & a_{1,j+1} & \cdots & a_{1n} \\ a_{21} & \cdots & a_{2,j-1} & b_2 & a_{2,j+1} & \cdots & a_{2n} \\ \cdots & & & & & & \\ a_{n1} & \cdots & a_{n,j-1} & b_n & a_{n,j+1} & \cdots & a_{nn} \end{pmatrix}.$$

The determinant of a matrix is the result of adding and subtracting different products of this matrix's elements. For example, the determinant of a 2×2 matrix

$$\begin{pmatrix} a_{11} & a_{12} \\ a_{21} & a_{22} \end{pmatrix}$$

is equal to $a_{11} \cdot a_{22} - a_{12} \cdot a_{21}$. In general, the determinant is a polynomial (hence, a differentiable function) of the coefficients. Therefore, the right-hand side of the Cramer's rules is a fraction of two differentiable functions and is, therefore, itself differentiable.

In particular, for the system (5.3a) − (5.3b),

$$D = \det \begin{pmatrix} y_1 & P(y_1) \\ y_2 & P(y_2) \end{pmatrix} = y_1 \cdot P(y_2) - y_2 \cdot P(y_1),$$

$$D_1 = \det \begin{pmatrix} P(C \cdot y_1) & P(y_1) \\ P(C \cdot y_2) & P(y_2) \end{pmatrix} = P(C \cdot y_1) \cdot P(y_2) - P(C \cdot y_2) \cdot P(y_1),$$

$$D_2 = \det \begin{pmatrix} y_1 & P(C \cdot y_1) \\ y_2 & P(C \cdot y_2) \end{pmatrix} = y_1 \cdot P(C \cdot y_2) - y_2 \cdot P(C \cdot y_1),$$

and Cramer's rule leads to the following expressions:

$$\lambda(C) = \frac{P(C \cdot y_1) \cdot P(y_2) - P(C \cdot y_2) \cdot P(y_1)}{y_1 \cdot P(y_2) - y_2 \cdot P(y_1)};$$

$$\mu(C) = \frac{y_1 \cdot P(C \cdot y_2) - y_2 \cdot P(C \cdot y_1)}{y_1 \cdot P(y_2) - y_2 \cdot P(y_1)}.$$

Since the function $P(y)$ is differentiable, these expressions show that the functions $\lambda(C)$ and $\mu(C)$ are differentiable as well.

Third step: deducing a differential equation. Now that we have shown that all three functions $P(y)$, $\lambda(C)$, and $\mu(C)$ form the functional equation (5.2) are differentiable, we can follow the path we followed in Lesson 4:

- differentiate both sides of this functional equation by C;

- take $C = 1$ and thus, get the desired differential equation.

When we differentiate both sides of (5.2) by C, we get the following expression:

$$y \cdot P'(C \cdot y) = \frac{d\lambda(C)}{C} \cdot y + \frac{d\mu(C)}{dC} \cdot P(y),$$

where P' denotes the first derivative of the function $P(y)$.

When we substitute $C = 1$ into this formula, we get the following differential equation:

$$y \cdot P'(y) = \lambda \cdot y + \mu \cdot P(y), \tag{5.4}$$

where we denoted

$$\lambda = \frac{d\lambda(C)}{dC}_{|C=1}; \quad \mu = \frac{d\mu(C)}{dC}_{|C=1}.$$

Comment. If this was the only differential equation that we will have to deal with in this text, we would just as well give its solution. However, since we will encounter several differential equations of this type, in the following lesson, we will show how to solve the differential equations of this type; as a particular case, we will describe the explicit solution of the equation (5.4).

5.4. Conclusion: How to solve constrained optimization problems?

Now that we know what are the possible optimal penalty functions, let us describe how exactly we can use penalty functions to actually solve constrained optimization problems. We will describe the resulting methodology on the example of the most widely used penalty function $P(y) = y \cdot \ln(y)$.

Derivation of a methodology. The main idea of using penalty function is that we reduce a constrained optimization problem $J(x) \to \max$ under the constraint $g(x) \geq 0$ to the unconstrained optimization problem $\tilde{J}(x) \to \max$, where

$$\tilde{J}(x) = J(x) + \lambda \cdot g(x) + \mu \cdot P(g(x)). \tag{5.5}$$

The function $P(y) = y \cdot \ln(y)$ is only defined for $y \geq 0$.

> Strictly speaking, since $\ln(y)$ is only defined for $y > 0$, this function is also defined only for $y > 0$. However, since for $y \to 0$, this function has a well-defined limit: we have $y \cdot \ln(y) \to 0$, we can extend this function to the point $y = 0$ by assuming that $0 \cdot \ln(0) = 0$. This extension is actively used in probability and information theory, because it is used in the definition of an *entropy* or an *information*.

Since $P(y)$ is only defined for $y \geq 0$, the function $\tilde{J}(x)$ is only defined when $y = g(x) \geq 0$. Thus, the point x_{opt} at which the unconstrained maximum of the new objective function $\tilde{J}(x)$ is attained, automatically satisfies the constraint $g(x) \geq 0$.

That this point x_{opt} is a point of an unconstrained maximum means that for every other point x, we have $\tilde{J}(x_{\text{opt}}) \geq \tilde{J}(x)$. In particular, this is true for all points x for which $g(x) \geq 0$. Therefore, the point x_{opt} at which the *unconstrained* maximum of the function $\tilde{J}(x)$ is attained is at the same time the solution to the following *constrained* optimization problem: $\tilde{J}(x) \to \max$ under the condition that $g(x) \geq 0$.

Therefore, the solution x_{opt} is close to the solution of the original constrained optimization problem $J(x) \to \max$ under the constraint $g(x) \geq 0$ when the new objective function (5.5) is close to the original objective function $J(x)$. This, in its turn, is true when the coefficients λ and μ at the new terms in $\tilde{J}(x)$ are close to 0.

Therefore, in order to get the solution to the original problem, we must take the values λ and μ as close to 0 as possible.

- For λ, we can simply take $\lambda = 0$.

- For μ, we cannot do that, because this will defeat the whole purpose of having a penalty function. Since we cannot take $\mu = 0$, we must take μ smaller and smaller and thus, get better and better solutions to the original constrained optimization problem.

So, we must take $\mu_n \to 0$. In order to describe the final methodology, we must now select the *sign* of μ_n.

The main idea of a *penalty function* is to *penalize* for getting close to the constraint boundary $g(x) = 0$. Penalizing means *decreasing* the value of the maximized function. In other words, we would like to choose a sign of μ for which the addition $\mu \cdot P(g(x))$ is negative for small values of $g(x)$.

When $y = g(x) > 0$ and $g(x)$ is close to 0, we have $y \ll 1$; hence, $\ln(y) < 0$, and $P(g(x)) = y \cdot \log(y) < 0$. Hence, to get the value $\mu \cdot P(g(x))$ negative, we must use positive values of μ.

Now, we are ready to describe the desired methodology.

Methodology itself. To solve the constrained optimization problem $J(x) \to$ max under the constraints $g_1(x) = 0, \ldots, g_m(x) = 0$, $m \geq 0$, $h_1(x) \geq 0, \ldots, g_p(x) \geq 0$, we must do the following:

- Select a positive number μ, and solve the following unconstrained optimization problem:

$$J(x) + \lambda_1 \cdot g_1(x) + \ldots + \lambda_m \cdot g_m(x) +$$

$$\mu(h_1(x) \cdot \ln(h_1(x)) + \ldots + h_p(x) \cdot \ln(h_p(x))) \to \max,$$

where, if $m > 0$, the values λ_i must be determined from the constraints $g_i(x) = 0$, $1 \leq i \leq m$.

- Decrease the value μ and repeat the solution, the decrease and repeat again, until the solutions that correspond to different μ become close enough (i.e., until these solution differ by a certain pre-defined value δ).

Exercise

5.3 Use the penalty functions methods to solve the following problem: find the closest point to $x = (2, 1)$ in the area $x_1^2 - x_2^2 \leq 1$.

6

SOLVING GENERAL LINEAR DIFFERENTIAL EQUATIONS WITH CONSTANT COEFFICIENTS: AN APPLICATION TO CONSTRAINED OPTIMIZATION

In Lesson 5, we started to show how continuous mathematics can help in algorithm design. In particular, we showed that the problem of finding the best penalty function leads to a linear differential equation. Since other algorithm design problems lead to similar differential equations, we need a general method for solving such differential equations. Such a method is described in this lesson.

6.1. Differential equations vs. functional equations

We would like to be able to solve differential equations of a certain type that would include the equation (5.4) $(y \cdot P'(y) = \lambda \cdot y + \mu \cdot P(y))$ as a particular case.

So far, in this course, we have learned how to solve *linear functional* equations. When we were describing different ideas of solving functional equations, we assumed that:

■ the reader was more or less acquainted with differentiation and resulting simple differential equations, but

■ the reader has probably never solved a functional equation before.

Therefore, we have mentioned, several times, a natural analogy between functional and differential equations, and used natural methods known for *differen-*

tial equations as a heuristic to develop similar techniques for *functional* equations.

At this point, however, the tables have been turned:

- we have learned a lot about how to solve *functional* equations, but

- we still have only a very limited knowledge of how to solve *differential* equations.

It is, therefore, natural to try to use the same analogy and use methods that we have described for *functional* equations as a source of heuristics for solving *differential* equations.

Historical comment. In this text,

- we start with functional equations, and then

- we go to differential equations.

Historically, however, it was the other way around:

- first, researchers learned how to solve differential equations, and

- after that, they learned how to solve functional equations.

The main reason for this difference is this:

- in the history of mathematical methods, the main incentive of analyzing different types of equations is the analysis of *real-life* phenomena,

- while in this text, we are interested in applications to *computer science*.

Let us elaborate a little bit on this difference.

- Most *real-life* processes are continuous. To describe such a process, we need to describe the values $x(t)$ of all related parameters x at different moments of time t. A natural way of describing how exactly these values change is to describe how the values $x(t + \Delta t)$ in the "next" moment of time $t + \Delta t$ depend on the values in the "previous" moment of time t. The smaller the interval of time Δt, the better the description. The ideal description corresponds to "infinitely small" values of Δt, i.e., to a *differential equation*. Therefore, differential equations are a natural way of describing real-life phenomena.

In some cases, of course, a system does not change continuously; its state changes only at certain moments of time. To describe such rare systems, functional equations were proposed and analyzed.

■ When we analyze *computers*, the situation is exactly the opposite. A typical computer operates in discrete time, its state changes discretely, and therefore, the most natural formalism for describing computer processes is *functional equations*.

Sometimes, we need *functional equations* as well (as we needed them, e.g., to describe program debugging).

Let us now describe what differential equations we need to solve and how we plan to solve them.

6.2. The class of differential equations that we will be solving: Linear differential equations with constant coefficients

We are discussing linear differential equations. All the methods described in Lessons 1 and 2 were for solving *linear* functional equations, i.e., functional equations in which both the left-hand side and the right-hand side were linearly depending on the unknown function $f(n)$. So, if we want to use these methods as an analogy, we must focus on *linear* differential equations, i.e., on equations in which both sides linearly depend on the unknown function $x(t)$.

Luckily, the equation (5.4) is an example of a *linear* differential equation, because its both sides linearly depend on the unknown function $P(y)$.

We will be solving linear differential equations with constant coefficients. Methods described in Lessons 1 and 2 only allow us to solve linear functional equations in which the coefficients at the values of an unknown function (i.e., at $f(n)$, $f(n-1)$, etc.) are *constants*.

It is therefore reasonable to expect that similar methods will only solve differential equations with *constant* coefficients at the unknown function $x(t)$, at its first derivative $x'(t)$, etc.

A minor problem with this restriction. There is a minor problem with this restriction: in the equation (5.4), the coefficient at the first derivative $P'(y)$ of the unknown function $P(y)$ is *not* a constant: it is y. So, first, let us show how we can reduce the equation (5.4) to an equation with constant coefficients. We will describe this reduction method in some detail, because it will be useful for other differential equations as well.

6.3. Reducing equations originating from scale invariance (like equation (5.4)) to equations with constant coefficients

The only non-constant term in the equation (5.4) is the term

$$y \cdot P'(y) = y \cdot \frac{dP}{dy}.$$

We want to represent this term as a term with a constant coefficient, i.e., as simply a first derivative.

Of course, if we keep the same variable y, we cannot do that; so, the only way to do the desired reduction is to find another variable t (depending on y), in terms of which the equation (5.4) will become an equation with constant coefficients. In other words, we would like to find a new variable $t = t(y)$ for which

$$\frac{dP}{dt} = y \cdot \frac{dP}{dy}. \tag{6.1}$$

We want to find the dependence $t(y)$ from this equation. To find this dependence, we must, first, re-write (6.1) as an equation that only contains t and y and does not contain P at all. This is the easiest part: the only way P enters into both sides of the equation (6.1) is by a factor dP in both sides of this equation. So, to get rid of P, we can simply divide both sides of the equation

(6.1) by the this factor dP. As a result, we get the following equation:

$$\frac{1}{dt} = \frac{y}{dy}. \tag{6.2}$$

In this equation, variables are already separated, so we are almost ready to integrate both sides and come up with an explicit relation between y and t. The only problem, e.g., with the left-hand, side of (6.2) is that we have to first inverse it to integrate. So, let us inverse both sides of this equation. As a result of this inversion, we get the following equation:

$$dt = \frac{dy}{y}.$$

Integrating both sides of this equation, we conclude that

$$\int dt = \int \frac{dy}{y},$$

i.e., that $t = \ln(y)$.

So, if we move to a new variable $t = \ln(y)$ (in which $y = \exp(t)$), and consider the new function $x(t) = P(\exp(t))$, then the term $y \cdot P'(y)$ becomes the new term $x'(t)$, and the equation (5.4) becomes a linear equation with constant coefficients:

$$x'(t) = \lambda \cdot \exp(t) + \mu \cdot x(t). \tag{6.3}$$

This method works for arbitrary equations originating from the requirement of scale invariance.

Exercise

6.1 Reduce the equation $y \cdot P'(y) = P(y) + 3y$ to an equation with constant coefficients.

6.4. Solving linear differential equations with constant coefficients: a plan

In order to solve general linear differential equations with constant coefficients, we will follow the same sequence of steps as we followed when we learned how to solve general linear functional equations with constant coefficients:

- First, we will consider the simplest equations without the right-hand side, of the type $x'(t) - \alpha \cdot x(t) = 0$.

- Second, we will consider general *non-degenerate* equations without a right-hand side.

- We will specifically handle the case when the roots are complex numbers.

- Then, we will consider general equations without a right-hand side.

- Finally, we will describe how to solve equations with a right-hand side.

6.5. Simplest linear differential equations without a right-hand side

Description of the equations. Let us start with the simplest equation with the right-hand side, i.e., with an equation of the type

$$x'(t) - \alpha \cdot x(t) = 0. \qquad (6.4)$$

The main idea of solving these equations. To solve this equation, let us express $x'(t)$ as dx/dt. Then, we get

$$\frac{dx}{dt} - \alpha \cdot x = 0.$$

We would like to separate terms that contain x and t. To do that, let us first move $\alpha \cdot x$ into the right-hand side:

$$\frac{dx}{dt} = \alpha \cdot x.$$

We want to have all terms that contain x in one side, and all terms that contain t in the other side.

- To get all terms that contain x into the left-hand side, i.e., to delete x from the right-hand side, let us divide both sides by x.

- The only term that contains t is a term dt in the left-hand side. To move this term to the right-hand side, we can multiply both sides by dt.

After these two transformations, we get the following equation:

$$\frac{dx}{x} = \alpha \cdot dt.$$

Integrating both sides of this equation, we conclude that

$$\int \frac{dx}{x} = \int \alpha \cdot dt,$$

i.e., that

$$\ln(x) = \alpha \cdot t + C \qquad (6.5)$$

for some constant C. To reconstruct x from $\ln(x)$, we can apply the exponential function, because by definition of a natural logarithm, $x = \exp(\ln(x))$. Therefore, if we apply exp to both sides of the equation (6.5), we conclude that $x(t) = \exp(\alpha \cdot t + C)$. An exponent of the sum is equal to the product of the exponents, so we get the final solution $x(t) = \exp(C) \cdot \exp(\alpha \cdot t)$, or

$$x(t) = C_1 \cdot \exp(\alpha \cdot t),$$

where $C_1 = \exp(C)$ is an arbitrary constant.

One can easily check that this function is indeed the solution of the original differential equation. In particular, for $\alpha = 1$, we get $x(t) = C_1 \exp(t)$ as a function for which $x'(t) = x(t)$. The fact that $\exp(x)$ is the solution of this equation can be easily remembered by the following joke:

> *A new patient is brought to the lunatic asylum. People who are already there produce all kind of irritating noise, so the newcomer yells: "Quiet, or I'll differentiate you all!" Everybody get quite except for one guy who pays not attention. "Why are you not afraid of me?" – asks the newcomer. "Because I am* $\exp(x)$*," replies the noisy guy.*

The constant C_1 can be determined, if we know the *initial condition* $x(t_0) = x_0$ for some known values x_0 and t_0. Substituting $t = t_0$ into the above expression, we conclude that $C_1 \cdot \exp(\alpha \cdot t_0) = x_0$ and hence, $C_1 = x_0/\exp(\alpha \cdot t_0) = x_0 \cdot \exp(-\alpha \cdot t_0)$. Hence, $x(t) = x_0 \cdot \exp(-\alpha \cdot t_0) \cdot \exp(\alpha \cdot t) = x_0 \cdot \exp(\alpha \cdot (t - t_0))$. As a result, we get the following algorithm:

Algorithm. Suppose that we have an equation (6.4) (with no free terms), and we want to find a solution of this equation that satisfies a given initial condition, i.e., which has the given value of $x(t_0) = x_0$. This solution is given by the following formula

$$x(t) = x_0 \cdot \exp(\alpha \cdot (t - t_0)). \tag{6.6}$$

Comment. The simple equation $x' = \alpha \cdot x$ (and its solution $x(t) = C_1 \cdot \exp(\alpha \cdot t)$) is indeed a very good first approximation for describing change in many real-life systems: it describes the growth of the population, the growth of the number of computer science publications, the growth of the internet, the increase of the computer speed, etc. With $\alpha < 0$, this equation also describes *decrease*: e.g., the amount of a radioactive substance left after a certain time.

The reader should beware that this equation is, usually, only a *simple first approximation*. If we take this equation too seriously, i.e., if we consider it to be the true equation that describes the process, then we run into all kinds of alarmist predictions: that in a few decades, due to population growth, there will be no standing room on Earth, that pollution will kill us all, etc. Of course, we need to worry about overpopulation and pollution, but luckily, gloomy predictions made in the 60s (based on simple exponential models) did not come true.

Exercise

6.2 Find a solution of the equation $x'(t) = 2x$ for which $x(0) = 3$.

6.6. General non-degenerate linear differential equations without a right-hand side

Idea. A general linear differential equation with constant coefficients and 0 right-hand side is an equation of the following type:

$$\frac{d^p x}{dt^p} + a_1 \cdot \frac{d^{p-1} x}{dx^{p-1}} + a_2 \cdot \frac{d^{p-2}}{dt^{p-2}} + \ldots + a_p \cdot x = 0. \tag{6.7}$$

This equation can be re-written as

$$Q(\frac{d}{dt})x = 0, \tag{6.8}$$

where we denoted

$$Q(z) = z^p + a_1 \cdot z^{p-1} + a_2 \cdot z^{p-2} + \ldots + a_p.$$

Every polynomial $Q(z)$ of p-th degree has exactly p roots $\alpha_1, \ldots, \alpha_p$, i.e., values for which $Q(\alpha_i) = 0$. (Roots may be complex numbers. In the degenerate case, if we have a double root, we count it twice, etc.). The polynomial can be represented as a product

$$Q(z) = (z - \alpha_1) \cdot (z - \alpha_2) \cdot \ldots \cdot (z - \alpha_p).$$

Therefore, the equation (6.8) can be represented as follows:

$$(\frac{d}{dt} - \alpha_1) \cdot (\frac{d}{dt} - \alpha_2) \cdot \ldots \cdot (\frac{d}{dt} - \alpha_p)x = 0. \tag{6.9}$$

From this representation, it is clear that the each j from 1 to p, every solution of the equation

$$(\frac{d}{dt} - \alpha_j)x = 0,$$

(i.e., the function $x_j(t) = \exp(\alpha_j \cdot t)$) is a solution to (6.9) as well: indeed, if, e.g.,

$$(\frac{d}{dt} - \alpha_p)x = 0,$$

then

$$(\frac{d}{dt} - \alpha_1) \cdot (\frac{d}{dt} - \alpha_2) \cdot \ldots \cdot (\frac{d}{dt} - \alpha_p)x =$$

$$(\frac{d}{dt} - \alpha_1) \cdot (\frac{d}{dt} - \alpha_2) \cdot \ldots \cdot (\frac{d}{dt} - \alpha_{p-1})[(\frac{d}{dt} - \alpha_p)x] = 0.$$

So, for every j, the function $x_j(t)$ is a solution. Thus, an arbitrary linear combination

$$x(t) = C_1 \cdot \exp(\alpha_1 \cdot t) + \ldots + C_p \cdot \exp(\alpha_p \cdot t)$$

is a solution to the original differential equation.

In the *non-degenerate case*, when all p roots are different, this is a general solution to the given differential equation. As a result, we get the following algorithm:

Algorithm: non-degenerate case. Suppose that we have an equation of the type (6.7)–(6.8) (with no free terms), and we want to find a solution of this equation that satisfies given initial conditions, e.g., which has the given values $x(t_0) = x_0$, $x'(t_0) = x'_0$, ..., $x^{(p-1)}(t_0) = x_0^{(p-1)}$. To find the corresponding solution $x(t)$, we do the following:

- First, we find all p roots $\alpha_1, \ldots, \alpha_p$ of the corresponding polynomial equation $Q(z) = 0$.

- From these roots, we construct a *general* solution

$$x(t) = C_1 \cdot \exp(\alpha_1 \cdot t) + \ldots + C_p \cdot \exp(\alpha_p \cdot t).$$

- Finally, we find the coefficients C_j by substituting this expression into the initial conditions and by solving the resulting system of p linear equations with p unknowns $C_1, \ldots, C_p$:

$$C_1 \cdot \exp(\alpha_1 \cdot t_0) + \ldots + C_p \cdot \exp(\alpha_p \cdot t_0) = x_0;$$

$$C_1 \cdot \alpha_1 \cdot \exp(\alpha_1 \cdot t_0) + \ldots + C_p \cdot \alpha_p \cdot \exp(\alpha_p \cdot t_0) = x_0';$$

$$\ldots$$

$$C_1 \cdot \alpha_1^{p-1} \cdot \exp(\alpha_1 \cdot t_0) + \ldots + C_p \cdot \alpha_p^{p-1} \cdot \exp(\alpha_p \cdot t_0) = x_0^{(p-1)}.$$

Example. Let us consider an equation $x'' - 3x' + 2x = 0$, with the initial conditions $x(0) = 1$ and $x'(0) = 0$ (i.e., $t_0 = 1$, $x_0 = 1$ and $x_0' = 0$).

- In this case, the polynomial equation $Q(z) = 0$ has the form $z^2 - 3z + 2 = 0$. This polynomial equation has two different roots $\alpha_1 = 2$ and $\alpha_2 = 1$.

- Therefore, a general solution of the corresponding differential equation can be written as $x(t) = C_1 \cdot \exp(2t) + C_2 \cdot \exp(t)$.

- To find the values of the coefficients C_1 and C_2, we substitute this general expression for $x(t)$ into the equations that describe the initial conditions. As a result, we get the following system of two linear equations with two unknowns C_1 and C_2 (these equations are much simpler than the general ones because for $t_0 = 0$, $\exp(\alpha \cdot t_0) = 1$):

$$C_1 + C_2 = 1;$$

$$2C_1 + C_2 = 0.$$

From the second equation, we conclude that $C_2 = -2C_1$; substituting this expression into the first equation, we get $C_1 - 2C_1 = 1$ and hence, $C_1 = -1$ and $C_2 = -2C_1 = 2$.

So, the desired solution is $x(t) = -\exp(2t) + 2\exp(t)$.

Exercise

6.3 Find a solution of the equation $x''(t) = -x' + 2x$ with the initial conditions $x(0) = 0$ and $x'(0) = 1$.

6.7. Case of complex roots

The problem: in case of complex roots, we get a complex-valued solution. We started with an equation with real-valued coefficients. Its solution is a real-valued function. However, a polynomial $Q(z)$ with real-valued coefficients can have complex roots α_p. In this case, our algorithm leads to a complex-valued expression.

It is, therefore, desirable to re-formulate the resulting solution in real-valued terms.

Idea. If one of the roots α_j of the polynomial is a complex number $\alpha_j = p_j + i \cdot q_j$, then, $\alpha_j \cdot t = p_j \cdot t + i \cdot q_j \cdot t$ and $\exp(\alpha_j \cdot t) = \exp(p_j \cdot t) \cdot \exp(i \cdot q_j \cdot t)$. Using the de Moivre formula (that we have already used in Lesson 2) $\exp(i\theta) = \cos(\theta) + i \cdot \sin(\theta)$, we conclude that

$$\exp(\alpha_j \cdot t) = \exp(p_j \cdot t) \cdot \cos(q_j \cdot t) + i \cdot \exp(p_j \cdot t) \cdot \sin(q_j \cdot t).$$

It is known that with every complex root $\alpha_j = p_j + i \cdot q_j$, a *complex conjugate* value $\alpha_j^* = p_j - i \cdot q_j$ is also the root of the same polynomial. The corresponding specific solution has the form

$$\exp(\alpha_j^* \cdot t) = \exp(p_j \cdot t) \cdot \cos(-q_j \cdot t) + i \cdot \exp(p_j \cdot t) \cdot \sin(-q_j \cdot t) =$$

(since $\cos(z)$ is an even function and $\sin(z)$ is an odd function)

$$= \exp(p_j \cdot t) \cdot \cos(q_j \cdot t) - i \cdot \exp(p_j \cdot t) \cdot \sin(q_j \cdot t).$$

Hence, a real-valued linear combination of the terms corresponding to these two roots is a linear combination of the terms

$$\exp(p_j \cdot t) \cdot \cos(q_j \cdot t) \quad \text{and} \quad \exp(p_j \cdot t) \cdot \sin(q_j \cdot t).$$

As a result, if we want only real-valued expressions, we can modify our algorithm in the following manner:

Algorithm: non-degenerate case. Suppose that we have an equation of the type (6.7)–(6.8) (with no free terms), and we want to find a solution of this equation that satisfies given initial conditions, e.g., which has the given values $x(t_0) = x_0$, $x'(t_0) = x'_0$, ..., $x^{(p-1)}(t_0) = x_0^{(p-1)}$. To find the corresponding solution $x(t)$, we do the following:

■ First, we find all p roots $\alpha_1, \ldots, \alpha_p$ of the corresponding polynomial $Q(z) = 0$.

■ Based on these roots, we form *specific* solutions $x_1(t), \ldots, x_p(t)$ to the original differential equation:

  – If j-th root α_j is a real number, we take $x_j(t) = \exp(\alpha_j \cdot t)$.

  – If j-th root α_j is a complex number $\alpha_j = p_j + q_j \cdot i$, then for this root and for the complex conjugate root $\alpha_k = \alpha_j^*$, we construct two specific solutions $x_j(t) = \exp(p_j \cdot t) \cdot \cos(q_j \cdot t)$ and $x_k(t) = \exp(p_j \cdot t) \cdot \sin(q_j \cdot t)$.

■ From these specific solutions, we construct a *general* solution

$$x(t) = C_1 \cdot x_1(t) + \ldots + C_p \cdot x_p(t).$$

■ Finally, we find the coefficients C_j by substituting the expression for the general solution into the equations that describe the initial conditions, and by solving the resulting system of p linear equations with p unknowns $C_1, \ldots, C_p$:

$$C_1 \cdot x_1(t_0) + \ldots + C_p \cdot x_p(0) = x_0,$$
$$C_1 \cdot x'_1(t_0) + \ldots + C_p \cdot x'_p(0) = x'_0,$$
$$\ldots$$
$$C_1 \cdot x_1^{(p-1)}(t_0) + \ldots + C_p \cdot x_p^{(p-1)}(0) = x_0^{(p-1)}.$$

Example 1. The functions that correspond to complex roots are proportional to sine and cosine and therefore, oscillate. Therefore, such solutions describe *oscillations*. The simplest oscillator (e.g., a spring or a pendulum) is a system in which, if you deviate from the equilibrium, a force f appears that brings you back. If we take 0 as an equilibrium point, then we get an equation of the type $f = f(x)$. By definition of an equilibrium, it is a state in which there no changes and therefore, no forces, so in the equilibrium, the force is 0: $f(0) = 0$. Therefore, if we expand the unknown function $f(x)$ into Taylor series:

$$f(x) = f(0) + f'(0) \cdot x + \frac{f''(0)}{2} \cdot x^2 + \ldots,$$

the first term disappears, and the first non-zero term is linear. In the first approximation, we can neglect the other terms and assume that $f(x) = kx$ for some constant k. The increase in the coordinate ($x > 0$) should cause the force bringing the system back ($f < 0$); therefore, we must have $k < 0$.

Due to second Newton's law, the acceleration x'' caused by this force is equal to $x'' = (k/m) \cdot x$. The simplest possible equation of this type is the equation $x'' = -x$ that corresponds to $k/m = -1$. Let us apply the above algorithm to see how an oscillator described by this equation will react to the initial deviation $x(0) = 1$ and $x'(0) = 1$.

- For this equation, $Q(z) = z^2 + 1$. The equation $z^2 + 1 = 0$ has two complex conjugate roots: $\alpha_1 = i$ and $\alpha_2 = -i$.

- Based on these roots, we construct two specific solutions
$$x_1(t) = \cdot(t) \quad \text{and} \quad x_2(t) = \sin(t).$$

- From these specific solutions, we construct a general solution
$$x(t) = C_1 \cdot \cos(t) + C_2 \cdot \sin(t).$$

- Finally, we determine the coefficients

- Finally, we determine the coefficients C_1 and C_2 by substituting the expression for the general solution into the equations that describe the initial conditions:
$$C_1 \cdot 1 + C_2 \cdot 0 = 1,$$
$$C_1 \cdot 0 + C_2 = 0.$$
From these two equations, we conclude that $C_1 = 1$ and $C_2 = 0$.

Thus, the desired solution is $x(t) = \cos(t)$.

Example 2. An oscillation from Example 1, once started, goes on forever. In real life, there is usually an additional friction force $f_{\text{fr}} = -\lambda \cdot x'$ that damps these oscillations. Let us see, e.g., what will happen if we add the friction force $f_{rmfr} = -2x'$ to the oscillator force $f_{osc} = -2x$ and the body mass $m = 1$. In this case, the total force is $f = -2x' - 2x$, and the total acceleration is $x'' = f/m = -2x' - 2x$.

So, in mathematical terms, we are interested in solving the differential equation $x'' + 2x' + 2x = 0$, with the initial conditions $x(0) = 1$, $x'(0) = 0$. For this problem, the above algorithm leads to the following solution:

- The corresponding polynomial $Q(z) = z^2 + 2z + 2$ has two complex conjugate roots

$$\alpha_{1,2} = \frac{-2 \pm \sqrt{4-8}}{2} = -1 \pm i.$$

- Based on these roots, we construct two specific solutions

$$x_1(t) = \exp(-t) \cdot \cos(t) \quad \text{and} \quad x_2(t) = \exp(-t) \cdot \sin(t).$$

- From these specific solutions, we construct a general solution

$$x(t) = C_1 \cdot \exp(-t) \cdot \cos(t) + C_2 \exp(-t) \cdot \sin(t).$$

- Finally, we determine the coefficients C_1 and C_2 by substituting the expression for the general solution into the equations that describe the initial conditions:

$$C_1 \cdot 1 + C_2 \cdot 0 = 1,$$

$$-C_1 + C_2 = 0.$$

From these two equations, we conclude that $C_1 = 1$ and $C_2 = 1$.

Thus, the desired solution is $x(t) = \exp(-t) \cdot \cos(t) + \exp(-t) \cdot \sin(t)$.

Exercise

6.4 Find a solution of the equation $x''(t) = 2x' - 5x$ with the initial conditions $x(0) = 0$ and $x'(0) = 1$.

6.8. Case of multiple roots

The problem. In the above sections, we described the solution for the *non-degenerate* linear differential equations without a right-hand side, non-degenerate meaning that all the roots of the corresponding polynomial $Q(z)$ were assumed to be different.

How can we solve this equation in the *degenerate* case, when we have a double or, in general, a multiple root?

Idea. We will handle degenerate *differential* equations in a manner that is very similar to handling degenerate *functional* equations in Lesson 2. Namely, just like in Lesson 2, we will take into consideration the fact that a polynomial with a *double* root $\alpha_1 = \alpha_2 = \alpha$ can be viewed as a *limit* case of polynomials with two close roots α and $\alpha + \varepsilon$ when $\varepsilon \to 0$. (Similarly, every equation with a *multiple* root can be described as a limit of non-degenerate equations in which all roots are different.) So, in order to find the solutions to the original *degenerate* equation (with multiple roots) we can:

- find the solution $x_\varepsilon(t)$ to the approximate *non-degenerate* equation (in which all roots are different), and then

- take the limit

$$x(t) = \lim_{\varepsilon \to 0} x_\varepsilon(t)$$

of these "approximate" solutions $x_\varepsilon(t)$ as the desired solution to the original (degenerate) equation.

To find out how this idea can be converted into an algorithm, let us first consider the case of the *double* root α. The corresponding non-degenerate approximate equation has *two* different roots α and $\alpha + \varepsilon$ and therefore, two specific solutions $\exp(\alpha \cdot t)$ and $\exp((\alpha + \varepsilon) \cdot t)$. Due to linearity, an arbitrary linear combination of these two solutions is also a solution; in particular, the difference $\exp((\alpha + \varepsilon) \cdot t) - \exp(\alpha \cdot t)$ between the specific solutions of the equation is also a solution. If we tend $\varepsilon \to 0$, then the limit expression is a solution to the original functional equation.

This difference, by itself, tends to 0, and therefore, does not lead to any meaningful solution of the original equation. However, due to linearity, the result of multiplying this difference by any number is also a solution. We can, therefore, multiply it by, say, $1/\varepsilon$, to prevent it from having a 0 limit. As a result, we can conclude that the following limit function is a solution to the original equation:

$$x_{\text{new}}(t) = \lim_{\varepsilon \to 0} \frac{\exp((\alpha + \varepsilon) \cdot t) - \exp(\alpha \cdot t)}{\varepsilon}.$$

The expression in the right-hand side is exactly the definition of the derivative:

$$x_{\text{new}}(t) = g'(\alpha) = \lim_{\varepsilon \to 0} \frac{g(\alpha + \varepsilon) - g(\alpha)}{\varepsilon}$$

for $g(\alpha) = \exp(\alpha \cdot t)$. The derivative of the function $g(\alpha) = \exp(\alpha \cdot t)$ is easy to compute: $x_{\text{new}}(t) = g'(\alpha) = t \cdot \exp(\alpha \cdot t)$. So, we can conclude that for the case of a double root α, in addition to $\exp(\alpha \cdot t)$, another function, $t \cdot \exp(\alpha \cdot t)$, is also a specific solution.

Similarly, if we have a *triple* root, we can form a linear combination that is equal to the *second derivative* of $g(\alpha) = \exp(\alpha \cdot t)$, and thus, conclude that not only $\exp(\alpha \cdot t)$, but also $t \cdot \exp(\alpha \cdot t)$ and $t^2 \cdot \exp(\alpha \cdot t)$ are solutions. If we have a root of multiplicity m, then $\exp(\alpha \cdot t), t \cdot \exp(\alpha \cdot t), \ldots, t^{m-1} \cdot \exp(\alpha \cdot t)$ are specific solutions.

Now, we are ready to describe the modified algorithm:

General algorithm for solving linear differential equations with constant coefficients and a zero right-hand side. Suppose that we have an equation (6.7)–(6.8) (with no free terms), and we want to find a solution of this equation that satisfies given initial conditions, i.e., which has the given values $x(t_0) = x_0$, $x'(t_0) = x_0'$, ..., $x^{(p-1)}(t_0) = x_0^{(p-1)}$. To find the corresponding solution $x(t)$, we do the following:

- First, we find all p roots $\alpha_1, \ldots, \alpha_p$ of the corresponding polynomial $Q(z) = 0$. Each root leads to *specific* solutions of the original functional equation:

    - For each *single* root α_j, we form a specific solution $\exp(\alpha_j \cdot t)$.

    - For each *double* root α_j, we form *two* specific solutions $\exp(\alpha_j \cdot t)$ and $t \cdot \exp(\alpha_j \cdot t)$.

    - For each *triple* root α_j, we form *three* specific solutions $\exp(\alpha_j \cdot t)$, $t \cdot \exp(\alpha_j \cdot t)$, and $t^2 \cdot \exp(\alpha_j \cdot t)$.

    - ...

    - For each root α_j of multiplicity m, we form m specific solutions $\exp(\alpha_j \cdot t), t \cdot \exp(\alpha_j \cdot t), \ldots, t^{m-1} \cdot \exp(\alpha_j \cdot t)$.

    It is known that if we count each double root twice, each triple root three times, etc., then the total number of roots is equal to the degree of the original polynomial. Hence, we get exactly p specific solutions $x_1(t), \ldots, x_p(t)$.

- From these specific solutions, we construct a *general* solution

$$x(t) = C_1 \cdot x_1(t) + \ldots + C_p \cdot x_p(t).$$

- Finally, we find the coefficients C_j by substituting the expression for the general solution into the equations that describe the initial conditions, and by solving the resulting system of p linear equations with p unknowns $C_1, \ldots, C_p$:

$$C_1 \cdot x_1(t_0) + \ldots + C_p \cdot x_p(0) = x_0,$$
$$C_1 \cdot x_1'(t_0) + \ldots + C_p \cdot x_p'(0) = x_0',$$

$$\ldots$$

$$C_1 \cdot x_1^{(p-1)}(t_0) + \ldots + C_p \cdot x_p^{(p-1)}(0) = x_0^{(p-1)}.$$

Comments.

- One can prove that this system of linear equations is always non-degenerate, and that therefore, this algorithm will always find the desired solution.

- When a double root α is a purely imaginary number, we get specific solutions of the type $t \cdot \cos(t)$ and $t \cdot \sin(t)$. These solutions represent oscillations whose amplitude grows with time. In physics, such oscillations are called *resonance*.

Example. As an example, let us solve the equation $x'' = 2x' - x$ with the initial conditions $x(0) = 1$ and $x'(0) = 0$. For this equation, our algorithm leads to the following solution:

- First, we solve the corresponding polynomial equation $Q(z) = z^2 - 2z + 1 = 0$. This equation has a double root $\alpha_1 = \alpha_2 = 1$.

- Since $\alpha_j = 1$ is a double root, we form two specific solutions

$$x_1(t) = \exp(\alpha_1 \cdot t) = \exp(t) \quad \text{and} \quad x_2(t) = t \cdot \exp(\alpha_1 \cdot t) = t \cdot \exp(t).$$

- From these specific solutions, we construct a *general* solution

$$x(t) = C_1 \cdot x_1(t) + C_2 \cdot x_2(t).$$

- Finally, we find the coefficients C_j by substituting the expression for the general solution into the equations that describe the initial conditions, and by solving the resulting system of two linear equations with two unknowns C_1 and C_2:

$$C_1 \cdot 1 + C_2 \cdot 0 = 1,$$
$$C_1 + C_2 = 0.$$

From this system, we get $C_1 = 1$ and $C_2 = -1$.

Hence, the desired solution is $x(t) = \exp(t) - t \cdot \exp(t)$.

Exercise

6.5 Solve the differential equation $x'' = 4x' - 4x$ with the initial conditions $x(0) = 0$ and $x'(0) = 1$.

6.9. Equations with non-zero right-hand sides

Equations with non-zero right-hand sides are important. Our whole interest in differential equations was caused by the fact that the problem of choosing a penalty function has lead us to a differential equation. This equation, however, had a non-zero right-hand side. Thus, for us, equations with non-zero right-hand sides are important.

How can we solve such equations?

Idea. The main idea of solving these equations is to reduce them to equations that we already know how to solve, i.e., to equations without any right-hand side. In other words, if we have an equation of the type

$$Q(\frac{d}{dt})x = y$$

with $y \neq 0$, we try to find a differential operator

$$R(\frac{d}{dt})$$

that will annul the right-hand side, i.e., for which

$$R(\frac{d}{dt})y = 0.$$

If we find such an operator R, then, by applying this operator R to both sides of the original equation

$$Q(\frac{d}{dt})x = y,$$

we get a new equation

$$R(\frac{d}{dt})Q(\frac{d}{dt})x = 0,$$

which we already know how to solve.

The only remaining question is: how to find this operator $R(z)$? The answer is easy: we know, from the above-described algorithm, which equations annul which functions. This description can be easily reversed. For example:

- We know that the equation $x' - \alpha \cdot x = 0$ with a single root α has a solution $C \cdot \exp(\alpha \cdot t)$. Thus, if we need to annul the expression $C \cdot \exp(\alpha \cdot t)$, we can take $R(z) = z - \alpha$.

- To annul a linear combination $C_1 \cdot \exp(\alpha_1 \cdot t) + C_2 \cdot \exp(\alpha_2 \cdot t)$, we can take an operator $R(z)$ that has two roots α_1 and α_2, e.g., $R(z) = (z - \alpha_1)(z - \alpha_2)$. Indeed, according to the above algorithm, the corresponding equation

$$R(\frac{d}{dt})x = 0$$

  has two specific solutions $x_1(t) = \exp(\alpha_1 \cdot t)$ and $x_2(t) = \exp(\alpha_2 \cdot t)$, and an arbitrary linear combination of these two specific solutions is also a solution (i.e., it is also annulled by the operator $R(z) = (z - \alpha_1)(z - \alpha_2)$).

- To annul a function $t \cdot \exp(\alpha \cdot t)$, we need an expression for which α is a double root, i.e., an expression of the type $R(z) = (z - \alpha)^2$.

- etc.

Example. Let us illustrate this idea on the example of the equation (6.3) $x' = \lambda \cdot \exp(t) + \mu \cdot x$ that come from the optimal choice of a penalty function. This equation has the form

$$Q(\frac{d}{dt})x = y$$

for $Q(z) = z - \mu$ and $y = \lambda \cdot \exp(t)$.

The right-hand side y of this equation has the form $C \cdot \exp(\alpha \cdot t)$ for $\alpha = 1$. Thus, according to the above idea, to annul this right-hand side, we can take $R(z) = z - 1$.

If we apply the corresponding operator $R(d/dt)$ to both sides of the equation $(d/dt - \mu)x = y$, we get a new equation

$$(\frac{d}{dt} - 1)(\frac{d}{dt} - \mu)x = 0$$

with a zero right-hand side.

We can now apply the algorithm from the previous section to this equation:

■ For this new equation, $Q(z) = (z-1)(z-\mu)$. Thus, the equation $Q(z) = 0$ has two roots $\alpha_1 = 1$ and $\alpha_2 = \mu$. Depending on whether these roots are equal or not (i.e., whether $\mu = 1$ or not), we get two different solutions:

– If $\mu \neq 1$, then the general solution of the new equation is

$$x(t) = C_1 \cdot \exp(\mu \cdot t) + C_2 \cdot \exp(t).$$

Substituting this expression into the equation (6.3), we conclude that

$$C_1 \cdot \exp(\mu \cdot t) + C_2 \cdot \exp(t) =$$

$$\mu \cdot C_1 \cdot \exp(\mu \cdot t) + \mu \cdot C_2 \cdot \exp(t) + \lambda \cdot \exp(t).$$

The coefficients at $\exp(\mu \cdot t)$ at both sides coincide. From the equality of the coefficients at $\exp(t)$, we conclude that $C_2 = \mu \cdot C_2 + \lambda$, i.e., that $C_2(1 - \mu) = \lambda$ and $C_2 = \lambda/(1 - \mu)$. Hence, for the case $\mu \neq 1$, the desired solution is

$$x(t) = C_1 \cdot \exp(\mu \cdot t) + \frac{\lambda}{\mu - 1} \exp(t).$$

– If $\mu = 1$, then the general solution to the new equation is

$$x(t) = C_1 \exp(t) + C_2 \cdot t \cdot \exp(t).$$

Substituting this expression into the equation (6.3), and taking into consideration that $\mu = 1$, we conclude that

$$C_1 \cdot \exp(t) + C_2 \cdot \exp(t) + C_2 \cdot t \cdot \exp(t) =$$

$$C_1 \cdot \exp(t) + C_2 \cdot t \cdot \exp(t) + \lambda \cdot \exp(t).$$

The coefficients at $t \cdot \exp(t)$ at both sides coincide. From the equality of the coefficients at $\exp(t)$, we conclude that $C_1 + C_2 = C_1 + \lambda$, i.e., that $C_2 = \lambda$. Hence, for the case $\mu = 1$, the desired solution is

$$x(t) = C_1 \cdot \exp(t) + \lambda \cdot t \cdot \exp(t).$$

From the viewpoint of the differential equation (6.3), we are done. However, let us recall that the original differential equation (5.4) that we started with was not an equation with constant coefficients. To transform the original equation with the unknown function $P(y)$ into this form, we introduced a new variable $t = \ln(y)$ and a new function $x(t) = P(\exp(t))$. So, to get the solution of the original differential equation, we must re-formulate our solution $x(t)$ in terms of the old variables $P(y)$. Substituting $t = \ln(y)$ into the above formulas, we get $P(y) = x(\ln(y))$:

- If $\mu \neq 1$, then

$$P(y) = C_1 \cdot \exp(\mu \cdot \ln(y)) + \frac{\lambda}{\mu - 1} \cdot \exp(\ln(y)).$$

- If $\mu = 1$, then

$$P(y) = C_1 \cdot \exp(\ln(y)) + \lambda \cdot \ln(y) \cdot \exp(\ln(y)).$$

The resulting expressions for $P(y)$ can be further simplified if we take into consideration that, by definition of a logarithm, $\exp(\ln(y)) = y$, and therefore, $\exp(\mu \ln(y)) = y^\mu$. Hence, the final expression for $P(y)$ is as follows:

- If $\mu \neq 1$, then

$$P(y) = C_1 \cdot y^\mu + \frac{\lambda}{\mu - 1} \cdot y.$$

- If $\mu = 1$, then

$$P(y) = C_1 \cdot y + \lambda \cdot \ln(y) \cdot y.$$

Exercise

6.6 Solve the differential equation $x'' + x = \cos(t)$ with the initial conditions $x(0) = 0$ and $x'(0) = 0$.

Comment. This equation describes a situation, in which the external perturbations occur at the system's own frequency. This is a typical case of resonance, the one that can cause, e.g., a well-designed bridge to vibrate and collapse.

7

SIMULATED ANNEALING: "SMOOTH" (LOCAL) DISCRETE OPTIMIZATION

In many real-life problems, we know what will happen if we make a decision, and what we want. Such problems are naturally formalized as optimization problems. In Lessons 5 and 6, we showed that when an optimization problem is continuous in nature, continuous mathematics is useful in designing an algorithm for this problem. In this lesson (and in the follow-up Lesson 8), we will show that continuous mathematics can help in designing algorithms for discrete optimization as well. Namely, in Lesson 7, we will show this on the example of simulated annealing, which is a good method for "almost smooth" discrete objective functions. In Lesson 8, the same idea will be applied to genetic algorithms, which optimize "non-smooth" objective functions.

7.1. Discrete optimization: Formulation of the problem

In many problems, we know exactly what will happen if we make a decision, and we know exactly what we want. Such problems are naturally formalized as *optimization problems*.

In Lessons 5 and 6, we showed that when an optimization problem is *continuous* (i.e., if the set of possible alternatives is continuous), then continuous mathematics can help in designing an algorithm.

Discrete optimization is important. In many problems, the set of possible alternatives is not continuous but *discrete* (usually, *finite*). The natural formalization of these problems leads to *discrete optimization* problems. For example, when we design a computing center, we must choose which computers and modems to buy, which software to buy, who to buy outside connection from, etc. Each of these decisions has only a finitely many alternatives.

Discrete optimization problems are often non-trivial. One may think that if there are only finitely many alternatives, then we can simply try them all, and no special algorithm is needed. In the simplest real-life problems, this is indeed true. However, in more complicated problems, the number of possible alternatives becomes so large that we are physically unable to simply try them all, we need an intelligent algorithm for solving the corresponding optimization problem.

For example, when designing a computing center, we face many choices; each of the choices is simple, there may be about 10 or so serious alternatives, and it is easy to test them all. However, the productivity of the computing center depends not only on the quality of the individual components, but also on how well these components agree with each other. Therefore, when designing a computing center, we must consider all possible *combinations* of alternatives. This number of combination can be enormously high. For example, if we have 10 features (a moderate number), and 10 alternatives for each feature (also quite a moderate number), then we have $10^{10} = 10$ billion possible combinations, much more than one can easily handle on a typical PC-type computer.

So, this problem *is* non-trivial.

What we are planning to do. Similar to the previous applications of continuous mathematics, this new application will follow the same pattern:

- First, we have a well-defined problem (in our case, it will be the problem of discrete optimization).

- Second, we have a class of possible algorithms that can be, in principle, used to solve this problem. Some of these algorithms are good, some are not.

- Finally, we do not know which algorithm we should choose.

We will show that continuous mathematics can help in choosing the best algorithm.

This application of continuous mathematics will be somewhat different for different discrete optimization problems, depending on how close these problems are to continuous optimization. So, before we describe this application in detail, let us briefly describe possible types of discrete optimization problems.

Two types of discrete optimization problems. In the last two lessons, we dealt with the optimization problems that are the easiest to solve: continuous optimization problems with differentiable objective function $J(x)$.

Why are these problems relatively easy to solve? From the practical viewpoint, the mathematical notion of differentiability means that if we change x a little bit, then, crudely speaking, the value of the objective function also changes only a little bit. (These local changes can be described by the derivatives of the objective function.) Therefore, if we have "almost" found the maximum, i.e., if we have found the point x for which $J(x)$ is almost equal to the desired maximum, then it makes sense to look for points in the small neighborhood of x for the desired maximum. The possibility to restrict the search to a small neighborhood saves lots of computer time.

A similar property of "smoothness" (small changes in x cause small changes in the value of the objective function $J(x)$) can occur for discrete optimization problems as well. For such "smooth" discrete optimization problems, we can use similar "neighborhood search" techniques. These techniques will be described and analyzed in this lesson.

In the following lesson, we will consider more complicated discrete optimization problems, in which small changes in x can cause drastic changes in the value of the objective function $J(x)$. For such "non-smooth" discrete optimization problems, local search makes no big sense, so different optimization algorithms are needed.

Let us start with an example of a "smooth" discrete optimization problem.

7.2. Knapsack: a (toy) example of a discrete optimization problem

An informal description. In the *knapsack* problem, we have n items of known costs $c_1, \ldots, c_n$ and weights $w_1, \ldots, w_n$, and we know the maximum weight W the knapsack can hold. Our *goal* is to fill the knapsack in such a way that the total cost of taken items is the largest possible.

Towards mathematical formulation of the problem. How can we formulate this problem in mathematical terms? Our goal is to determine, for each i from 1 to n, whether i-th object will be taken or not. In other words, for each i, we need a *Boolean* ("true"–"false") value. In most of the computers, "true" is represented by 1, and "false" by 0. Therefore, we can describe the desired solution as a sequence of values $t_1, \ldots, t_n$ (t for *taken*), so that:

- $t_i = 1$ means that we take i-th object;

- $t_i = 0$ means we do not take i-th object.

To express the total weight of the taken items in these terms, we must, for each object i, add its weight if the object is taken and do not add anything (i.e., add 0) if the object is not taken. This "conditional" addition is equivalent to adding the term $w_i \cdot t_i$:

- If $t_i = 1$, this terms is equal to w_i.

- If $t_0 = 0$, this terms is equal to 0.

Thus, the total weight of taken objects is equal to the following sum:

$$w_1 \cdot t_1 + \ldots + w_n \cdot t_n.$$

Similarly, the total cost of the taken objects is equal to

$$c_1 \cdot t_1 + \ldots + c_n \cdot t_n.$$

Knapsack: precise mathematical formulation. We can now formulate the desired discrete optimization problem:

$$c_1 \cdot t_1 + \ldots + c_n \cdot t_n \to \max$$

under the constraints that $t_i \in \{0, 1\}$ and

$$w_1 \cdot t_1 + \ldots + w_n \cdot t_n \leq W.$$

This problem is also called Ali-Baba problem. Another name for knapsack problem is *Ali-Baba* problem, after the name of the person, who, in a well-known tale from the Arabic Nights, has found his way into a cave where the forty thieves kept their treasure, and he could only carry away a certain amount.

(For simplicity, we will assume that the thieves were really accurate, and each object in this case is equipped with a label stating its cost and weight, like in a supermarket.)

This problem is non-trivial. In each instance of the knapsack problem, the total number of possible solutions $x = (t_1, \ldots, t_n)$ is finite. Since each variable t_i takes two possible values, we thus get $2 \times 2 \times \ldots \times 2 = 2^n$ possible combinations.

In principle, it may seem like it is possible to simply try all the combinations x and find out which of them leads to the largest value of $J(x)$. However, in reality, even for moderate number of objects (e.g., for $n = 300$), we need more computational time to test all these possible solutions than the lifetime of the Universe.

Thus, simple as it is, knapsack is an example of a *non-trivial* discrete optimization problem. It is worth mentioning (although we do not want to concentrate on this right now) that knapsack is an example of a so-called *NP-hard* problem, i.e., a discrete optimization problem that is as hard-to-solve as such problems can be.

When all objects are small, this discrete optimization problem is "smooth". In general, "smooth" means that a small change in x causes a small change in the value of the objective function $J(x)$. In our problem:

- x is a vector $(t_1, \ldots, t_n)$ of 0-1 values, and

- the objective function is $J(x) = \sum w_i \cdot t_i$.

Since x consists of the values t_i, a change in x means a change in some (or all) values. It is reasonable to call a change *small* if only a few values t_i are changed. The resulting change in $J(x)$ consists of the sum of the (few) corresponding terms w_i.

In principle, we may have a large object, with a large value of w_i. However, in many applications of the knapsack problem, all the objects are relatively small in the sense that:

- the weight w_i of each object is small as compared to the total admissible weight W, and

- the cost c_i of each object is small in comparison with the total cost that we can carry.

In such applications, small changes in x (i.e., changes that affect a few objects i) lead to a relatively small change in the values of the objective function. In other words, these applications are indeed "smooth".

Let us start describing algorithms for such "smooth" discrete optimization problems.

7.3. Local maximization algorithms: main idea

Full search is often impossible. In the knapsack example, the total number of possible alternatives x grows exponentially fast (as 2^n) with the size of the problem (i.e., with n), so that even for moderate values of n, the full search of all possible alternatives x becomes practically impossible.

Full search is often impossible for other discrete optimization problems as well.

The main idea of local minimization: local search (search in the neighborhood) instead of the global search. If for a given problem, full search is impossible, we must somehow restrict our search. Since we are considering "smooth" optimization problems, in which the notion of "closeness" makes sense, it is reasonable to restrict this search to a *neighborhood* of some point x, i.e., to all the points that are close to x.

Neighborhood must contain only a few points. The very need for restricting the search came from the fact that we had too many (exponentially many) points to look at. Thus, for this restriction to work, we must be sure that the newly defined "neighborhood" only contains a few points.

Let us check this for the knapsack example.

Neighborhoods for knapsack: Hamming metric. When we argued that knapsack is a "smooth" discrete optimization problem, we used the following natural notion of closeness: two vectors $x = (t_1, \ldots, t_n)$ and $x' = (t'_1, \ldots, t'_n)$

are close if they differ in a few points only, i.e., if there are only a few indices i for which $t_i \neq t'_i$.

If we formalize "a few" as meaning "1 or smaller", "2 or smaller", etc., we get a formal definition of neighborhood. This definition is known in communication theory, where due to noise, some transmitted bits may arrive corrupted: the number of indices i for which $t_i \neq t'_i$ is called the *Hamming distance* between the vectors x and x'.

Are the corresponding Hamming neighborhoods really small? Let us check. To define a neighborhood, we must fix some threshold distance d. The larger the threshold distance, the larger the neighborhood.

The smallest neighborhood corresponds to the threshold distance 1. For this threshold, a neighborhood of a point $x = (t_1, \ldots, t_n)$ is defined as a set of all points $x' = (t'_1, \ldots, t'_n)$ that differ from x in at most one place. This neighborhood contains:

- the original point x;

- n points
$$(1 - t_1, t_2, \ldots, t_n), (t_1, 1 - t_2, t_3, \ldots, t_n), \ldots,$$
$$(t_1, \ldots, t_{i-1}, 1 - t_i, t_{i+1}, \ldots, t_n), \ldots, (t_1, t_2, \ldots, t_{n-1}, 1 - t_n)$$
that are obtained from x by changing the value of exactly one index.

Totally, this neighborhood contains $1 + n$ elements. For reasonable n, this is a reasonable number (and much smaller than 2^n).

The second-in-size neighborhood corresponds to the threshold distance 2. For this threshold, a neighborhood of a point $x = (t_1, \ldots, t_n)$ is defined as a set of all points $x' = (t'_1, \ldots, t'_n)$ that differ from x in at most two places. This neighborhood contains:

- the original point x;

- n points that differ from x in exactly one place; and

- $n(n-1)/2$ points $x' = (t_1, \ldots, t_{i-1}, 1 - t_i, t_{i+1}, \ldots, t_{j-1}, 1 - t_j, t_{j+1}, \ldots, t_n)$ that differ from x in exactly two places: i and j.

Totally, this neighborhood contains $1 + n + n(n-1)/2$ elements. For reasonable n, this number is of order n^2 and is therefore, still quite reasonable (and still much smaller than 2^n).

In general, for distance d, the total number of points in a neighborhood is bounded by $\approx n^d$. For small d, this number is still reasonable and still $\ll 2^n$. So, for small d, Hamming neighborhoods are quite suitable for a local search.

Let us now return to the description of an algorithm.

Algorithm. This is an iterative algorithm in which, at any given moment of time, we have some candidate x for a solution. We start with some point $x = x^{(0)}$ (selected either based on some intuition, or, if no such intuition is available, simply at random). If the original optimization problem contains some constraints, then we try to choose this initial point $x^{(0)}$ that satisfies these constraints.

Then, on each iteration, we do the following:

- For each point y from the neighborhood of the current candidate x_{cur}, we check the constraints (if any). For all the points y that satisfy the constraints, we compute the value $J(y)$ of the objective function.

- Among all the neighboring points y that satisfy the given constraints, we choose a point x_{new} with the largest value of $J(y)$. This point is then selected as a new candidate, and the process starts anew.

The process stops when $x_{\text{new}} = x_{\text{cur}}$ (i.e., when no improvement has been achieved). The current candidate is then returned as the desired solution.

Comment. Since each point belongs to its own neighborhood, on each iteration, the value $J(x_{\text{new}})$ for the newly selected candidate will be at least as large as the value $J(x_{\text{cur}})$ for the current one. Thus, from iteration to iteration, we can only increase the value of the objective function.

An example where this algorithm work perfectly well. Let us illustrate this algorithm on an example of the following particular case of the knapsack problem: We have 4 objects, with costs $c_1 = 1$, $c_2 = 3$, $c_3 = 5$, and $c_4 = 2$, weights $w_1 = 4$, $w_2 = 6$, $w_3 = 7$, and $w_4 = 3$, and the total weight $W = 10$. As neighborhoods, we will take Hamming neighborhoods that correspond to a unit threshold.

For this problem, we can enumerate all $2^4 = 16$ possible combinations $x = (t_1, t_2, t_3, t_4)$, and conclude that the optimal solution is to take objects No. 3 and 4 (i.e., $t_1 = 0$, $t_2 = 0$, $t_3 = 1$, $t_4 = 1$). Let us show that the local minimization technique yields the same solution.

Starting value. We will start with the choice that definitely satisfies our constraint $\sum w_i \cdot t_i \leq W$: the choice $x^{(0)} = (0,0,0,0)$ in which we do not take anything at all. For this choice, $J(x^{(0)}) = \sum c_i \cdot t_i = 0$.

First iteration. The neighborhood of this point $x^{(0)}$ contains, in addition to this point itself, four other points $y_1 = (1,0,0,0)$, $y_2 = (0,1,0,0)$, $y_3 = (0,0,1,0)$, and $y_4 = (0,0,0,1)$. It is easy to check that all four points satisfy the constraint. Therefore, we compute $J(y)$ for all these points: $J(y_1) = 1 \cdot c_1 + 0 \cdot c_2 + 0 \cdot c_3 + 0 \cdot c_4 = 1$, $J(y_2) = 3$, $J(y_3) = 5$, and $J(y_4) = 2$. This value is the largest for $x_{\text{new}} = y_3$, so, we will take $y_3 = (0,0,1,0)$ as the new current candidate x_{cur}.

Second iteration. The neighborhood of the new candidate $x = (0,0,1,0)$ contains, in addition to the candidate itself, four other points (obtained by changing one of the values t_i): $y_1 = (1,0,1,0)$, $y_2 = (0,1,1,0)$, $y_3 = (0,0,0,0)$, and $y_1 = (0,0,1,1)$. For y_1, the total weight of taken objects is $w_1 + w_3 = 4 + 7 = 11 > 10$, so, y_1 does not satisfy the desired constraint. Similarly, y_2 does not satisfy this constraint. Thus, we only need to compare the value of $J(y)$ for the two points y_3 and y_4 that satisfy the constraint. For these points, $J(y_3) = 0$, and $J(y_4) = 7$. If we add the current point, with $J(x_{\text{cur}}) = 5$, to this list, we conclude that $J(y)$ is the largest for $x_{\text{new}} = y_4$. Thus, we take $y_4 = (0,0,1,1)$ as the new current candidate.

Third iteration. The neighborhood of the new candidate $x = (0,0,1,1)$ consists of this point x itself plus the following four points: $y_1 = (1,0,1,1)$, $y_2 = (0,1,1,1)$, $y_3 = (0,0,0,1)$, and $y_4 = (0,0,1,0)$. Of these points, only y_3 and y_4 satisfy the constraint, but for them, $J(y_3) = 5$ and $J(y_4) = 2$ are smaller than the value of $J(x)$ for the current candidate x. Since this iteration brought no improvement, the algorithm stops.

The current candidate $x = (0,0,1,1)$ is then returned as the desired solution. This solution is exactly what we were looking for.

An example in which the algorithm is not working well. In the above example, we start with a vector $(0,0,0,0)$ and compute the desired maximum.

However, this same example, with a different starting point, can show that the local maximization algorithm does not always lead to the desired maximum.

Indeed, let us take a new starting point $x^{(0)} = (1, 1, 0, 0)$. This point satisfies the constraint; for this point, $J(x) = 1 + 3 = 4$. The neighborhood of this point consists of this point itself plus four more points $y_1 = (0, 1, 0, 0)$, $y_2 = (1, 0, 0, 0)$, $y_3 = (1, 1, 1, 0)$, and $y_4 = (1, 1, 0, 1)$. From these point, only the points y_1 and y_2 satisfy the desired constraint. Both of these points, however, are worse than the current maximum: $J(y_1) = 3 < 4$ and $J(y_2) = 1 < 4$. Hence, the algorithm returns the same point $x^{(0)}$ as the new candidate. Since no improvement was achieved, the algorithm stops, and $x^{(0)} = (1, 1, 0, 0)$ is returned as the desired solution.

This is not, however, the desired maximum, because for the global maximum $(0, 0, 1, 1)$, the value of the objective function is larger: $J((0, 0, 1, 1)) = 7 > J((1, 1, 0, 0)) = 4$. So, in this case, the local maximization algorithm does not perform that well.

Why is this algorithm not working well. In the last example, we have a *local* maximum, in which the value $J(x)$ is larger than all the values in the neighborhood, but still not as large as possible.

This is a major problem, and for solving this problem, a new method of *simulated annealing* has been invented, that will be described in the next section.

Exercise

7.1 Apply the local maximization method to any other particular case of the knapsack problem.

7.4. Simulated annealing: idea

We must be able to "jump" out of a local maximum. Suppose that we started the local maximization process at some initial point $x^{(0)}$, and ended up in a local maximum. In this case, according to the above algorithm, we stop and give up. It is, therefore, desirable to modify the local maximization algorithm in such a way that after falling into the local maximum, we will be able to get out of it and try again.

In other words:

- If we are at a point that can be locally improved, we should follow the same local maximization algorithm.

- However, if we have reached a local maximum x (i.e., a point in which the value $J(x)$ of the objective function is greater than or equal to the values $J(y)$ in all neighboring points y), then we should not always stop in x, we should sometimes "jump" to one of the neighboring points y that satisfy the constraints.

By "jumping" to a new point, we *worsen* the situation (decrease the value of the objective function), in hope that this *decrease* will eventually pay off with a larger *increase* on the further iterations.

Since we want the algorithm to stop, we must use probabilities. Usually, we do not know whether we are in a local or in the global maximum. Therefore, we cannot make this "jump" deterministic, because otherwise, we would jump out of any maximum and never stop at all.

Hence, we must arrange this jump step in such a way that we jump only with a certain probability, and with the remaining probability, we stay in the local maximum. In other words, to decide whether we stay or jump, we should use some *random* physical process like a roulette.

Let us now start describing the jump process.

Where shall we jump to? Idea. If after looking at all the points $N(x)$ from the neighborhood of x (that satisfy the given constraints), we found out that we are at a local maximum x, then what point y from this set $N(x)$ shall we jump to?

Since the desired jump is a random step, this question can be reformulated as follows: with what probability $p(y)$ shall we jump to each point $y \in N(x)$? The probability of *staying* in the point x can also be expressed in the same way, as $p(x)$.

To determine these probabilities, let us use common sense.

- If for some point y, the value of $J(y)$ is very *small*, then we may lose all we have gained by jumping into the corresponding point y. Thus, even if we do jump into this point y, we should do it with a *small probability $p(y)$*.

- On the other hand, if for some other point y', the corresponding value $J(y')$ is reasonably *large*, we do not lose so much by jumping into y, so it makes more sense to do it. Hence, in this case, the corresponding *probability $p(y')$* can be *larger*.

In other words, the larger the value $J(y)$, the larger the probability $p(y)$.

Where to jump to? First guess. The above commonsense idea leads to the following natural formula: take

$$p(y) = f(J(y))$$

for some increasing function $f(z)$.

Problem with the first guess formula. There is one problem with this idea: as a result of the jump step, we must jump to one of the points $y \in N(x)$ (it could be the same point x, in which case "jumping to x" means that we actually stay at x). We cannot jump to two different points at the same time, and we must jump to one of these points. Thus, the sum of probabilities of jumping to different points $y \in N(x)$ must be equal to 1:

$$\sum_{y \in N(x)} p(y) = 1.$$

If we simply take $p(y)$ from the first guess formula, we cannot guarantee that this equality will be always satisfied.

Where to jump to? Final formula. A natural way to satisfy the desired equality $\sum p(y) = 1$ (a way typical for probability theory), is to *normalize* the probabilities, i.e., to replace the values $\widetilde{p}(y) = f(J(y))$ by the normalized values

$$p(y) = \frac{\widetilde{p}(y)}{\sum \widetilde{p}(z)} = \frac{f(J(y))}{\sum f(J(z))},$$

where the sum is taken over all points z from the set $N(x)$.

Computer implementation of the final formula requires random simulation. The above formula enables us to compute the probabilities

$p_1, p_2, \ldots, p_N$ of choosing all possible neighbors $y_1, y_2, \ldots, y_N$ that satisfy the constraints.

In order to implement the random step, we must be able to choose each of these neighbors y_i with the probability p_i. How can we do that?

Main idea of random simulation. Most programming languages have a built-in random number generator that generates numbers uniformly distributed on an interval $[0, 1]$. *Uniformly distributed* means that:

- If we divide the interval $[0, 1]$ into two equal parts $[0, 1/2]$ and $[1/2, 1]$, then the probability of this random number to be in each half is exactly $1/2$ (i.e., if we run this random number generator long enough, the fraction of numbers that turn out to be in $[0, 1/2]$ goes to $1/2$).

- If we divide the interval $[0, 1]$ into four equal parts $[0, 1/4]$, $[1/4, 1/2]$, $[1/2, 3/4]$, and $[3/4, 1]$, then the probability of this random number to be in each part is exactly $1/4$.

- etc.

In all these cases, the probability of a random number to be in each of the chosen subintervals ($[0, 1/2]$, $[1/2, 1]$, $[0, 1/4]$, $[1/4, 1/2]$, etc.) is equal to the width of this subinterval. In general, whatever subinterval $[a, b] \subseteq [0, 1]$ we choose, the probability that the random number is in this sub-interval is equal to the width $b - a$ of this subinterval.

Thus, to simulate the desired choice of a neighbor, we can do the following.

- The fact that $1 = p_1 + p_2 + \ldots + p_N$ means that we can subdivide the unit interval $[0, 1]$ into N sub-intervals of widths $p_1, p_2, \ldots, p_N$:

    - the sub-interval $[0, p_1]$ of width p_1;

    - the sub-interval $[p_1, p_1 + p_2]$ of width p_2;

    - ...

    - the sub-interval $[p_1 + p_2 + \ldots + p_{i-1}, p_1 + p_2 + \ldots + p_{i-1} + p_i]$ of width p_i;

    - ...

    - the sub-interval $[p_1 + p_2 + \ldots + p_{N-1}, p_1 + p_2 + \ldots + p_{N-1} + p_N]$ of width p_N.

■   Then, we run the standard random number generator. Since the width of
    i-th interval is exactly p_i, and the random number generator simulates the
    uniform distribution, the probability of the resulting number α to be in
    i-th interval is exactly p_i.

Algorithm for random simulation. Thus, we can make the desired selection
as follows:

■   First, we compute the values $p_i = p(y_i)$ (by using the above formula).

■   Second, we compute the partial sums $s_0 = 0$, $s_1 = s_0 + p_1$, $s_2 = s_1 + p_2$,
    ..., $s_i = s_{i-1} + p_i$, ..., $s_N = s_{N-1} + p_N = 1$.

■   Third, we run a standard random number generator that generates a ran-
    dom number α that is uniformly distributed on the interval $[0, 1]$.

■   Finally, by comparing α with the partial sums s_i, we find the index i for
    which $s_i \leq \alpha \leq s_{i+1}$, and jump to i-th neighbor y_i.

Now, we are ready to describe the final algorithm for discrete optimization:

Algorithm for discrete optimization. This algorithm requires that we fix
some increasing function $f(z)$.

This is an iterative algorithm in which, at any given moment of time, we have
some candidate x for a solution. We start with some point $x = x^{(0)}$ (selected
either based on some intuition, or, if no such intuition is available, simply at
random). If the original optimization problem contains some constraints, then
we try to choose the initial point $x^{(0)}$ that satisfies these constraints.

Then, on each iteration, we do the following:

■   For each point y from the neighborhood of the current candidate x_{cur}, we
    check the constraints (if any). For all the points y from this neighborhood
    that satisfy the constraints, we compute the value $J(y)$ of the objective
    function.

■   Among all the neighboring points y that satisfy the given constraints, we
    choose a point x_{new} with the largest value of $J(y)$. Then:

    – If $x_{\text{new}} \neq x_{\text{cur}}$, we take x_{new} as the new candidate, and start the
      process anew.

– If $x_{\text{new}} = x_{\text{cur}}$, then, from all the points y (from the neighborhood of x) that satisfy the constraints, we select a point y with a probability

$$p(y) = \frac{f(J(y))}{\sum f(J(z))}, \qquad (7.1)$$

where the sum is taken over all z from the neighborhood of x that satisfy the constaints.

* If the newly chosen candidate is still equal to x_{cur}, we stop.
* Otherwise, we start the process anew.

Why is this algorithm called annealing? A physical analogy. The above algorithm is called *simulated annealing*. Where did this name come from? What exactly is annealing and how is it related to this algorithm?

Annealing is a rather easy-to-describe physical process. Usually, at any give temperature and pressure, every substance has one and only one most stable state (i.e., the state with the smallest possible potential energy). Typically, this state is either a gas, or a liquid, or a crystal.

For example, at room temperatures, the most stable state of the iron is the crystal form (like in meteorites). This state of iron is not the best for most applications. For example, if we design a cutting instrument (in ancient days, a sword; in modern days, a manufacturing tool), we would prefer it to have different properties on the cutting edge and in the body itself. Luckily, for iron, there are states with such properties. These states are not *global* mimima of the potential function, but they are *local* mimuma and therefore, pretty stable. In physics, states corresponding to local mimima are called *meta-stable*.

For example, the famous Damascus steel from which the world's best swords were made is an example of a meta-stable state. Here is when annealing comes into picture. If we want to use a discarded piece of such a sword, we want to bring it back to the stable state. How can we do that?

■ One possibility would be to *melt* the sword; then, when the melted iron cools down, it will automatically get into the stable state. However, melting requires lots of heat, and this heat is mostly wasted, because after melting, we cool the iron back. Can we avoid this waste of energy?

■ Yes, we can, and this is what annealing is about. Since the sword is in a meta-stable state, i.e., in a state that is a *local* mimimum, all we have to do to push it into the *global* minimum is:

- to push it out of the local mimimum state by heating it a little bit;
- and then slowly cool it back so that hopefully, it will go into the state corresponding to the global minimum.

From the physical viewpoint, the heat means random fluctuations, so heating means that we apply a random "push" to get the state out of the local minimum. This heating and a consequent cooling is called *annealing*.

In some cases, one heating–cooling cycle is sufficient. In some other cases, we need to have several cycles.

From this description of physical annealing, one can see that the above-described optimization algorithm is, indeed, a direct computer analogie of physical annealing.

Exercise

7.2 Program the simulated annealing algorithm for knapsack (for simplicity, take $f(z) = z$). Check whether your algorithm finds the optimal solution for the example given at the end of Section 7.3.

> *Comment:* Do not be discouraged if it does not. This is, after all, an algorithm that uses random numbers, so it may sometimes work, and sometimes it may not work.

7.5. Simulated annealing: what function $f(z)$ should we choose?

The choice of $f(z)$ is a serious problem. In our description of simulated annealing, we did not specify how exactly we can choose the function $f(z)$.

One way to choose it is to simply take the function $f(z)$ from the physical annealing, but since the analogy between physical annealing and simulated annealing is qualitative rather quantitative, there does not seem to be much reason in such direct copying.

Numerical experiments show that the choice of the function $f(z)$ *is* important:

- a bad choice can keep the value in the local maximum (or jump around without improving the value of the objective function), while

- a good choice can lead to the desired global maximum fast.

We want the best $f(z)$: in what sense the best? We would like to choose a function $f(z)$ for which the simulatied annealing method works the best.

What do we mean by "the best"? It is not so difficult to come up with different criteria for choosing a function $f(z)$:

- We may want to choose the function $f(z)$ for which the average value of the objective function $J(x)$ in the resulting point x is the largest (i.e., for which the *quality of the answer* is, on average, the best).

- We may also want to choose the function $f(z)$ for which the *average computation time* is the smallest (average in the same of some reasonable probability distribution on the set of all problems).

- We may also want to minimize the *computer memory*, or any other characteristic of the computation or of the resulting solution.

At first glance, the problem of choosing $f(x)$ may seem hopeless. All these criteria sound quite meaningful; the trouble is that, unfortunately, we cannot do anything about them. We do not know how to estimate these meaningful numerical criteria even for a single function $f(z)$.

At first glance, the situation seems hopeless: since we cannot estimate these numerical criteria even for a single function $f(z)$, how can we undertake an even more ambitious task of finding the *optimal* function $f(z)$?

There is hope. Our successful discovery of the optimal software models (in Lessons 3 and 4), and of the optimal penalty function (in Lessons 5 and 6), in situations that, at first glance, may have seemed similarly hopeless, makes us want to try and apply similar methods here.

Going slightly ahead: this try does work, we do get the optimal function $f(z)$. To get to this result, let us follow the pattern established in Lessons 4 and 5.

It is reasonable to consider differentiable functions $f(z)$. From the purely mathematical viewpoint, this requirement is not really needed, but, as before, it makes the computations much easier.

A *plausbile* argument in favor of differentiability comes from the very motivation of using the function $f(z)$ itself. This argument was that if we increase $J(y)$, then our fear of decreasing the value of the objective function $J(x)$ decreases, and the hope of achieving the global maximum increases. Thus, we took the probability of jumping to y as an increasing function $p(y) \sim f(J(y))$ of $J(y)$. It seems natural to follow the same logic further: if we change the value $J(y)$ slightly, the hopes and fears also change slightly, so it makes sense to describe probabilities in such a way that *small* changes in $J(y)$ lead to *small* changes in probability. In other words, it is reasonable to assume that the function $f(z)$ is differentiable.

This is only a *plausible* argument, in no way a convincing proof, but since, as we said, we do not really need differentiability, we feel that it is OK to have this plausible argument.

The function $f(z)$ must be non-negative. Our goal is to find probabilities. Probabilities are always non-negative numbers, so the function $f(z)$ must also take only non-negative values.

We need to choose a family of functions, not a single function. All we want from the function $f(z)$ is the probabilities. These probabilities are computed according to the formula (7.1). From this expression (7,1), one can easily see that if we multiply all the values of this function $f(z)$ by an arbitrary constant C, i.e., if we consider a new function $\widetilde{f}(z) = C \cdot f(z)$, then this new function will lead (after the normalization involved in (7.1)), to exactly the same values of the probabilities:

$$\widetilde{p}(y) = \frac{\widetilde{f}(J(y))}{\sum \widetilde{f}(J(z))} = \frac{C \cdot f(J(y))}{\sum C \cdot f(J(z))} = \frac{f(J(y))}{\sum f(J(z))} = p(y).$$

Thus, whether we choose $f(z)$ or $\widetilde{f}(z) = C \cdot f(z)$, does not matter. So, what we are really choosing is not a *single* function $f(z)$, but a *family* of functions. This argument leads to the following definition:

Definition 7.1.

■   *Let $f(z)$ be a differentiable strictly increasing function from real numbers to non-negative real numbers. By a family that corresponds to this function $f(z)$, we mean a family of all functions of the type $\widetilde{f}(z) = C \cdot f(z)$, where $C > 0$ is an arbitrary positive real number.*

- *Two families are considered* equal *if they coincide, i.e., consist of the same functions.*

Denotation. *Let's denote the set of all possible families by* Φ.

In this lesson, we will consider optimality criteria on the set Φ of all families.

Natural symmetries. Similarly to the problem of choosing the optimal software reliability model and of choosing the optimal penalty function, we have natural *symmetries* here.

Namely, our ultimate goal is to find the solution to the discrete optimization problem $J(x) \to$ max under some constraints. This discrete optimization problem is, usually, a formalization of a real-life problem, in which we want to find an alternative x_{opt} that is the best. Our main goal is to find this best alternative (i.e., in the mathematical terms, the point x_{opt} at which this maximum is attained); we are, usually, not that much interested in the actual value of $J(x_{opt})$ for this optimal point x_{opt} (and once in a while we are interested in this value, we can always compute it by applying J to x_{opt}).

Now, if we add a constant to all the values of the function $J(x)$, i.e., if we consider a new function $\widetilde{J}(x) = J(x) + s$ for some constant s, then the *absolute value* of the objective function at different points will change, but the relation between the values at different points remain the same: if we had $J(x) > J(x')$, then we will still have $\widetilde{J}(x) = J(x) + s > J(x') + s = \widetilde{J}(x')$, etc. Hence, the value x_{opt} where maximum is attained will remain the same.

Thus, whether we maximize the original objective function $J(x)$ or the new objective function $\widetilde{J}(x) = J(x) + s$, this is the same optimization problem. It is therefore desirable to require that the relative quality of two different functions $f(z)$ and $g(z)$ should not change with this shift: If for the original function $J(x)$, $f(z)$ was better than $g(z)$, then $f(z)$ should be better than $g(z)$ for the "shifted" objective functions as well. How can we express this requirement?

If we "shift" the objective function, i.e., change from $J(x)$ to $\widetilde{J}(x) = J(x) + s$, we thus replace the original expression $f(J(x))$ in the formula for probability with a new expression $f(\widetilde{J}(x)) = f(J(x) + s)$. We can re-formulate this expression as $\widetilde{f}(J(x))$, where we denoted $\widetilde{f}(z) = f(z + s)$. Thus, *shifting the objective function* leads to exactly the same change in probabilities as keeping the same objective function but *switching to a new function* $\widetilde{f}(z) = f(z + s)$. Thus, the above invariance requirement can be reformulated as follows:

If $f(z)$ is better than $g(z)$, then the "shifted" function $\tilde{f}(z) = f(z + s)$ is better than the similarly shifted function $\tilde{g}(z) = g(z + s)$.

In other words, we want the optimality criterion (that enables us to compare different functions $f(z)$) to be *shift-invariant* (in the sense described in Lesson 4). Let us recall the corresponding definitions:

Definition 7.2. *Let s be a real number.*

■ *By a s-shift of a function $f(z)$ we mean a function $\tilde{f}(z) = f(z + s)$.*

■ *By a s-shift of a family of functions F we mean the family consisting of s-shifts of all functions from F.*

Denotation. *s-shift of a family F will be denoted by $S_s(F)$.*

Definition 7.3. *We say that an optimality criterion on Φ is shift-invariant if for every two families F and G and for every real number s, the following two conditions are true:*

i) *if F is better than G in the sense of this criterion (i.e., $F \succ G$), then $S_s(F) \succ S_s(G)$;*

ii) *if F is equivalent to G in the sense of this criterion (i.e., $F \sim G$), then $S_s(F) \sim S_s(G)$.*

It turns out that this requirement describes a reasonably small class of possibly optimal families:

Theorem 7.1. *If a family F is optimal in the sense of some optimality criterion that is final and shift-invariant, then this family corresponds to $f(z) = \exp(\beta \cdot z)$ for some $\beta > 0$.*

Comment.

■   This result is in good accordance with the experience of numerical methods, that show that this function $f(z)$ is indeed the most widely used and the most successful type of simulated annealing. (Interestingly, this is also the same function that comes from annealing in physics.)

■ Similarly to the results from Lessons 4 and 5, this result does not necessarily mean that in every possible situation, one of these functions $f(z)$ will be better than any other possible choice.

  – In every *specific* application, if we have a good understanding of the function $J(x)$ and of the constraints, then other functions $f(z)$ may turn out to be better than the functions described here.

  – However, if we want to choose the best method for *general* use, so that this method will be applicable to any situation, then it is natural to require shift invariance and thus, for such a general application, one of these functions will be indeed the best.

7.6. Proof

The outline of the proof. In proving this result, we will follow the sequence of steps similar to the ones used in the proofs of Theorems 4.2 and 5.1:

■ first, we will find a functional equation describing the desired optimal function $f(z)$;

■ then, we will show that the functions used in this equations are indeed differentiable;

■ after that, we will deduce the differential equation for this desired function $f(z)$;

■ finally, we will solve this differential equation and get the desired expression for the optimal function $f(z)$.

First step: deducing a functional equation. According to Proposition 4.1, the optimal family $\{C \cdot f(z)\}$ must be shift-invariant. Therefore, for every s, there must exist a value $C(s)$ such that

$$f(z + s) = C(s) \cdot f(z). \tag{7.2}$$

This is the desired functional equation.

Second step: proving differentiability. From the equation (7.2), we can conclude that $C(s) = f(z + s)/f(z)$. We have assumed that the function $f(z)$ is differentiable. Therefore, the function $C(s)$ is also differentiable.

Third step: deducing a differential equation. Now, we can follow the path we followed in Lessons 4 and 5:

- differentiate both sides of this functional equation by s;
- take $s = 0$ and thus, get the desired differential equation.

When we differentiate both sides of (7.2) by s, we get the following expression:

$$f'(z + s) = \frac{dC(s)}{ds} \cdot f(z).$$

When we substitute $s = 0$ into this formula, we get a differential equation: $f'(z) = C \cdot f(z)$.

Final step: solving the differential equation. We already know, from Lesson 6, how to solve such equations. As a result, we get $f(z) = C \cdot \exp(\beta \cdot z)$ for some β. The theorem is proven.

Comment. In this lesson, we used the invariance with respect to *shifting* all the values of the objective function $J(x)$, i.e., with respect to choosing a different starting point for $J(x)$. In principle, the values x_{opt} does not change also if we choose a different *unit* for measuring the values of the objective function, i.e., if we go from $J(x)$ to $\widetilde{J}(x) = \lambda \cdot J(x)$. In one of the Appendices, we use this additional invariance to further improve the simulated annealing algorithm.

8

GENETIC ALGORITHMS: "NON-SMOOTH" DISCRETE OPTIMIZATION

In Lesson 7, we described an algorithm (called simulated annealing) that solves "almost smooth" discrete optimization problems, i.e., problems in which a "small" change in the point x leads to a small change in the value of the objective function $J(x)$. In this lesson, we consider "non-smooth" discrete optimization problems. For such problems, a different class of algorithms has been developed: genetic algorithms that simulate evolution in nature.

8.1. "Non-smooth" discrete optimization problems

Discrete optimization problems: a brief reminder. Many important problems are formalized as problems of *discrete optimization* $J(x) \to \max$ under certain constraints. In many of these problems, the set of all possible alternatives is simply finite.

- *Theoretically*, it is possible to enumerate all possible alternatives x and find the one with the largest value of the objective function $J(x)$.

- However, *in reality*, the number of possible solutions is often astronomically large, so that it is not practically possible to enumerate all possible alternatives. It is therefore desirable to find more "intelligent" discrete optimization techniques.

"Non-smooth" discrete optimization problems and why they are important. Some of the discrete optimization problems have an extra property

153

of "smoothness": that a small change in the point x (e.g., changing a single bit in the binary representation of x) leads to a small change in the value of the objective function $J(x)$.

> For example, a knapsack problem is "smooth" if all the objects are relatively small.

For such problems, if we are already close to the maximum, it makes sense to consider new points y that are close for the current one: for these points, the value $J(y)$ is close to $J(x)$ and, therefore, also close to the maximum. This idea lies behind the methods of *local optimization* and *simulated annealing* described in Lesson 7.

However, not all discrete optimization problems have this "smoothness" property.

> For example, a knapsack may have several relatively large objects, so that taking or not taking one of these objects may drastically change the resulting value of the objective function.

"Non-smooth" discrete optimization problems are difficult to solve. For such problems, the closeness of y to x does not guarantee that $J(y)$ is close to $J(x)$. Therefore, if we are currently at a certain point x, but we are not satisfied with this x, then:

- on one hand, there is no special reason for testing points y that are *close* to x; but

- on the other hand, testing *all* points y, whether they are close to x or not, means the full search, and we already know that full search is often not feasible.

So, what do we do?

Idea.

- The local maximization algorithm described in Lesson 7 was based on the possibility to take a point x that is reasonably close to a solution, and to transform into several possible reasonable candidate solutions.

- For "non-smooth" discrete optimization problems, we cannot follow the same idea, because we cannot transform a *single* point x (that is close to the solution) into another point y that is also possibly close to the solution. Since we cannot transform a *single* point, a natural idea is to try to take *two* points x, x' and transform them into a new candidate y.

From the biological viewpoint, this idea may sound familiar. The idea is to take two different objects and produce a new one as an "offspring". This is a direct analogy of what happens in nature, when two organisms meet and a new offspring is born as a result.

It is therefore natural to try to use this biological analogy with reproduction process to develop reasonable algorithms for solving "non-smooth" discrete optimization problems. This analogy is indeed successfully used. The corresponding algorithms, called *genetic* algorithms, will be described in this lesson.

It is even more reasonable to use the biological analogy if we recall that according to the evolution theory, reproduction is, actually, often very selective, and the whole purpose of this selection is to produce the *fittest* offspring, the offspring with the largest potential for survival in the (often hostile) environment. Aiming at "the fittest" means that nature is, actually, solving an *optimization* problem.

Moreover, the very fact that a "family-type" reproduction (in which two parents produce an offspring) is, at present, the prevalent way of biological reproduction, means that over billions of years of highly competitive evolution this particular way of reproduction turned out to be the best possible in the sense that it leads to the fittest species.

Before we describe genetic algorithms in detail, let us briefly recall the corresponding biological processes.

8.2. Survival of the fittest and evolution in nature as an example of optimization

Survival of the fittest as optimization. In nature, organisms that are best suited for their environment have the best chances to survive and to produce offspring. Organisms in nature compete for food, space, for other resources, and for mates. The stronger and/or smarter organisms have a better chance of finding a mate and reproducing than the weaker and/or dumber organisms. This is called *survival of the fittest*.

As a result, less fit organisms are eventually replaced by the fitter and fitter ones. This process of constant improvement is called *evolution*.

The genes as a description of the organism. How can we describe an organism? Our main goal is to describe the evolution. From this viewpoint, we are not interested in minor details of an individual organism, we are mainly interested in major features that are essential for survival.

These features are determined by the organism's *genetic code*. In these terms, survival of the fittest means survival of the fittest genes.

What is a genetic code? Almost all of the information about an organism is contained in its *genetic code*. A genetic code is contained in a molecule called *DNA*. This molecule has the shape of a double helix and consists of two closely connected strands of nucleic acids. There are four types of acids, that are denoted by A, G, C, and T. The acid on the first strand uniquely determines the corresponding acid on the second strand:

- A in the first strand is always teamed up with T on the second strand;

- G in the first strand is always teamed up with C on the second strand;

- C in the first strand is always teamed up with G on the second strand;

- T in the first strand is always teamed up with A on the second strand.

As a result, depending on the acid in the first strand, there are four possible pairs: A–T, G–C, C–G, and T–A.

So, a genetic code is uniquely determined by a sequence of acids in the first string.

For example, a code can consist of the letters AGGTC...

In other words, a code is a word (sequence of symbols) in a *genetic alphabet* consisting of four symbols.

Comment.

- In biology, for convenience of analysis and description, this sequence is usually divided into pieces that are responsible for different features of the organism; these pieces are called *genes*.

- In this text, our goal is not to go deeply into biological analysis, but simply to reproduce the very *basic* features of genetic code and evolution. Therefore, in this text, we will simply treat the genetic code as a single long sequence.

Reproduction from genetic viewpoint. In a reproduction process, the DNAs that contain the parents' genetic codes combine to produce the child's DNA. Since two DNAs are made into one DNA, some defects of each of the parents do not go into the child's genes and thus, the child can survive better than the parents.

How is the DNA of a child formed? The corresponding process is called *crossover*.

- At first, both DNAs are in their double helix form.

- Then each parent's DNA un-twists and splits into two strands.

- The first strand of the mother's DNA and the first strand of the father's DNA line up together.

- These two strands:
    - get "glued" to each other,
    - twist randomly around each other, and then
    - "un-tangle".

    As a result, we get a strand that is partially formed by acids from the father's DNA, and partially by acids from the mother's DNA.

This resulting strand is the child's DNA.

In other words, the child's DNA is formed as follows:

- the first several acids of the child's DNA are the same as the corresponding acids from one parent's DNA;

- the next few acids are the same as the corresponding acids from the other parent;

- etc.

Mutation: an additional mechanism that leads to survival of the fittest. Another way in which DNA changes is by *mutation*, i.e., by the changing of one or a few genes. Mutations are caused by chemical substances or radioactive elements that influence the DNA molecule.

This influence is usually small and therefore, it usually only affects one nucleic acid.

What we are planning to do. Now that we have recalled how optimization is done in nature, we will describe algorithms that simulate this evolution. Then, we will show that continuous mathematics can help to obtain the best possible version of a genetic algorithm.

Historical comment. The idea of using simulated biological evolution to solve optimization problems dates back to the 60s (H. J. Bremmerman, M. Rogson, S. Salaff, and others). Genetic algorithms, in their present form, were invented by John H. Holland from the University of Michigan at Ann Arbor. Since then, many improving modification have been proposed, and the resulting algorithms are among the most promising optimization and machine learning techniques: see, e.g.,

> D. E. Goldberg, "Genetic and Evolutionary Algorithms Come of Age" (*Communications of the ACM*, March 1994, Vol. 37, No. 3, pp. 113–119)

and references therein, in particular,

- D. E. Goldberg, *Genetic algorithms in search, optimization, and machine learning* (Addison-Wesley, Reading, MA, 1989).

- Y. Davidor, *Genetic algorithms and robotics: A heuristic strategy for optimization* (World Scientific, Singapore, 1991).

In this lesson, we will only describe *simple* genetic algorithms, algorithms that are close to the original Holland's idea.

8.3. Genetic algorithms: main idea

Main objects.

- In biological evolution, an organism is characterized by its genetic code, i.e., by a *word* in the 4-letter alphabet.

- Inside the computer, all the numbers, objects, etc., are represented by using a *binary code*, i.e., every object is represented as a sequence of 0's and 1's.

It is, therefore, natural to take this binary representation x as the desired computer description of different objects.

Fitness function.

- Biological evolution results in the survival of the fittest organisms. If we characterize the fitness of an object x by a number $f(x)$, then we can say that biological evolution leads to maximizing the value of the fitness function $f(x) \rightarrow$ max. Usually, the way to characterize fitness by numbers satisfies the natural condition that an organism that is not fit at all is characterized by the fitness value $f(x) = 0$, and all the other organisms have a positive value of fitness.

- Our goal is to maximize the given objective function $J(x)$.

In view of this analogy, in genetic algorithms, the objective function is usually called a *fitness function*. It is usually assumed that the fitness function takes only non-negative values (i.e., that $J(x) \geq 0$ for all x).

Selection of the fittest "parents".

- In nature, for an organism to participate in the reproduction process, it must first survive. In nature, there are many random unpredictable forces that can influence the organism's survival. If we start with two organisms, one better fit that the other, then it is quite possible that, in *some* situations, the better fit will perish while the weaker one will survive. However, in *most* situations, the fittest one will survive. In other words, the fitter the organism, the larger its probability of survival.

■ Similarly, in computer simulations, we will simulate survival, and only let the surviving objects mate.

The larger the value of the fitness function $J(x)$, the larger the probability $p(x)$ that the object x will survive and therefore, be able to mate. How to describe the probability of being selected for reproduction?

Survival probabilities: first guess. The problem of designing appropriate survival probabilities $p(x)$ is similar to the problem of assigning "jump" probabilities $p(y)$ that we described in Lesson 7: in both cases, the larger the value of the objective (fitness) function, the larger the probability should be. It is, therefore, natural to take the probability $p(x)$ of choosing an object x to be an increasing function of $J(x)$: $p(x) = f(J(x))$ for some increasing function $f(z)$.

Problem with the first guess formula. Similarly to the simulated annealing case, there is a problem with this idea: Our goal is to select an organism that will mate. Thus, the total probability of selecting one of the existing organisms must be equal to 1: $\sum p(x) = 1$. If we simply take $p(x)$ from the first guess formula, we cannot guarantee that this equality will be always satisfied.

Survival probabilities: final formula. Similarly to Lesson 7, a natural way to satisfy the desired equality $\sum p(x) = 1$ (a way typical for probability theory), is to *normalize* the probabilities, i.e., to replace the values $\tilde{p}(x) = f(J(x))$ by the normalized values

$$p(x) = \frac{\tilde{p}(x)}{\sum \tilde{p}(y)} = \frac{f(J(x))}{\sum f(J(y))},$$

where the sum is taken over all existing organisms x.

Survival probabilities: example. Let us illustrate this formula on an example of a knapsack problem that we considered in Lesson 7. In this example, we are looking for a 4-bit sequence $x = (t_1, t_2, t_2, t_4)$, $t_i \in \{0, 1\}$, for which $J(x) = t_1 + 3t_2 + 5t_3 + 2t_4 \to$ max under the constraint that $4t_1 + 6t_2 + 7t_3 + 3t_4 \leq 10$. Let us assume that at some point, we have three organisms $x_1 = 0000$, $x_2 = 0100$, and $x_3 = 0010$ that satisfy the above constraint. For the simple choice of $f(z) = z$, we have $\tilde{p}(x_1) = J(x_1) = 0$, $\tilde{p}(x_1) = J(x_2) = 3$, and $\tilde{p}(x_1) = J(x_1) = 5$. The sum of these values is equal to $0 + 3 + 5 = 8$, so the selection probabilities are $p(x_1) = 0/8 = 0$, $p(x_2) = 3/8 = 0.375$, and $p(x_3) = 5/8 = 0.625$.

Crossover: algorithm. We can use the above formula to pick the first parent x; we can, similarly, pick the second parent x'. Now, we must describe their "offspring". The process of computing the offspring is called *crossover*.

In genetic algorithms, we use the exact same crossover as in the biological genetic code. Let us assume that the codes that we are dealing with are binary sequences of length L: $x = x[1]x[2]\ldots x[L]$. To describe the child c of the two parents x and x', we first choose the number of places p at which the "combined" DNA breaks. Then:

- We pick p breaking places $b_1, \ldots, b_p$ at random.

  For that, we do the following:
  * We run another standard random number generator to pick p random integers $r_1, \ldots, r_p$ between 1 and L[1]. If any of the two integers r_i are equal, we repeat the procedure until we get them all different.
  * After that, we sort the values $r_1, \ldots, r_p$ into the desired ordered sequence $b_1 < b_2 < \ldots < b_p$. We also define $b_{p+1} = L$.

- Then, we form bits $c[1], c[2], \ldots, c[L]$ of the child's "DNA" as follows:
  - For all j from 1 to b_1, we copy the bits $x[j]$ from the first parent's DNA x into the child's DNA c: $c[j] \leftarrow x[j]$.
  - For all j from $b_1 + 1$ to b_2, we copy the bits $x'[j]$ from the second parent's DNA x' into the child's DNA c: $c[j] \leftarrow x'[j]$.
  - For all j from $b_2 + 1$ to b_3, we copy the bits $x[j]$ from the first parent's DNA x into the child's DNA c: $c[j] \leftarrow x[j]$.
  - etc.

Crossover: example. To illustrate the crossover algorithm, let us consider a simple example. In this example, we start with two sequences $x = 01010001$ and $x' = 10100110$; $L = 8$ and $p = 2$. Then:

---

[1] In most programming languages (including C and C++) there is no direct way to get a number from 1 to L, but there is a way to get a number from 0 to a given integer with equal probabilities. So, to get a number from 1 to L, we generate a random integer from 0 to $L - 1$ and then add 1 to it.

    – First, we find the breaking places. For that, we do the following:

        * First, we run a standard random generator to pick 2 random integers between 1 and 8. In this example, we got $r_1 = 6$ and $r_2 = 3$.

        * After that, we sort the values 5 and 3 into the ordered sequence $b_1 = 3 < b_2 = 6$.

    – Then, we form the bits of the child's DNA as follows:

        * For all j from 1 to $b_1 = 3$, we copy the bits $x[j]$ from the first parent's DNA x into the child's DNA c. Thus, we get $c[1] = 0$, $c[2] = 1$, and $c[3] = 0$.

        * For all j from $b_1 + 1 = 4$ to $b_2 = 6$, we copy the bits $x'[j]$ from the second parent's DNA x' into the child's DNA c. Thus, we get $c[4] = 0$, $c[5] = 0$, and $c[6] = 1$.

        * Finally, for all j from $b_2 + 1 = 7$ to $b_3 = L = 8$, we copy the bits $x[j]$ from the first parent's DNA x into the child's DNA c. Thus, we get $c[7] = 0$ and $c[8] = 1$.

The resulting sequence is $c = 01000101$.

Mutation: algorithm. The sequence generated by a crossover may not necessarily be what the algorithm returns as child's DNA, because after the crossover, we may want to add some *mutation*.

■   From the biological viewpoint, a mutation is a local change, usually, a change in a single codon (element of the sequence that defines the genetic code). Since mutation is a random process, the location of a mutated codon is completely random.

■   It is, therefore, natural to define a mutation step in a genetic algorithm in a similar manner, as a change of a single bit at a random place.

To describe the corresponding step, we must select a *mutation probability* p_{mut}. Then, we do the following:

■   With probability $1 - p_{mut}$, we leave the child unchanged.

■   With probability p_{mut}, we select a random integer b from 1 to L, and change b-th bit in the child's binary sequence.

To make a change with a known probability p_{mut}, we can follow the same idea as we used in Lesson 7:

- We run a standard random number generator to generate a number α that is uniformly distributed on the interval $[0, 1]$.

- Then, we do the change only if $\alpha \leq p_{mut}$.

Since α is uniformly distributed, the probability of α to be in the interval $[0, p_{mut}]$ is equal to the length of this interval, i.e., to the desired number p_{mut}.

Mutation: example. Let us trace the mutation algorithm on the same example that we used to trace crossover. Let us assume that $p_{mut} = 0.1$. After the crossover, we got $c = 01000101$. To check whether mutation is needed, we run a random number generator. When we ran this generator, we got $\alpha = 0.1356$. Since $0.1356 > 0.1$, we return the unchanged binary string c as the child's DNA.

This returning of the unchange value will happen in $1 - p_{mut} = 90\%$ of all cases. To illustrate the case of actual mutation, we ran the random number generator several times until we got a number that was smaller than 0.1: $\alpha = 0.0863 < 0.1$. If this is the output of the random number generator, then we pick a random number b from 1 to $L = 8$ and change b-th bit. Our random number generator returned $b = 5$, so, we change 5-th bit of c (from 0 to 1), and get a new "mutated" child $c = 01001101$.

Now that we have described all components of the genetic algorithms, let us combine these descriptions into the description of the algorithm itself:

Genetic algorithm: description. This algorithm is designed to solve discrete optimization problems of the type $J(x) \rightarrow \max$ under some constraints. To apply this algorithm, we assume that the objects x among which we maximize are represented in the computer as binary (0–1) sequences of a certain length L.

In the genetic algorithm, we will simulate evolution generation after generation. To apply genetic algorithm, we must specify the following:

- The n number of organisms in each generation. This number is called the *population size*.

- An increasing function $f(z)$ from real numbers to real numbers. This function is called *fitness scaling*.

■   A number p_{mut} from the interval $[0, 1]$. This number is called a *probability of mutations*.

The algorithm itself runs as follows:

■   First, we choose n random elements $x_1^{(1)}, \ldots, x_n^{(1)}$ of length L. These elements will be called *organisms of the first generation*.

   – If we are solving an unconstrained optimization problem, then we simply generate all L bits of all n elements at random.

   – If we are solving a constrained optimization problem, then after generating a random object x, we check whether this object satisfies the constraints, and if not, generate a new random objects until the constraints are satisfied.

■   The algorithm consists of transforming the first generation into the second one, the second one into the third one, etc. Let us describe how the k-th generation $x_1^{(k)}, \ldots, x_n^{(k)}$ is transformed into the next generation $x_1^{(k+1)}, \ldots, x_n^{(k+1)}$. The elements of the next generation are generated one by one, so it is sufficient to show how each of them is generated. Before we generate these elements, we need to compute the survival probabilities $p_1, \ldots, p_n$:

   – First, we compute the fitness values $F(x_i^{(k)})$ for all organisms from k-th generation.

   – Then, we apply the fitness scaling function $f(z)$ to these fitness values, and get "pre-probabilities" $\widetilde{p}_i = f(F(x_i^{(k)}))$.

   – Finally, we compute the sum $s = \sum \widetilde{p}_i$ of these values, and the desired survival probabilities $p_i = \widetilde{p}_i / s$.

To generate one of the elements $x_i^{(k+1)}$ of the next generation, we do the following:

   – First, we pick an element $x_i^{(k)}$ with probability p_i as the first parent x (in Lesson 7, we showed how this can be done).

   – Second, we pick another element $x_j^{(k)}$ with probability p_j as the second parent x'.

   – We apply the above crossover procedure to compute a child c.

   – Finally, with probability p_{mut}, we change a random bit in the child's binary sequence.

If the resulting binary sequence satisfies the desired constraints, we proclaim it the organism from the next generation. Otherwise, we repeat the same reproduction procedure again and again until the constraints are satisfied.

- On each generation k, we look for the "record" value

$$\max(J(x_1^{(k)}), \ldots, J(x_n^{(k)})).$$

When the next generation does not lead to a tangible improvement, we stop and produce the "best" organism generated so far as the solution to the desired optimization problem.

The main steps of this algorithm can be represented as a following diagram:

$$\text{given } x_1^{(k)}, \ldots, x_n^{(k)}$$
$$\downarrow$$
$$\text{compute } F(x_i^{(k)})$$
$$\downarrow$$
$$\text{compute } \tilde{p}_i = f(F(x_i^{(k)}))$$
$$\downarrow$$
$$\text{compute } p_i = \tilde{p}_i / \sum \tilde{p}_i$$
$$\downarrow$$
$$\text{choose } x_i^{(k)} \text{ with probability } p_i$$
$$\downarrow$$
$$\text{apply crossover and get } x_1^{(k+1)}, \ldots, x_n^{(k+1)}$$

Exercise

8.1 Program the genetic algorithm and apply it to the knapsack problem. For simplicity, use $f(z) = z$.

8.4. Genetic algorithms: what fitness scaling $f(z)$ should we choose?

The choice of the fitness scaling function is important. At first glance, it may seem that we can always use the simplest possible fitness scaling function $f(z) = z$. However, in many cases, this is a bad choice:

- First of all, it may happen that the function $J(x)$ that we want to maximize takes negative values for some x. In this case, we cannot use $f(z) = z$, because this choice of fitness scaling would lead to meaningless negative values of probabilities.

- Even if $J(x) \geq 0$ for all x, the simplest fitness scaling function is often not the best:

  - In the beginning, the genetic algorithm often leads to the appearance of a few "super-organisms" with relatively large values of the objective function (relatively large but still far from the desired maximum), which all have basically the same genes. If we use $f(z) = z$, then the probability of any other organism to reproduce is practically 0. Therefore, these "super-organisms" dominate the selection process. However, since they all have approximately the same genes, they will never evolve and the evolution process will stop. To be more precise, the evolution will go on (e.g., due to mutations), but since we have much fewer choices than we may think based on the population size, the evolution drastically slows down.

    To prevent this slow-down, we need a fitness scaling function $f(z)$ that would "help" the underdeveloped organisms compete. In other words, we need a function $f(z)$ for which the initial ratio $z_1/z_2 \gg 1$ would lead to a decreased ratio $f(z_1)/f(z_2) \ll z_1/z_2$.

  - At the end, when the population consists largely of the organisms x, for which $J(x)$ is close to the maximum, all the values $J(x)$ are almost equal and, as a result, if we use the fitness scaling function $f(z) = z$, the resulting selection probabilities of different organisms will be practically equal to each other. Therefore, the competition is practically absent, which again slows down the process.

    To prevent this slow-down, we need a fitness scaling function $f(z)$ that would promote the slightly better organisms. In other words, we need a function $f(z)$ for which the initial ratio $1 < z_1/z_2 \approx 1$ would lead to an increased ratio $f(z_1)/f(z_2) \gg z_1/z_2$.

Which fitness scaling functions are actually used in genetic algorithms. Genetic algorithms started with the simple choice of the function $f(z) = z$. In this case, to determine the probability of selecting x, we simply divide the fitness $J(x)$ of the organism x by the sum $\sum J(y)$ of all the fitnesses.

The next simplest increasing function is a linear function $f(z) = a \cdot z + b$, where a and b are constants. This scaling is useful when some of the values of the objective function $J(x)$ are negative. Then, if we know the lower bound $B < 0$ for the function $J(x)$, the choice of $a = 0$ and $b = -B$ will make all the values $f(J(x))$ non-negative.

Two non-linear scaling functions are also actively used:

- *Power law scaling* $f(z) = z^\alpha$ for some constant $\alpha > 0$.

  - Power scaling with $\alpha < 1$ solves the problem of the "super-organisms" at the beginning of the process. It solves it because it decreases the difference between the fitnesses of a "super-organism" and the fitnesses of other organisms.

    For example, if $J(x_1) = 1$ and $J(x_2) = 8$, the ratio $J(x_2)/J(x_1) = 8$. If we use $f(z) = z^{1/3}$, then $f(J(x_2))/f(J(x_1)) = 8^{1/3}/1^{1/3} = (8/1)^{1/3} = 2 \ll 8$.

  - Power scaling with $\alpha > 1$ solves the problem of "no competition" at the end of the process. solves this problem because it amplifies the difference between two almost equal fit organisms.

    For example, if $J(x_1) = 1$ and $J(x_2) = 2$, the ratio $J(x_2)/J(x_1) = 2$. If we use $f(z) = z^3$, then $f(J(x_2))/f(J(x_1)) = 2^3/1^3 = (2/1)^3 = 8$.

  Power law scaling is mainly used in *machine vision*.

- *Exponential scaling.* In exponential scaling, the scaling function is $f(z) = \exp(\beta \cdot z)$ for some constant $\beta > 0$. Depending on the choice of β, we can get the same results as the power law fitness scaling function. Exponential scaling is most commonly used in *robotics*.

Numerical experiments confirm that the choice of the fitness scaling function $f(z)$ *is* important:

- a bad choice can keep the value in the local maximum (or jump around without improving the value of the objective function), while

- a good choice can lead to the desired global maximum fast.

Choosing the best fitness scaling function is a difficult problem. We would like to choose a fitness scaling function $f(z)$ for which the genetic algorithm works the best. Similarly to simulated annealing, we can, in principle, formulate several possible meaning of the term "the best":

- We may want to choose the function $f(z)$ for which the *average computation time* is the smallest (average in the same of some reasonable probability distribution on the set of all problems).

- We may also want to choose the function $f(z)$ for which the average value of the objective function $J(x)$ in the resulting point x is the largest (i.e., for which the *quality of the answer* is, on average, the best).

- etc.

However, as for simulated annealing, we do not know how to compute the quality of a *single* fitness scaling function, not to talk about choosing the best one.

Therefore, we will use an *indirect* approach to optimization similar to the one used in Lesson 7.

We need to choose a family of functions, not a single function. The only purpose of using the fitness scaling function $f(z)$ is to generate probabilities. From the expression for probabilities, one can easily see that if we multiply all the values of the function $f(z)$ by an arbitrary constant C, i.e., if we consider a new function $\tilde{f}(z) = C \cdot f(z)$, then this new function will lead to exactly the same values of probabilities. Thus, what we are really choosing is not a single function $f(z)$, but a *family* of functions $\{C \cdot f(z)\}$ that correspond to all possible positive real numbers $C > 0$.

We can, therefore, repeat a definition from Lesson 7:

Definition 8.1.

- Let $f(z)$ be a differentiable strictly increasing function from real numbers to real numbers. By a family that corresponds to this function $f(z)$, we mean a family of all functions of the type $\tilde{f}(z) = C \cdot f(z)$, where $C > 0$ is an arbitrary positive real number.

- *Two families are considered equal if they coincide, i.e., consist of the same functions.*

Comment. In this lesson, we will consider optimality criteria on the set Φ of all possible families.

Natural symmetries. Similarly to Lesson 7, we can consider two possible transformations that, intuitively, should not change the relative quality of different families:

- First, we can, as in Lesson 7, *change the starting point* for the quantity that describes the objective function $J(x)$. In mathematical terms, this means that we move from the original objective function $J(x)$ to a new *shifted* function $\widetilde{J}(x) = J(x) + s$.

  We have already noticed in Lesson 7 that applying the old function $f(z)$ (in our new case, the old fitness scaling function $f(z)$) to the new objective function $\widetilde{J}(x) = J(x) + s$ is equivalent to applying, to the old objective function $J(x)$, of a new fitness scaling function $\widetilde{f}(z)$ defined as $f(z + s)$:

  $$\widetilde{f}(J(x)) = f(\widetilde{J}(x)).$$

  Therefore, the requirement that

  > *the relative quality of two different fitness scaling functions should not depend on the choice of the starting point for describing $J(x)$*

  can be re-formulated as follows:

  > *an optimality criterion should be shift-invariant.*

- Second, we can also *change the units* for the quantity that describes the objective function $J(x)$. For example, if our goal is to minimize the distance, then we could switch from miles to kilometers in describing the quality $J(x)$ of each possible route. In mathematical terms, switching to a new unit that is λ times smaller means that we move from the original objective function $J(x)$ to a new *scaled* function $\widetilde{J}(x) = \lambda \cdot J(x)$.

  Applying the old fitness scaling function $f(z)$) to the new objective function $\widetilde{J}(x) = \lambda \cdot J(x)$ is equivalent to applying, to the *old* objective function $J(x)$, of a new fitness scaling function $\widetilde{f}(z)$ defined as $f(\lambda \cdot z)$:

  $$\widetilde{f}(J(x)) = f(\widetilde{J}(x)).$$

  Therefore, the requirement that

the relative quality of two different fitness scaling functions should not depend on the choice of the units in which we express $J(x)$

can be re-formulated as follows:

an optimality criterion should be scale-invariant (in the sense described in Lesson 4).

First result: fitness scaling functions that are optimal with respect to a shift-invariant optimality criterion. If the optimality criterion is shift-invariant, then we in exactly the same situation as described and solved in Lesson 7. Therefore, in this case, Theorem 7.1 leads to the following corollary:

Corollary. *If a family F of fitness scaling functions is optimal in the sense of some optimality criterion that is final and shift-invariant, then this family corresponds to $f(z) = \exp(\beta \cdot z)$ for some $\beta > 0$.*

Comment. This result explains why exponential scaling is indeed widely used: it is optimal with respect to many different criteria.

Second result: fitness scaling functions that are optimal with respect to a scale-invariant optimality criterion. Let us recall the corresponding definitions from Lesson 4:

Definition 8.2. *Let $\lambda > 0$ be a positive real number.*

- *By a λ-rescaling of a function $f(z)$ we mean a function $\tilde{f}(z) = f(\lambda \cdot z)$.*

- *By a λ-rescaling of a family of functions F we mean the family consisting of λ-rescalings of all functions from F.*

Denotation. *λ-rescaling of a family F will be denoted by $R_\lambda(F)$.*

Definition 8.3. *We say that an optimality criterion on Φ is shift-invariant if for every two families F and G and for every positive real number λ, the following two conditions are true:*

i) if F is better than G in the sense of this criterion (i.e., $F \succ G$), then $R_\lambda(F) \succ R_\lambda(G)$;

ii) if F is equivalent to G in the sense of this criterion (i.e., $F \sim G$), then $R_\lambda(F) \sim R_\lambda(G)$.

It turns out that this requirement describes a reasonably small class of possibly optimal families:

Theorem 8.1. *If a family F is optimal in the sense of some optimality criterion that is final and scale-invariant, then this family corresponds to $f(z) = z^\alpha$ for some real number α.*

Comment.

- This result explains why power scaling is indeed so widely used: because it is optimal with respect to many reasonable criteria.

- Similarly to the results from Lessons 4, 5, and 7, this result does not necessarily mean that in every possible situation, one of these fitness scaling functions $f(z)$ will be better than any other possible choice.

  - In every *specific* application, if we have a good understanding of the function $J(x)$ and of the constraints, then other functions $f(z)$ may turn out to be better than the functions described here.

  - However, if we want to choose the best method for *general* use, so that this method will be applicable to any situation, then it is natural to require shift invariance or scale-invariance and thus, for such a general application, one of these functions will be indeed the best.

Exercise

8.2 Prove Theorem 8.1.

   Hint: Its proof is very similar to the proof of Theorem 7.1.

For those readers who want to check whether their proof is correct, and for those who did not succeed in proving this theorem, the actual proof is given in the appendix to this lesson.

Appendix: proof of theorem 8.1

The outline of the proof. In proving this result, we will follow the sequence of steps similar to the ones used in the proofs of Theorems 4.2, 5.1, and 7.1:

- first, we will find a functional equation describing the desired optimal fitness scaling $f(z)$;

- then, we will show that the functions used in this equations are indeed differentiable;

- after that, we will deduce the differential equation for this desired function $f(z)$;

- finally, we will solve this differential equation and get the desired expression for the optimal function $f(z)$.

First step: deducing a functional equation. According to Proposition 4.2, the optimal family $\{C \cdot f(z)\}$ must be scale-invariant. Therefore, for every $\lambda > 0$, there must exist a value $C(\lambda)$ such that

$$f(\lambda \cdot z) = C(\lambda) \cdot f(z).$$

This is the desired functional equation.

Second step: proving differentiability. From the functional equation, we can conclude that $C(\lambda) = f(\lambda \cdot z)/f(z)$. We have assumed that the function $f(z)$ is differentiable. Therefore, the function $C(\lambda)$ is also differentiable.

Third step: deducing a differential equation. Now, we can follow the path we followed in Lessons 4, 5, and 7:

- differentiate both sides of this functional equation by λ;

- take $\lambda = 1$ and thus, get the desired differential equation.

When we differentiate both sides of the functional equation by λ, we get the following expression:

$$z \cdot f'(\lambda \cdot z) = \frac{dC(\lambda)}{d\lambda} \cdot f(z).$$

When we substitute $\lambda = 1$ into this formula, we get a differential equation: $z \cdot f'(z) = C \cdot f(z)$.

Final step: solving the differential equation. We already know, from Lesson 6, how to solve such equations.

- First, we reduce this equation to a differential equation with constant coefficients, by introducing a new variable $t = \ln(z)$. In terms of this new variable, $z = \exp(t)$), $f(z) = F(t)$, where we denoted $F(t) = f(\exp(t))$, and the differential equation takes the form

$$\frac{dF}{dt} = C \cdot F(t).$$

- Second, we apply a known method from Lesson 6 to solve this equation. As a result, we get the following solution: $F(t) = C \cdot \exp(\alpha \cdot t)$ for some real number α.

- Finally, we substitute $t = \ln(z)$ into this solution, and get

$$f(z) = F(\ln(t)) = C \cdot \exp(\alpha \cdot \ln(z)) = C \cdot (\exp(\ln(x))^\alpha = C \cdot z^\alpha.$$

The theorem is proven.

9

RISC COMPUTER
ARCHITECTURE AND INTERNET
GROWTH: TWO APPLICATIONS
OF EXTRAPOLATION

In some problems, we know the consequences of every possible alternative. For such problems, finding the best alternative is naturally formalized as an optimization problem. In many problems, however, we only have a partial information about the consequences of different alternatives. To solve such problems, we first need to extrapolate this partial information to a complete description. In this lesson (and in several follow-up lessons), we provide several examples of important computer science problems in which such an extrapolation is necessary, and we show how continuous mathematics helps such extrapolation.

In this lesson, we start with two important computer hardware problems that require simple extrapolation: RISC (Restricted Instruction Set) computer architecture and Internet growth.

9.1. Extrapolation and interpolation are needed

Not all practical problems can be directly formalized as optimization problems: we need reconstruction first.

- In some problems, we have a *complete* information about the consequences of every possible alternative x. For such problems, for every alternative x, we can compute the corresponding value $J(x)$ of the objective function that characterizes the quality of this alternative x. In this case, the problem of finding the best alternative is naturally formalized as an *optimization* problem $J(x) \rightarrow \max$.

- In many problems, however, we only have a *partial* information about the consequences of different alternatives; in other words, we only know the values of $J(x_i)$ for a few alternatives $x_1, \ldots, x_s$. In such situations, in order to formalize the problem of choosing the best alternative as an optimization problem, we first need to *reconstruct* the missing values of the objective function $J(x)$.

How do we call a reconstruction? In *computer science* (especially in Artificial Intelligence), reconstruction is usually known as *learning*:

- We know the *patterns*, i.e., the pairs $(x_1, J(x_1))$, $(x_2, J(x_2))$, ..., $(x_s, J(x_s))$.

- Based on these patterns, we want to determine the function $J(x)$.

This is how we un-cover nature's laws: we make several experiments and try to find a pattern. This is often how we learn a new material: we solve several exercises, and after working through several exercises, we get the idea. Thus, "learning" seems to be an appropriate word for reconstruction. However:

- In Artificial Intelligence, this word is usually reserved for the *complicated* reconstruction problems, problems that require a certain *intelligence*.

- On the other hand, in many real-life situations, a simple reconstruction algorithm is sufficient.

Since in this text, we want to cover simple reconstruction algorithms as well, we will not use the computer science term "learning" (unless we are really solving a complicated reconstruction problem); instead, we would use mathematical terms for reconstruction.

In *mathematics*, there are two terms for reconstruction: *extrapolation* and *interpolation*. Both terms describe the same reconstruction process:

- We know the values $J(x_1), \ldots, J(x_s)$.

- We want to determine the value $J(x)$ for some $x \neq x_i$.

The difference between these two terms is very clear and precise in the situations where alternatives x are characterized by a single number:

- When the point x (at which we want to reconstruct the value of the function $J(x)$) is in between the smallest $\min x_i$ and the largest $\max x_i$ of the values x_i, i.e., when the point x is *internal* to the interval $[\min x_i, \max x_i]$, the problem of reconstructing $J(x)$ is called *interpolation*.

- When the point x is outside the interval $[\min x_i, \max x_i]$ (i.e., *external* to the interval), then the problem of reconstructing $J(x)$ is called *extrapolation*.

Typically, an alternative x is characterized by *several* numbers. In this case, we can also distinguish between the cases when a new point x is *internal* and *external* to the known points x_i, but there is not longer a clear distinction.

Usually, there are much more points outside than inside, so, most of the time, reconstruction is an *extrapolation*, not an interpolation. As a result, if we want a generic word for both types of reconstruction, we use the word *extrapolation* (that describes the most frequent type of reconstruction).

What we are planning to do. In this lesson (and in two follow-up lessons), we provide several examples of important computer science problems in which such an extrapolation (reconstruction) is necessary, and we show how continuous mathematics helps such extrapolation. In this lesson, we start with two problems that require simple extrapolation: RISC architecture and Internet growth.

9.2. RISC architecture

CISC: traditional approach to computer architecture. Computers are getting faster and faster and still, as anyone who ever used a computer knows, there are still some problems in which a computer is painstakingly slow. So, the computer designers try to make the computers even faster. How can we speed up a computer?

Ideally, we should make all operations faster, and this is exactly what computer engineers are constantly doing: new technology is making computers faster and faster. The technology level determines the speed with which hardware supported operations are run; this hardware speed determines the computations speed.

■   Some elementary operations are directly hardware supported, so they run
     as fast as the hardware allows.

■   Some other elementary computer operations are not directly hardware sup-
     ported. To perform such an operation, we need several hardware steps.
     These operations are, therefore, much slower than the ones that are di-
     rectly hardware supported. Hence, to speed up the computations, it is
     desirable to hardware support these operations.

For example, in the first computers, of all arithmetic operations with integers,
only addition and subtraction were directly hardware supported, while multiplication had to be implemented as a sequence of additions. Thus, in the first
computers, multiplication was much slower than the hardware allowed it to be.
To speed up the computations, designers managed to add multiplication to the
list of hardware supported operations, thus speeding it up. Next, the designers
incorporated a hardware support for operations with floating point numbers,
etc.

As a result, we get a computer in which many different operations are hardware
supported. The resulting instruction set is pretty complex; so, such computers
are called *Complex Instruction Set Computers*, or *CISC*, for short.

RISC architecture: a better approach. Adding a new operation to the
list of hardware supported ones (i.e., adding a new *instruction*) speeds up this
operation. The computers designers hope that speeding up the operation would
lead to the speed up of the computations in general. In many cases, this is
indeed the case, but in some other cases, the addition of a new instruction
actually *slows down* the computer. Why?

To perform an instruction, the computer has to decode it and to activate the
corresponding hardware operation. How much time does this decoding take?

■   If we had only *one* instruction in our instruction set, we would simply
     perform it without any need for decoding.

■   If we have *two* instructions in our instruction set, then we will need (at
     least) one bit to describe the choice of an instruction. Decoding would
     simply mean that we read this bit and, depending on whether its value is
     0 or 1, activate one of the two hardware supported instructions.

■   If we have $256 = 2^8$ instructions (a typical set of CISC instructions), we
     need 8 bits to store the instruction. Therefore, to decode such an instruc-

tion, we need to read these eight bits, and, depending on the particular combination of bits, activate one of the 256 hardware supported operations.

The more instructions we have, the more bits we need to store an instruction; hence, the longer it takes to *decode* an instruction, and therefore, the longer it takes to *perform* each of these instructions. So, when we add new instructions to the original instruction set:

- On one hand, by hardware supporting new operations, we *speed up* these operations.

- On the other hand, by increasing the size of the instruction set, we make it more difficult (and thus, slower) to decode an instruction, and thus, we *slow down* all instructions.

Originally, when we added frequently used operations such as multiplication, etc., to the list of hardware supported operations, the overall effect was positive. However, nowadays, when we add a rarely used operation to the instruction set, we speed up this rarely used operation but we slow down all more frequently used ones, and, as a result, we may slow down the computer instead of speeding it up.

When this adverse effect of adding new instruction was realized (in the early 80s), computer designers started to *restrict* the list of hardware supported instructions instead of the previous trend of supporting as many of them as possible. The resulting computers are called *Restricted Instruction Set Computers*, or *RISC*, for short.

How to choose the number of instructions? A problem. If we have already agreed on the number h of hardware supported operations, then the question "Which of the possible operations should we support?" has an easy answer: we should hardware support (and thus, speed up) the most frequent operations.

The frequency of different operations can be easily determined by collecting data from existing computer systems. So, the main question is: how many operations should we hardware support? To decide on that, we must know the instruction execution speed $S(h)$ (i.e., number of executed instructions per second) for different values of h.

How to choose the number of instructions? A (toy) example. Let us have a simple (toy) example of choosing h. In this simplified example, we have four operations that can, in principle, be hardware supported; we will denote these operations o_1, o_2, o_3, and o_4.

- The frequencies of these operations are, accordingly, $f_1 = 0.6$, $f_2 = 0.25$, $f_3 = 0.1$, and $f_4 = 0.05$.

- If we do not directly support each operation, then it must be implemented as a sequence of 10 hardware supported ones.

- The speeds $S(h)$ are, correspondingly, equal to $S(1) = 1$, $S(2) = 0.9$, $S(3) = 0.8$, and $S(4) = 0.5$.

In this example, we have four possible choices: we can support from 1 to 4 operations. To find out how many operations we should support, let us compute the average operation speed for each of the four choices:

- If we only support one operations ($h = 1$), then we should support the most frequently used operation, i.e., o_1. Thus, for this operation, we need 1 instruction, while operations $o_2 - o_4$ require ten instructions each. So, to implement one operations, in 0.6 of all cases, we need 1 instruction, and in the remaining $0.25 + 0.1 + 0.05 = 0.4$ cases, we need 10 instructions. The average number of instructions per operation is thus $1 \cdot 0.6 + 0.4 \cdot 10 = 4.6$. The resulting computer performance, measured by the number of operations per second, can be obtained if we divide the number of instructions per second ($S(1) = 1.0$) by the number of instructions per operation (4.6). As a result, we get $1/4.6 \approx 0.22$ operations per second.

- If we support two operations, then the average number of instructions per operation is equal to $(0.6 + 0.25) \cdot 1 + (0.1 + 0.05) \cdot 10 = 2.35$. Thus, the average number of operations per second is equal to $S(2)/2.35 = 0.9/2.35 \approx 0.39$.

- If we support three operations, then the average number of instructions per operation is equal to $(0.6 + 0.25 + 0.1) \cdot 1 + 0.05 \cdot 10 = 1.45$. Thus, the average number of operations per second is equal to $S(3)/1.45 = 0.8/1.45 \approx 0.55$.

- Finally, if we support all four operations, then we get exactly one instructions per operation. Thus, the average number of operations per second is equal to the number of instructions per second, i.e., to $S(4) = 0.5$.

Of these options, the best performance (i.e., the largest number of operations per second) is achieved when we support three operations: then, we have 0.55 operations per second. So, in this toy example, the best option is to support the three most frequent operations.

Extrapolation is needed. To decide how many operations we need to support, we must know the values $S(h)$ for different h.

- *In principle*, we can determine all these values experimentally, by designing chips that correspond to all possible values $h = 1, 2, 3, \ldots, 256$, and by experimentally measuring the instruction-per-second performance of each of these chips.

- However, *in practice*, designing a chip is very costly and time-consuming, so it is not realistic to design 256 different chips for the only purpose of designing a single one of them. It is therefore desirable:

  - to design a few chips (say, two or three) that correspond to a few values h_i;
  - to measure the values $S(h_i)$ for these h_i, and then
  - to apply some extrapolation procedure for estimating the values $S(h)$ for different h.

How can we extrapolate? A general idea. In general, how do we extrapolate? We have already described a particular case of extrapolation in Lessons 3 and 4, when we discussed program testing. Usually, we have some idea of how the function will behave. In mathematical terms, this "idea" means that although we do not know the desired function, but we do know (or at least we do assume that we know) a *family* of functions to which the desired function belongs. To select a function from this family, we need to fix the values of one or several parameters. If we have experimental data, we can determine the values of these parameters, and get the desired function.

If, for example, we know that $J(0) = 0$ and $J(1) = 1$, then the extrapolation result depends on the choice of the family:

- If the family consists of linear functions $J(x) = a \cdot x + b$, then from $J(0) = 0$ and $J(1) = 1$, we conclude that $a = 1$, $b = 0$, and $J(x) = x$.

- If the family consists of quadratic functions $J(x) = a \cdot x^2$, then from $J(0) = 0$ and $J(1) = 1$, we conclude that $a = 1$, and $J(x) = x^2$.

■   If the family consists of sine functions $J(x) = \sin(k \cdot x)$, $0 \leq k \leq 2$, then from $J(0) = 0$ and $J(1) = 1$, we conclude that $k = \pi/2$, and $J(x) = \sin(\pi \cdot x/2)$.

The resulting three functions are very different. This simple example shows that the result of extrapolation drastically depends on the choice of the family. Therefore, if we choose a wrong family, extrapolation results will be very misleading.

So, the question is: for this particular problem related to RISC architectures, how do we choose the best approximation family?

Choosing a family of functions: first step. We want to describe a class of functions $S(h)$ that describe how the instruction speed S depends on the total number h of hardware supported operations (instructions).

As we have already mentioned, the instruction speed S depends not only on the number of instructions, but also on the hardware technology used in their implementation. As the technology becomes better, the number of operations increases:

■   If we manage to get a new technology that it twice faster, then, for every h, the number of instructions per second will be not $S(h)$, but $2S(h)$.

■   In general, if the new technology is C times faster (for some real number $C > 1$), then the new dependence of S on h take the form $\widetilde{S}(h) = C \cdot S(h)$.

Thus, if $S(h)$ is a reasonable description of the desired RISC dependence on instruction speed on h, then, for every $C > 0$, the function $C \cdot S(h)$ is also a reasonable description of such a dependence. In other words, if $S(h)$ is a reasonable function, then the entire family of functions $\{C \cdot S(h)\}$ that correspond to different positive real numbers C, consists of reasonable functions.

For simplicity, let us consider such families $\{C \cdot S(h)\}$ for fixed functions $S(h)$. Which of these families should we choose?

An additional problem: how to count the number of instructions? In the previous text, we assumed that it is difficult to choose the function $S(h)$, but it is easy to estimate the number of instructions h that are hardware supported in a given computer. Alas, the situation is not so simple. For example, in most IBM computers (IBM mainframe and IBM PC), there are different versions of arithmetic operations that correspond to the different bit length of the corresponding integers: we can add integers of standard length (2 bytes), of half-length (1 byte), of double length (4 bytes), etc.

- All these operations are hardware supported, so we may say that we have as many different instructions as there are different word lengths.

- On the other hand, all these versions use the same adder, with minor variations added to accommodate different word lengths, so, we may also say that we have a single instruction with several minor versions.

The same is true for many other instructions who often come in different versions. Depending on what we consider to be one instruction, we get different number of instructions for one and the same processor.

Comment. This situation is somewhat similar to the one encountered in program testing, where the number of un-covered bugs depended on how we counted them. For RISC architecture, the dependence is amplified by the marketing concerns of the hardware manufacturers:

- In the old days, when CISC was the prevailing paradigm, manufacturers would brag about how many operations are hardware supported. From this viewpoint, it makes sense to count different versions of a single instruction as *several* different instructions.

- Now, when RISC is the buzzword, it makes more sense to brag about the restricted character of the instruction set. Therefore, it makes more sense to count several versions as a *single* instruction.

How can we describe the change in a way of counting in mathematical terms? Let λ denote the average number of new instructions that a single old instruction corresponds to. For example:

- If we previously counted all versions as a single instruction, and now decided to count them separately, then $\lambda > 1$.

- Vice versa, if we previously counted all versions as separate instructions, and now decided to count them as a single instruction, we get $\lambda < 1$.

In general, we can say that h old instructions means $\tilde{h} = \lambda \cdot h$ new instructions. Therefore, the dependence that, in old count, was equal to $S(h)$, in the new units, is described by a new function $\tilde{S}(h) = S(\lambda \cdot h)$.

Optimal choice of an extrapolation method for RISC design: precise formulation and result. It is reasonable to demand that the relative quality of different models $\{C \cdot S(h)\}$ and $\{C \cdot T(h)\}$ used for extrapolation should not depend on the how we count the number of instructions. Thus, the optimality criterion must be invariant with respect to *scaling* $S(h) \to S(\lambda \cdot h)$.

Formally:

> *Among all families of the type $\{C \cdot S(h)\}$ that correspond to different differentiable functions $S(h)$, we must choose a family that is the best in the sense of some final scale-invariant optimality criterion.*

We have already analyzed this problem in Lesson 8, and we know the solution to this problem: the optimal family consists of the power functions $S(h) = C \cdot h^\alpha$ for some constants C and α. Thus:

> *The best extrapolation model for RISC performance is $S(h) = C \cdot h^\alpha$.*

Comment. This model is indeed in good accordance with the experimental data; see, e.g., D. C. Gazis, "Brief time, long march: the forward drive of computer technology", In: D. Leebart, *Technology 2001: The future of computing and communications*, MIT Press, Cambridge, MA, 1991, pp. 41–76.

How to use this extrapolation model: idea. Our extrapolation model $S(h) = C \cdot h^\alpha$ has two parameters to be determined from the experimental data: C and h. To determine two parameters, we need to have at least two experimental points.

- If we know the values $S(h_1)$ and $S(h_2)$ for exactly two points h_1 and h_2, then, to determine the two values C and α, we must solve the system of two equations with two unknowns: $S(h_1) = C \cdot h_1^\alpha$ and $S(h_2) = C \cdot h_2^\alpha$. To solve this system, let us eliminate one of the unknowns. The easiest is to eliminate C: we can do it by simply dividing both sides of the first equation by the second one. As a result, we get the following equation:

$$\frac{S(h_1)}{S(h_2)} = \left(\frac{h_1}{h_2}\right)^{\alpha}.$$

To find α, we apply logarithm to both sides of this equation. If we take into consideration that $\ln(a^{\alpha}) = \alpha \cdot \ln(a)$, we conclude that $\ln(S(h_1)/S(h_2)) = \alpha \cdot (\ln(h_1/h_2))$, and hence, $\alpha = (\ln(S(h_1)/S(h_2))/(\ln(h_1)/h_2))$. Then, from the equation $S(h_1) = C \cdot h_1^{\alpha}$, we can determine C as $S(h_1)/h_1^{\alpha}$.

- If we know the values $S(h_i)$ at more than 2 points, we must use the least squares method described in Lesson 3.

How to use this extrapolation model: algorithm. For a given technology, to determine the dependence of the computer performance S (i.e., the number of instructions per second) on the total number h of hardware implemented instructions, we must do the following:

- Based on the known empirical data, we sort all possible operations according to the frequency of their use.

- Second, we select two numbers $h_1 \neq h_2$, and design two chips:
    - a chip that implements h_1 most frequent operations; and
    - a chip that implements h_2 most frequent operators.

  For these two chips, we measure the performances $S(h_1)$ and $S(h_2)$.

- Based on the measured values $S(h_i)$, we compute

$$\alpha = \frac{\ln(S(h_1)/S(h_2))}{\ln(h_1/h_2)}.$$

- Then, we compute $C = S(h_1)/h_1^{\alpha}$.

- Finally, we return $S(h) = C \cdot h^{\alpha}$ as the desired model.

Example. If for $h_1 = 16$, we have $S(h_1) = 1$, and for $h_2 = 32$, we have $S(h_2) = 0.5$, then the above formulas lead to $\alpha = \ln(0.5)/\ln(2) = -1$ and $C = S(h_1)/h_1^{\alpha} = 1/(16^{-1}) = 16$. Thus, the desired extrapolation model is $S(h) = 16 \cdot h^{-1} = 16/h$.

Further reading.

- P. J. Denning, "RISC architecture", *American Scientist*, 1993, No. 1, pp. 7–10.

- D. A. Patterson and J. L. Hennessy, *Computer architecture: A quantitative approach*, Morgan Kaufmann, San Mateo, CA, 1990.

- D. Tabak, *RISC systems*, Wiley, N.Y., 1990.

Exercises

9.1 Describe the extrapolation model $S(h)$ based on the following data: for $h_1 = 2$, $S(h_1) = 4$; for $h_1 = 4$, $S(h_2) = 1$.

9.2 Use this model to find the optimal number of implemented operations $h = 1, 2, 3, 4$. Use the frequencies (and the number of instructions per non-supported operation) from the above toy example.

9.3. How to predict the growth of the Internet?

The problem. Internet is becoming more and more important to computer science: The Internet-related communication, search, and remote computing tasks form an increasing portion of computer usage. The increasing importance of the Internet leads to the necessity to constantly upgrade its hardware and software. And here lies a problem. Upgrading the Internet is an expensive and long-term task. Just like the current state of the Internet is the result of the decisions made decades ago, the decisions that we are about to make will affect the state of the Internet for years and decades to come. If we make a wrong choice now, we may face very negative consequences in the future:

- If we under-finance the Internet, then in the future, the connections will be clogged, and the growing number of users will only encounter the frustrating long lines.

- If we over-finance the Internet, i.e., if we overextend it to the level that is not needed in the nearest future, we have thus tied up, in the under-used connections, the scarce budget resources that could have been used on something useful.

It is therefore very important to be able to predict the growth of the Internet.

Growth equation. At any given moment of time t, the state of the Internet is characterized either by the total load L (measured in bytes per second)measurement,of Internet load, or by any other meaningful characteristic, such as the total number of users that use the Internet. Although for some of these characteristics (like the number of users), the corresponding parameter L only takes *discrete* values, the discreteness of adding 1 to a few million is so relatively small that for all practical reasons, we can consider L to be a *continuously* changing variable. It is also reasonable to assume that the changes are reasonably *gradual*, i.e., that the function $L(t)$ is *differentiable*.

Within these assumptions, to describe how this function $L(t)$ changes with time, we must describe, for every given moment of time t, the value of the derivative dL/dt. This derivative is usually called the *growth rate*.

If we had a *constant* growth rate $dL/dt = C$, then $L(t)$ would be a *linear* function $L(t) = C \cdot t + C_0$. Since the real-life growth of Internet is *not linear*, the growth rate is *not constant*, it must depend on the current load. Hence, we must consider models of the type

$$\frac{dL}{dt} = g(L). \tag{9.1}$$

This equation describes the growth and is therefore called the *growth equation*. The function $g(L)$ is called a *growth function*.

Extrapolation is needed.

- At any given moment of time t_0, we know the *past* values of the Internet load: $L(t)$, $t < t_0$. Based on these values, we can estimate the values of the time derivative dL/dt and thus, reconstruct the value of the growth function $g(L)$ for all *past* load values.

- However, what we are really interested in is prediction. Therefore, we are interested in the values of the growth function $g(L)$ for the *future* loads L. The only way to get these values is by using some *extrapolation*.

To perform a reasonable extrapolation, we must select a *model* of the growth function. In other words, we want to select a *family* from which we will choose functions that are consistent with the experimental data.

First guess: 1-parametric family of functions. What can we say about the desired family of potential growth functions? One of the reasons why in all previous lessons, we ended up with a *family* of functions (as opposed to a *single* function) is that there usually was a possibility to *change the measuring units*. This possibility did not change the situation, but changed the numerical values of the corresponding quantities. It is, therefore, natural to use a similar idea here. The equation (9.1) involves two quantities: load L and time t, and for both quantities, the change of measuring units should not affect the extrapolation results.

Let us start with the units of *time*. When we say that a growth function $g(L)$ is a reasonable description of the Internet's growth, it means that the actual dependence $L(t)$ satisfies the differential equation (9.1). If we change a unit of time to a new unit that is C times larger, e.g., if, instead of the day, we measure time in weeks $(C = 7)$, months $(C = 30)$, or years $(C = 365)$, then instead of previous numerical values t, we get new values $\tilde{t} = t/C$. If we use a new time unit that is C times larger, then the numerical values of the time derivative become C times larger. Thus, in the new units,

$$\frac{dL}{d\tilde{t}} = C \cdot \frac{dL}{dt} = C \cdot g(L).$$

In other words, in the new units, the same growth process is described by a new growth function $\tilde{g}(L) = C \cdot g(L)$.

Thus, if $g(L)$ is a reasonable description of growth, then, for an arbitrary positive real number $C > 0$, the function $C \cdot g(L)$ should also be a reasonable description, because it describes exactly the same growth, only in the new units.

The situation is very similar to the one described in the previous section, and so, as a first guess, we will consider 1-parametric families $\{C \cdot g(L)\}$ that correspond to different differentiable functions $g(L)$.

Scale invariance. So far, we have used the idea that a mere change of units for measuring *time* should not affect the relative quality of different extrapolation techniques. Similarly, a change of units for measuring the *load* itself must not affect this relative quality. We may measure load in bits per second, in bytes per second, in Terabytes per year, etc. If we simply replace a unit for measuring L to a new unit that is λ times larger, then the numerical values of the load will be replaced by new numerical values $\tilde{L} = L/\lambda$. As a result, the new differential equation takes the form

$$\frac{d\tilde{L}}{dt} = \frac{1}{\lambda} \cdot \frac{dL}{dt} = \frac{1}{\lambda} \cdot g(L) = \frac{1}{\lambda} \cdot g(\lambda \cdot \tilde{L}).$$

Thus, this unit change leads to the following change in the growth function: from $g(L)$ to $\tilde{g}(L) = (1/\lambda) \cdot g(\lambda \cdot L)$. The transition from $g(L)$ to $\tilde{g}(L)$ involves two steps:

- first, we "scale" the *parameter* inside the function $g(L)$ (from L to $\lambda \cdot L$);

- second, we "scale" the *values* of the function $g(L)$ (by dividing these values by a constant λ).

When we move from *individual* growth functions $g(L)$ to *families* of functions $\{C \cdot g(L)\}$, this transition becomes even simpler, because multiplying all the values of a function does not change the family: the family of all functions of the type $C \cdot g(L)$ contains exactly the same functions as the family of all the functions $C \cdot (g(L)/\lambda)$. Thus, for families, the change of unit simply means that we replace a family $\{C \cdot g(L)\}$ by a new family $\{C \cdot g(\lambda \cdot L)\}$. This is exactly the scaling transformation that we have considered starting from Lesson 4.

Thus, the requirement that the optimality criterion should not depend on the choice of the unit for measuring L simply means that the *optimality criterion should be scale-invariant*.

Optimal choice of the Internet growth model: 1-parametric case. If we use 1-parametric models of the type $\{C \cdot g(L)\}$, the problem of choosing the optimal Internet growth model can be reformulated as follows:

> Among all families of the type $\{C \cdot g(L)\}$ that correspond to different differentiable functions $g(L)$, choose a family that is the best in the sense of some final scale-invariant optimality criterion.

This is the same problem that we have considered in Lesson 8 and in the previous section, and the solution to this problem is known: $g(L) = C \cdot L^{\alpha}$. If we substitute this function into the equation (9.1), we get a differential equation from which we can determine the desired dependence of the load L on time t:

$$\frac{dL}{dt} = C \cdot L^{\alpha}. \tag{9.2}$$

We can solve this equation by separating its variables (L and t). To separate these variables, we:

- move the term L^{α} to the left-hand side (by dividing both sides of (9.2) by this term); and

■   move the term dt to the right-hand side (by multiplying both sides by this term).

As a result, we get the equation

$$\frac{dL}{L^\alpha} = C \cdot dt.$$

Integrating both parts of this equation, we get

$$\int \frac{dL}{L^\alpha} = \int C \cdot dt. \tag{9.3}$$

The right-hand side of (9.3) is equal to $C \cdot t + C_1$. The expression for the left-hand side depends on α:

■   If $\alpha \neq 1$, then the left-hand side $\int dL \cdot L^{-\alpha}$ of (9.3) is equal to $L^{1-\alpha}/(1-\alpha)$. Equating this expression with $C \cdot t + C_1$, we conclude that $L^{1-\alpha} = \widetilde{C} \cdot t + \widetilde{C}_1$ (where $\widetilde{C} = C \cdot (1 - \alpha)$ and $\widetilde{C}_1 = C_1 \cdot (1 - \alpha)$), and

$$L(t) = (\widetilde{C} \cdot t + \widetilde{C}_1)^\beta \tag{9.4}$$

for $\beta = 1/(1 - \alpha)$.

■   If $\alpha = 1$, then the left-hand side $\int dL/L$ of (9.3) is equal to $\ln(L)$, and thus, from the equation $\ln(L) = C \cdot t + C_1$, we conclude that

$$L(t) = L(0) \cdot \exp(C \cdot t), \tag{9.5}$$

where we denoted $L(0) = \exp(C_1)$.

Thus, in this case, we either have an exponential growth, or a power-law growth.

1-parametric description is too crude. Up to now, the situation was similar to the RISC case: we considered 1-parametric models, and we chose a 1-parametric model that is the best with respect to some final scale-invariant criterion.

To confirm the resulting model, we must compare it with the experiments. And here lies a big difference with the RISC case:

■   For *RISC*, the 1-parametric models are in pretty *good accordance* with the experimental data.

- However, for the *Internet* growth, these models lead only to a very *crude* qualitative *description*.

So, to adequately describe the growth of the Internet, we need a better model.

Comment. The difference between the success and un-success of using the same simple (1-parametric) models to describe RISC and Internet can be easily explained:

- For RISC, every measurement requires lots of effort. Therefore, for each architecture level, at most a few different chips and made and compared. Thus, for each particular technology level, we need a model $S(h)$ that adequately explains a small number of experimental points $S(h_i)$. It is relatively easy to find a function that goes through a few points.

- Internet, on the contrary, is constantly monitored. There is a large amount of experimental data, and it is, therefore, very difficult to find a model that agrees with all this data.

In search of more adequate models: multi-parametric families. Since 1-parametric models do not describe the Internet growth well enough, we need better models. How can we find these better models? The only simplifying assumption that we have made so far was to consider 1-parametric models of the type $\{C \cdot g(L)\}$ for some function $g(L)$. Since *1-parametric* models, with a single parameter C, turned out to be not sufficient for our extrapolation purposes, a natural idea is to consider *multi-parametric* models.

- To describe a 1-parametric model, we fix a single differentiable function $g(L)$, and consider all functions of the type $C \cdot g(L)$ that correspond to different values of C.

- Similarly, for every integer $m > 1$, to describe an m-parametric model, we will fix m different differentiable functions $g_1(L), \ldots, g_m(L)$, and consider all functions of the type $g(L) = C_1 \cdot g_1(L) + \ldots + C_m \cdot g_m(L)$ that correspond to different values of $C_1, \ldots, C_m$. Such functions $g(L)$ are called *linear combinations* of the (fixed) functions $g_1(L), \ldots, g_m(L)$.

Optimal multi-parametric families. Justification of von Bertalanffy's general system theory. Let us describe this idea in precise mathematical terms.

Definition 9.1.

- Let m be a positive integer, and let $g_1(L), \ldots, g_m(L)$ be differentiable functions from real numbers to real numbers. By an m-dimensional family F that corresponds to these functions $g_i(L)$, we mean a family of all functions of the type $\tilde{g}(L) = C_1 \cdot g_1(L) + \ldots + C_m \cdot g_m(L)$, where $C_1, \ldots, C_m$ are arbitrary real numbers.

- Two families are considered equal if they coincide, i.e., consist of the same functions.

Comment. In this lesson, we will consider optimality criteria on the sets Φ_m of all possible m-dimensional families. Our definitions of the optimality criterion, final criterion, shift-invariant and unit-invariant criteria can be applied to these families. This leads to the following result:

Theorem 9.1. *If an m-dimensional family F is optimal in the sense of some optimality criterion that is final and scale-invariant, then every function $g(L)$ from the family F is equal to a linear combination of the functions of the type $(\ln^p(L) \cdot L^\alpha \cdot \sin(\beta \cdot \ln(L) + \varphi)$, where p is a non-negative integer, α, β and φ are real numbers.*

Comments.

- As we will see in a short time (see a comment after Theorem 9.2), this theorem is in good accordance with the actual Internet growth data (and with the semi-heuristic differential equation that has been proposed to describe this growth).

- This result is, therefore, one more case when all we achieve by a complicate use of continuous mathematics is justifying the formulas that have already been empirically discovered. Based on these examples, the reader may get a wrong impression that this is all continuous mathematics can do. The reason why only had these types of examples is that in this introductory course, we mainly illustrate the mathematical methods on *simple* problems, for which, usually, a solution is already known. In reality, continuous mathematics can not only explain the simple solution, but it can also, if the simple solution turn out to be too crude, provide us with a set of possible more complicated solutions. For example, the result of Theorem 9.1 can be applied not only to the description of the Internet growth, but also to other scale-invariant situations, such as the choice of a function $f(z)$

in genetic algorithms, RISC architecture, etc. For example, if a genetic algorithm with a fitness scaling function $f(z) = z^\alpha$ works well in the first approximation, but not ideally well, this theorem provides us with a family of more complicated fitness scaling functions that are worth trying. In other words:

- In simple case, when we more or less know the solution, mathematical methods mainly *justify* the existing solution, by showing that this solution is indeed the best.

- In more complicated cases, in which the solution is not yet know, mathematical methods *can* lead to *new* previously unknown solutions.

- *Mathematical comment.* If we use complex numbers, then the resulting optimal models are even easier to describe: every element of the optimal family is a linear combination of the functions of the type $(\ln L)^p \cdot L^\alpha$, where p is a non-negative integer, and α is an arbitrary complex number.

- *The proof will follow.* The proof of this theorem (and of the following Theorem 9.2) will be given in the next lesson. These theorems are proved along the lines of Theorems proven in the previous lessons: we show that the desired functions satisfy a system of functional equations, reduce this system to a system of *differential* equations, and solve the resulting system of differential equations.

- Theorem 9.1 is a general result that describes optimal families for an arbitrary value of m. The larger m, the more complicated the model, and the more difficult it is to use this model. Therefore, to simplify computations, we would like to use the smallest possible m for which the model is consistent with the experimental data. Since $m = 1$ turned out to be inadequate, it is natural to take $m = 2$. For $m = 2$, we can get a more detailed description of the growth functions from the optimal families:

Theorem 9.2. *Let a 2-dimensional family F be optimal in the sense of some optimality criterion that is final and scale-invariant. Then, every function $g(L)$ from the family F has one of the following forms:*

1. $g(L) = C_1 \cdot L^{\alpha_1} + C_2 \cdot L^{\alpha_2}$;

2. $g(L) = C_1 \cdot L^\alpha + C_2 \cdot L^\alpha \cdot \ln(L)$;

3. $g(L) = C \cdot L^\alpha \cdot \sin(\beta \cdot \ln(L) + \varphi)$.

Comments.

- The models that correspond to $m = 2$ are already in good accordance with the Internet growth. For example, the growth of the Bitnet is in good accordance with the second model, for $\alpha = 1$ (see, e.g., V. Gurbaxani, "Diffusion in computing networks: Bitnet", *Communications of the ACM*, 1990, Vol. 33, No. 12, pp. 65-ff). For this particular case, the differential equation (9.1) has an explicit solution

$$L = K \cdot A^{b^t} \qquad (9.6)$$

(see exercise 9.3).

- In our derivations, we did not use many specific features of the Internet. The only thing we used was scale invariance. It is, therefore, no wonder that the optimal models that we have just described are exactly the ones that were described, on a semi-heuristic basis, by Ludwig von Bertalanffy in his General System Theory (see, e.g., his books *Perspectives on general system theory* (G. Braziller, N.Y., 1975) and *General system theory* (G. Braziller, N.Y., 1984)).

  Bertalanffy mainly considered equations of the *first* type. These so-called *Bertalanffy equations* turned out to be very adequate for describing growth in biology (namely, the growth of individual organisms and of their organs), so adequate that they are routinely used by fisheries in England and Japan and by by the Food and Agriculture Organization (FAO) of the United Nations. The following particular cases of the Bertalanffy equation describe the simplest growth processes:

  - For $\alpha_1 = 1$, $C_1 > 0$, and $C_1 = 0$, we get the equation $dL/dt = C_1 \cdot L$ that describe an *exponential* growth $L(t) = C \cdot \exp(C_1 \cdot t)$.

  - For $\alpha_1 = 1$, $\alpha_2 = 2$, $C_1 > 0$, and $C_2 < 0$, we get the equation $dL/dt = C_1 \cdot L - |C_2| \cdot L^2$ that describes a so-called *logistic curve* that starts with an exponential growth but then flatters out. For this particular growth function, the growth equation also admits an explicit solution

$$L(t) = \frac{1}{K + A \cdot b^t}. \qquad (9.7)$$

  Equations of the *second* type were originally proposed by Gompertz (for $\alpha = 1$). These equations describe, e.g., such growth processes as *population dynamics* (see, e.g., A. G. Nobile, L. M. Riccardi, L. Sacerdote, "On a class of difference equations modeling growth processes", In: L. M. Riccardi and

A. C. Scott, *Biomathematics in 1980*, North-Holland, Amsterdam, 1982, pp. 217–244). Thus:

> Theorem 9.2 provides a precise mathematical justification for the (highly successful) semi-heuristic formulas of von Bertalanffy's general system theory.

In the equations of the *third* type, $g(L)$ is sometimes negative, so they describe a situation in which $L(t)$ sometimes increase and sometimes decreases.

Exercise

9.3 Derive the formulas (9.6) and (9.7). *Hints:*

- First, separate the variables.
- After that, the growth equation can be expressed as an equality of two integrals.
- Then:
    * For the formula (9.6), introduce a new variable $y = \ln(L)$. After that, you should get an integral that we already know how to compute.
    * For the formula (9.7), represent the expression

$$\frac{1}{aL - bL^2} = \frac{1}{L \cdot (a - bL)}$$

as a linear combination of the fractions $1/L$ and $1/(a-bL)$. After this representation, the integral will be easy to compute.

10

SYSTEMS OF DIFFERENTIAL EQUATIONS AND THEIR USE IN COMPUTER-RELATED EXTRAPOLATION PROBLEMS

In many computer science problems, it is necessary to extrapolate the partial knowledge. Different extrapolation methods lead to drastically different results, so, it is important to choose an appropriate extrapolation technique. In the previous lesson, we formulated the task of choosing the best extrapolation technique in precise mathematical terms. In this lesson, we show that this task can be reduced to a system of differential equations, and we explain how to solve such systems.

It is important to choose an appropriate extrapolation: a brief reminder. In the previous lesson, we mentioned that in many computer science problems, it is necessary to extrapolate the partial knowledge. Since different extrapolation methods often lead to drastically different result, it is extremely important to choose an appropriate extrapolation technique.

What we have done so far. In the previous lesson, for two real-life problems, we formulated the task of choosing the *best* extrapolation technique in precise mathematical terms:

- For the first simpler problem (related to RISC architecture), we reduced the corresponding optimization problem to a simple linear differential equation. Since we already know how to solve such equations (we learned this in Lesson 6), we applied the corresponding algorithm and got the desired solution.

- We also mentioned that for the second, more complicated problem (of predicting the Internet's growth), a similar approach leads not to a *single* differential equation, but to a *system* of differential equations.

So, to solve this problem, we must learn how to solve such systems. Before we do that, let us describe the reduction in detail.

10.1. The problem of choosing the best extrapolation technique can be reduced to a system of differential equations: case of scale invariance

The problem: a brief reminder. We must choose the best family of functions. Here, by a family, we mean the set of all functions of the type $g(L) = C_1 \cdot g_1(L) + \ldots + C_m \cdot g_m(L)$, where $g_1(L), \ldots, g_m(L)$ are fixed functions, and $C_1, \ldots, C_m$ are arbitrary real numbers. By "the best", we mean the best relative to some scale-invariant final optimality criterion.

The outline of the reduction. In this reduction, we will follow the sequence of steps similar to the ones used in the proofs of Theorems 4.2, 5.1, and 7.1:

- first, we will find a system of functional equations describing the functions $g_i(L)$ from the optimal family;

- then, we will show that the functions used in these equations are indeed differentiable;

- after that, we will deduce a system of differential equations for the desired functions $f_i(z)$.

First step: deducing a system of functional equations. Similarly to the proof of Proposition 4.2, we come to a conclusion that the optimal family F is scale-invariant. In particular, for every i the result $g_i(\lambda \cdot L)$ of scaling the function $g_i(L)$ must belong to the same family, i.e.,

$$g_i(\lambda \cdot L) = C_{i1}(\lambda) \cdot g_1(L) + C_{i2}(\lambda) \cdot g_2(L) + \ldots + C_{im}(\lambda) \cdot g_m(L) \quad (10.1)$$

for some constants C_{ij} (that may depend on λ). This is the desired system of functional equations.

Second step: proving differentiability. Let us prove that the functions $C_{ij}(\lambda)$ are differentiable. Indeed, if we take m different values L_k, $1 \leq k \leq m$, we get m linear equations for $C_{ij}(\lambda)$:

$$g_i(\lambda \cdot L_k) = C_{i1}(\lambda) \cdot g_1(L_k) + C_{i2}(\lambda) \cdot g_2(L_k) + \ldots + C_{im}(\lambda) \cdot g_m(L_k),$$

from which we can determine the values $C_{ij}(\lambda)$, $1 \leq j \leq m$, using Cramer's rule. Cramer's rule expresses every unknown as a ratio of two determinants, and these determinants polynomially depend on the coefficients. The coefficients either do not depend on λ at all ($g_j(L_k)$) or depend smoothly ($g_i(\lambda \cdot L_k)$) because g_i are smooth functions. Therefore these polynomials are also smooth functions, and so is their ratio $C_{ij}(\lambda)$.

Third step: deducing a system of differential equations. Now, we can follow the path we followed in Lessons 4, 5, and 7:

- differentiate both sides of each of the functional equations by λ;

- take $\lambda = 1$.

When we differentiate both sides of i-th functional equation by λ, we get the following system of differential equations:

$$L \cdot g_i'(L) = c_{i1} \cdot g_1(L) + c_{i2} \cdot g_2(L) + \ldots + c_{im} \cdot g_m(L), \qquad (10.2)$$

where, by c_{ij}, we denoted the derivative $C_{ij}'(1)$ of $C_{ij}'(\lambda)$ at $\lambda = 1$.

So the set of functions $g_i(L)$ satisfies a system of linear differential equations. How can we solve such systems?

10.2. How to solve systems of linear differential equations?

Let us generalize methods of solving single differential equations to systems of equations. A system of differential equations means that we have several differential equations that the unknown function (or functions) must satisfy. We do not yet know how to solve general systems, but we do know how to solve a particular class of systems:

- systems that consist of a single differential equation with a single unknown function, and

- a slightly more general systems of *independent* differential equations, in which each equation contains only one unknown function.

Methods of solving such equations were described in Lesson 6. It is therefore reasonable to try to generalize these methods to a more general situation of *systems* of differential equations.

Reducing systems of equations originating from scale invariance (like a system (10.2)) to systems with constant coefficients. In Lesson 6, our main object was a linear differential equation with *constant* coefficients. We also considered slightly different differential equations, of the type $y \cdot P'(y) = \lambda \cdot y + \mu \cdot P(y)$, that originate from scale invariance, and we showed that these equations can be reduced to equations with constant coefficients if, instead of the original variable y, we introduce a new variable $t = \ln(y)$ (so that $y = \exp(t)$). If we introduce this new variable, and consider the new function $x(t) = P(\exp(t))$, then $y \cdot P'(y)$ turns into

$$ y \cdot \frac{dP}{dy} = \frac{dP}{dy/y} = \frac{dP}{d(\ln(y))} = \frac{dx}{dt}, $$

and the original differential equation with non-constant coefficients is reduced to an equation with constant coefficients.

The only non-constant coefficients in the system (10.2) are exactly of the same type $(L \cdot g_i'(L))$, so, a similar transformation can reduce this system to a system with linear differential equations with constant coefficients. Namely, if we introduce a new variable $t = \ln(L)$ (so that $L = \exp(t)$), and consider the new functions $x_i(t) = g_i(\exp(t))$, then the system (10.2) takes the form

$$ x_i'(t) = c_{i1} \cdot x_1(t) + c_{i2} \cdot x_2(t) + \ldots + c_{im} \cdot x_m(t). \tag{10.3} $$

So, in order to solve the original system of equations (10.2), it is sufficient to be able to solve systems of linear differential equations with constant coefficients (i.e., systems of the type (10.3)). How can we solve such systems?

Solving systems of linear differential equations with constant coefficients: main idea. We will try to design a general solution to a system of differential equations as a natural generalization of known solutions of systems of independent equations.

In Lesson 6, to describe a general solution of a linear differential equation with constant coefficients, we did the following:

■   We started with the simplest linear differential equation $x'(t) = \alpha \cdot x(t)$. Its general solution has the form $x(t) = C \cdot \exp(\alpha \cdot t)$.

■ Then, we showed that an arbitrary linear differential equation with constant coefficients has a solution of this same type $x(t) = C \cdot \exp(\alpha \cdot t)$ for an appropriately chosen α. By substituting this solution into the differential equation, we got a polynomial equation $P(\alpha) = 0$ for describing the appropriate values of α.

■ For *non-degenerate* differential equations (i.e., equations in which all the roots of the polynomial equation $P(\alpha) = 0$ are different), a general solution to this equation can be represented as a linear combination of such "elementary solutions" $x(t) = C \cdot \exp(\alpha \cdot t)$ corresponding to different values of α.

■ Finally, we showed that if we treat the degenerate case (when the polynomial describing α has a double root) as a limit case of non-degenerate cases, then we get additional "elementary solutions" of the type $t \cdot \exp(\alpha \cdot t)$, $t^2 \cdot \exp(\alpha \cdot t)$, etc., and that a general solution can be represented as a linear combination.

For *systems*, we will try a similar approach:

First step: solving simplest possible systems. We start with the simplest possible systems: namely, systems that are *independent* (in the above-described sense), and in which all the equations are identical and the simplest possible. In other words, we consider systems that consist of m identical equations $x'_i(t) = \alpha \cdot x_i(t)$. It is easy to describe the general solution to this system: Since each function $x_i(t)$ only occurs in i-th equation, this function is a solution to i-th equation, i.e.,

$$x_i(t) = C_i \cdot \exp(\alpha \cdot t) \qquad (10.4).$$

Second step: showing that a general system has solutions of this type. Let us show that an arbitrary systems (10.3) of linear differential equations with linear coefficients has a solution of the type (10.4) for an appropriate value of α. Indeed, according to (10.4), all the functions $x_i(t)$ are proportional to the same basic function $e(t) = \exp(\alpha \cdot t)$: $x_i(t) = C_i \cdot e(t)$. If we substitute this expression for (10.4) into the system (10.3) and take into consideration that $e'(t) = \alpha \cdot e(t)$, we conclude that

$$C_i \cdot \alpha \cdot e(t) = c_{i1} \cdot C_1 \cdot e(t) + \ldots + c_{im} \cdot C_m \cdot e(t).$$

If we divide both sides of this equation by $e(t) > 0$, we get a simplified equation:

$$c_{i1} \cdot C_1 + \ldots + c_{im} \cdot C_m = \alpha \cdot C_i. \qquad (10.5)$$

This equation can be easily represented in terms of linear algebra: its left-hand side is a product of the matrix $c = \{c_{ij}\}$ and a vector $\vec{C} = (C_1, \ldots, C_m)$, and the right-hand side is simply the same vector C multiplied by α. So, the equation (10.5) has the following form:

$$c \cdot \vec{C} = \alpha \cdot \vec{C}. \tag{10.6}$$

This matrix equation is well known in linear algebra. For an arbitrary matrix c:

- the values α for which this equation has a non-zero solution $\vec{C} \neq 0$ are called *eigenvalues* of the matrix c, and

- the corresponding vector $\vec{C}$ is called the *eigenvector*.

It is known, from linear algebra, that every matrix c has an eigenvalue α and, therefore (by definition of an eigenvalue), this matrix has a corresponding eigenvector $\vec{C} = (C_1, \ldots, C_m)$. Therefore, every system of linear differential equations (10.3) with constant coefficients has a solution of the type $x_i(t) = C_i \cdot \exp(\alpha \cdot t)$.

Since we are interested not so much in *proving* the existence of solutions, but rather in actually *computing* these solutions, let us recall how eigenvalues are actually computed (we do not mean fancy and efficient algorithms, just the basics).

How to compute eigenvalues: in brief. The matrix equation (10.6) that describes eigenvalues can be reformulated as

$$(c - \alpha \cdot I) \cdot \vec{C} = 0, \tag{10.7}$$

where by I, we denoted a unit matrix

$$I = \begin{pmatrix} 1 & 0 & \cdots & 0 \\ 0 & 1 & \cdots & 0 \\ & \cdots & & \\ 0 & 0 & \cdots & 1 \end{pmatrix}.$$

It is known, from linear algebra, that for an arbitrary matrix a, the equation $a \cdot \vec{C} = 0$ has a non-zero solution $\vec{C} = 0$ if and only if this matrix a is *degenerate* in the sense that its determinant is equal to 0: $\det(a) = 0$. Therefore, α is an eigenvalue if and only if

$$P(\alpha) = 0, \tag{10.8}$$

where we denoted
$$P(\alpha) = \det(c - \alpha \cdot I) \qquad (10.9).$$

The determinant $\det(a)$ of an $m \times m$ matrix a is, by definition, a sum of products, each product containing m elements of the matrix. Therefore, $P(\alpha)$ is a *polynomial* of order m. This polynomial is called the *characteristic polynomial* of the matrix c, and the equation (10.9) is called the *characteristic equation* of the matrix c.

In these terms, for every matrix c, eigenvalues can be found as roots of the corresponding characteristic polynomial (or, equivalently, as solutions to the corresponding characteristic equation).

How to compute eigenvalues: an example. In particular, for a 2×2 matrix
$$c = \begin{pmatrix} c_{11} & c_{12} \\ c_{21} & c_{22} \end{pmatrix},$$
the characteristic polynomial $P(\alpha)$ is equal to the following determinant:
$$P(\alpha) = \det \begin{pmatrix} c_{11} - \alpha & c_{12} \\ c_{21} & c_{22} - \alpha \end{pmatrix} = (c_{11} - \alpha) \cdot (c_{22} - \alpha) - c_{12} \cdot c_{21}.$$

For example, for a matrix
$$c = \begin{pmatrix} 1 & 7 \\ 0.25 & 4 \end{pmatrix},$$
the characteristic polynomial $P(\alpha)$ is equal to the following determinant:
$$P(\alpha) = \det \begin{pmatrix} 1 - \alpha & 7 \\ 0.25 & 4 - \alpha \end{pmatrix} = (1 - \alpha) \cdot (4 - \alpha) - 7 \cdot 0.25 =$$
$$\alpha^2 - 5\alpha + 4 - 1.75 = \alpha^2 - 5\alpha + 2.25.$$

The corresponding polynomial equation $P(\alpha) = 0$ has two roots: $\alpha = 0.5$ and $\alpha = 4.5$ that are, therefore, the desired eigenvalues of the matrix c.

How to find solutions of the type (10.3): algorithm. Now that we recalled how to compute eigenvalues, we can use this knowledge to find all solutions of type (10.4). Namely, if we have a system of the type (10.3), we do the following:

- First, we find the eigenvalues α as solutions to the polynomial equation $P(\alpha) = \det(c - \alpha \cdot I) = 0$.

■   Second, for each eigenvalue α, we solve the corresponding system (10.7) and find the corresponding vector $\vec{C} = (C_1, \ldots, C_m)$.

■   The solution corresponding to each α has the form $x_i(t) = C_i \cdot \exp(\alpha \cdot t)$.

How to find solutions of the type (10.4): example. Let us consider the following system of linear differential equations:

$$x_1' = x_1 + 7 \cdot x_2; \tag{10.10a}$$

$$x_2' = 0.25 \cdot x_1 + 4 \cdot x_2. \tag{10.10b}$$

We have already found that the corresponding matrix has two different eigenvalues: $\alpha^{(1)} = 0.5$ and $\alpha^{(2)} = 4.5$. Let us find the corresponding solutions:

■   For $\alpha^{(1)} = 0.5$, the corresponding system of linear equations takes the form

$$\begin{pmatrix} 0.5 & 7 \\ 0.25 & 3.5 \end{pmatrix} \cdot \vec{C} = 0,$$

i.e.,

$$0.5 \cdot C_1 + 7 \cdot C_2 = 0;$$

$$0.25 \cdot C_1 + 3.5 \cdot C_2 = 0.$$

The second equation is equivalent to the first one, so, it is sufficient to find a solution to the first equation. If we set, e.g., $C_2 = 1$, we get $C_1 = -14$. Thus, we get the following solution to the original system of differential equations $x_1^{(1)}(t) = -14 \cdot \exp(0.5 \cdot t)$, $x_2^{(1)}(t) = \exp(0.5 \cdot t)$.

■   For $\alpha^{(2)} = 4.5$, the corresponding system of linear equations takes the form

$$\begin{pmatrix} -3.5 & 7 \\ 0.25 & -0.5 \end{pmatrix} \cdot \vec{C} = 0,$$

i.e., the form

$$-3.5 \cdot C_1 + 7 \cdot C_2 = 0;$$

$$0.25 \cdot C_1 - 0.5 \cdot C_2 = 0.$$

The second equation is equivalent to the first one, so, it is sufficient to find a solution to the first equation. If we set, e.g., $C_2 = 1$, we get $C_1 = 2$. Thus, we get the following solution to the original system of differential equations $x_1^{(2)}(t) = 2 \cdot \exp(4.5 \cdot t)$, $x_2^{(2)}(t) = \exp(4.5 \cdot t)$.

Substituting these expressions into the original system of differential equations
(10.10a) – (10.10b), we can easily check that these expressions indeed form the
solution.

How to find a general solution in non-degenerate case: idea. If,
after using the above algorithm, we get m linearly independent solutions
$x_i^{(1)}(t), \ldots, x_i^{(m)}(t)$ of the type (10.4) (i.e., if the original system is *non-
degenerate*), then we can describe a *general* solution as a linear combination
of these solutions:

$$x_i(t) = K_1 \cdot x_i^{(1)}(t) + \ldots + K_m \cdot x_i^{(m)}(t), \qquad (10.11)$$

where the coefficients $K_1, \ldots, K_m$ can be obtained from the initial conditions.
For example, if initial conditions have the form $x_i(t_0) = x_i^{(0)}$ for given values
$x_i^{(0)}$, $1 \leq i \leq m$, then we can find the values $K_1, \ldots, K_m$ by solving the following
system of linear equations:

$$K_1 \cdot x_1^{(1)}(t_0) + \ldots + K_m \cdot x_1^{(m)}(t_0) = x_1^{(0)};$$

$$K_1 \cdot x_2^{(1)}(t_0) + \ldots + K_m \cdot x_2^{(m)}(t_0) = x_2^{(0)};$$

$$\ldots$$

$$K_1 \cdot x_m^{(1)}(t_0) + \ldots + K_m \cdot x_m^{(m)}(t_0) = x_m^{(0)}.$$

How to find a general solution in non-degenerate case: example. If
we add the initial conditions $x_1(0) = 1$ and $x_2(0) = 0$ to the system (10.10a) –
(10.10b) (i.e., if we take $t_0 = 0$, $x_1^{(0)} = 1$, and $x_2^{(0)} = 0$), then we have the
following system of linear equations to determine the coefficients K_1 and K_2:

$$-14 \cdot K_1 + 2 \cdot K_2 = 1;$$

$$K_1 + K_2 = 0.$$

From the second equation, we conclude that $K_1 = -K_2$. Thus, the first equation implies that $12 \cdot K_2 = 1$, i.e., that $K_2 = 1/12$ and $K_1 = -1/12$. Hence,
the desired solution (10.11) takes the form:

$$x_1(t) = -\frac{1}{12} \cdot (-14) \cdot \exp(0.5 \cdot t) + \frac{1}{12} \cdot 2 \cdot \exp(4.5 \cdot t) =$$

$$-\frac{7}{6} \exp(0.5 \cdot t) + \frac{1}{6} \exp(4.5 \cdot t);$$

$$x_2(t) = -\frac{1}{12} \cdot \exp(0.5 \cdot t) + \frac{1}{12} \cdot \exp(4.5 \cdot t).$$

Case of complex numbers: idea. If one of the roots $\alpha^{(j)}$ of the characteristic polynomial is a *complex* number $\alpha^{(j)} = p + q \cdot i$, we get the solution of the type $x_i^{(j)}(t) = C_i \cdot \exp(\alpha_j \cdot t) = C_j \cdot \exp(p \cdot t + i \cdot q \cdot t)$ expressed in terms of complex numbers. Since the expressions are complex, the coefficients C_i may also turn out to be complex numbers. We can express this same solution *without* using complex numbers if we use the the de Moivre formula $\exp(a + ib) = \exp(a) \cdot (\cos(b) + i \cdot \sin(b))$.

Case of complex numbers: example. Let us illustrate this idea on the following 2×2 example:

$$x_1'(t) = -x_2(t);$$

$$x_2'(t) = x_1(t),$$

with the initial condition $x_1(0) = 1$ and $x_2(0) = 0$. In this example, the matrix c has the form

$$c = \begin{pmatrix} 0 & -1 \\ 1 & 0 \end{pmatrix}.$$

The corresponding characteristic equation has the form

$$\det \begin{pmatrix} -\alpha & -1 \\ 1 & -\alpha \end{pmatrix} = \alpha^2 + 1 = 0.$$

For this equation, $\alpha^2 = -1$ and therefore, this equation has two roots: $\alpha^{(1)} = i$ and $\alpha^{(2)} = -i$. For the first root, the equation (10.7) takes the form

$$(-i) \cdot C_1 - C_2 = 0,$$

so we can have a solution with $C_2 = 1$ and $C_1 = i$: $x_1^{(1)}(t) = i \cdot \exp(i \cdot t)$, $x_2^{(1)}(t) = \exp(i \cdot t)$. Similarly, for the second root, we get a solution $x_1^{(1)}(t) = (-i) \cdot \exp(-i \cdot t)$, $x_2^{(1)}(t) = \exp(-i \cdot t)$. The desired solution

$$x_i(t) = K_1 \cdot x_i^{(1)}(t) + K_2 \cdot x_i^{(2)}(t)$$

can be found from the initial conditions:

$$i \cdot K_1 - i \cdot K_2 = 1;$$

$$K_1 + K_2 = 0.$$

From the second equation, we get $K_2 = -K_1$; substituting this expression into the first equation, we get $2 \cdot i \cdot K_1 = 1$ and $K_1 = 1/(2 \cdot i) = -(1/2) \cdot i$. Hence, $K_2 = -K_1 = (1/2) \cdot i$, and

$$x_1(t) = -\frac{1}{2} \cdot i \cdot i \cdot \exp(i \cdot t) + \frac{1}{2} \cdot i \cdot (-i) \cdot \exp(-i \cdot t) = \frac{\exp(i \cdot t) + \exp(-i \cdot t)}{2};$$

$$x_2(t) = -\frac{1}{2} \cdot i \cdot \exp(i \cdot t) + \frac{1}{2} \cdot i \cdot \exp(-i \cdot t) = i \cdot \frac{-\exp(i \cdot t) + \exp(-i \cdot t)}{2}.$$

To re-formulate this expression in real terms, we use the de Moivre formula, according to which $\exp(i \cdot t) = \cos(t) + i \cdot \sin(t)$ and

$$\exp(-i \cdot t) = \cos(-t) + i \cdot \sin(-t) = \cos(t) - i \cdot \sin(t)$$

(since $\cos(t)$ is an even function, and $\sin(t)$ is an odd function). Substituting these expressions into the above complex formulas, we get the desired reformulation: $x_1(t) = \cos(t)$ and $x_2(t) = \sin(t)$.

General case: idea. To complete the description of how to solve systems of differential equations with constant coefficients, we must consider the *degenerate* cases, in which for an $m \times m$ systems, there are fewer than m particular solutions of type (10.4).

Similarly to Lesson 6, we can consider such a degenerate case, when α is a multiple root of a characteristic polynomial, as a *limit* of non-degenerate cases in which we have two (or more) different roots α and $\alpha + \varepsilon$. In Lesson 6, we have already learned that this idea leads to the expressions of the type $t \cdot (\alpha \cdot t)$ (for the double root), $t^2 \cdot (\alpha \cdot t)$ (for the triple root), etc. Thus, instead of solutions of type (10.4) (which are simply proportional to $\exp(\alpha \cdot t)$), we must consider more general particular solutions. For example, for the case of the double root, we must consider solutions of the type $x_i(t) = C_i \cdot \exp(\alpha \cdot t) + D_i \cdot t \cdot \exp(\alpha \cdot t)$, etc.

General case: example. Let us consider a simple example of degenerate system:

$$x_1'(t) = x_1(t) + x_2(t); \tag{10.12a}$$

$$x_2'(t) = x_2(t), \tag{10.12b}$$

with the initial conditions $x_1(0) = 0$ and $x_2(0) = 1$. For this system, the matrix c is equal to

$$c = \begin{pmatrix} 1 & 1 \\ 0 & 1 \end{pmatrix},$$

and therefore, the characteristic equation takes the form

$$P(\alpha) = \det \begin{pmatrix} 1 - \alpha & 1 \\ 0 & 1 - \alpha \end{pmatrix} = (1 - \alpha)^2 = 0.$$

This equation has a double root $\alpha^{(1)} = \alpha^{(2)} = 1$. Therefore, we must consider possible solutions of the type $x_i(t) = C_i \cdot \exp(t) + D_i \cdot t \cdot \exp(t)$. Substituting this expression into the original system $(10.12a) - (10.12b)$, we get the following equations:

$$C_1 \cdot \exp(t) + D_1 \cdot \exp(t) + D_1 \cdot t \cdot \exp(t) = (C_1 + C_2) \cdot \exp(t) + (D_1 + D_2) \cdot t \cdot \exp(t);$$

$$C_2 \cdot \exp(t) + D_2 \cdot \exp(t) + D_2 \cdot t \cdot \exp(t) = C_2 \cdot \exp(t) + D_2 \cdot t \cdot \exp(t).$$

Equating coefficients at $\exp(t)$ and at $t \cdot \exp(t)$ in both sides, we conclude that

$$C_1 + D_1 = C_1 + C_2;$$

$$D_1 = D_1 + D_2;$$

$$C_2 + D_2 = C_2;$$

$$D_2 = D_2.$$

The fourth equation is always true. The third equation implies that $D_2 = 0$. For this D_2, the second equation is also always true. The first equation leads to $D_1 = C_2$. Thus, we can choose arbitrary values of C_1 and C_2, and get $D_1 = C_2$ and $D_2 = 0$. Since we want two linearly independent solutions, we can consider two cases:

- $C_1 = 1$ and $C_2 = 0$ (in which case $D_1 = D_2 = 0$), and
- $C_1 = 0$ and $C_2 = 1$ (in which case $D_1 = 1$ and $D_2 = 0$).

The corresponding particular solutions are:

- $x_1^{(1)}(t) = \exp(t); \; x_2^{(1)}(t) = 0.$
- $x_1^{(2)}(t) = t \cdot \exp(t); \; x_2^{(2)}(t) = \exp(t).$

Since we have two linearly independent solutions, we can consider genera;ΦΦal solutions of the type $x_i(t) = K_1 \cdot x_i^{(1)}(t) + K_2 \cdot x_i^{(2)}(t)$. Substituting $t = 0$ into the equations that describe initial conditions, we arrive at the following system of equations:

$$K_1 \cdot 1 + K_2 \cdot 0 = 0;$$

$$K_1 \cdot 0 + K_2 \cdot 1 = 1.$$

The first equation leads to $K_1 = 0$, and hence, the second leads to $K_2 = 1$. Thus, $x_1(t) = t \cdot \exp(t)$ and $x_2(t) = \exp(t)$.

First conclusion: general form of the solution. From the above algorithm, we can make the following conclusion:

Proposition 10.1. *For every solution $x_1(t), \ldots, x_n(t)$ of a system of linear differential equations with constant coefficients, each of the components $x_i(t)$ of this solution is a linear combination of the functions of the type $\exp(\alpha \cdot t)$, $t \cdot \exp(\alpha \cdot t)$, ..., $t^p \cdot \exp(\alpha \cdot t)$, where p is a non-negative integer, and α is a complex number.*

If we use de Moivre formula to re-formulate each complex-valued expression in terms of real numbers, and take into consideration that $\cos(z) = \sin(z + \pi/2)$, the same conclusion becomes slightly more complicated:

Proposition 10.2. *For every solution $x_1(t), \ldots, x_n(t)$ of a system of linear differential equations with constant coefficients, each of the components $x_i(t)$ of this solution is a linear combination of the functions of the type*

$$t^p \cdot \exp(\alpha \cdot t) \cdot \sin(\beta \cdot t + \varphi),$$

where p is a non-negative integer, and α, β, and φ are real numbers.

Final remark: how to transform the solution of the system (10.2) with constant coefficients into the solution of the original system (10.1).

- We started this section with a system of differential equations (10.1) with non-constant coefficients, and

- we reduced it to a system (10.2) with *constant* coefficients by introducing a new variable $t = \ln(L)$.

- We already know how to solve the resulting system (10.2); by applying known methods, we get a solution $x_i(t)$ in terms of this new variable t.

- All we need to do to get the solution to the original system is to substitute $t = \ln(L)$ into the resulting solution $x_i(t)$.

In precise terms, we transform the solution $x_i(t)$ to the system (10.2) into the solution $g_i(L) = x_i(\ln(L))$ to the original system (10.1).

Corollary 1: general form of the solution to the system (10.1). As a result of this substitution, from Proposition 10.2 (that describes the general solution to the system (10.3)), we can deduce the general form of the solution to the original system (10.1). We will formulate this corollary in a second. Before formulating it, we want to mention that we will simplify one term in the resulting expression: namely, the term $\exp(\alpha \cdot \ln(L))$ can be simplified if we take into consideration that $\exp(\ln(L) = L$ (this is the definition of a natural logarithm), and therefore, $\exp(\alpha \cdot \ln(L)) = (\exp(\ln(L))^{\alpha} = L^{\alpha}$.

Corollary. *For every solution $g_1(L), \ldots, g_n(L)$ of a system of linear differential equations (10.1), each of the components $g_i(L)$ of this solution is a linear combination of the functions of the type $(\ln(L))^p \cdot L^{\alpha} \cdot \sin(\beta \cdot \ln(L) + \varphi)$, where p is a non-negative integer, and α, β, and φ are real numbers.*

Corollary 2: Proof of Theorem 9.1. Since in the previous section 10.1, we have shown that for *scale-invariant* optimality criteria, functions $g_i(L)$ from the optimal family satisfy the system (10.1), we can conclude that the *functions from the optimal family can be represented as linear combinations of functions of this type*. Thus, we have proven Theorem 9.1.

Corollary 2: Proof of Theorem 9.2. In this theorem, we consider 2-dimensional families, i.e., the case $m = 2$. In this case, the characteristic equation $P(\alpha) = 0$ is a quadratic equation. For quadratic equations, there are three possible cases:

- This equation can have two different real roots α_1 and α_2.

  In this case, according to our algorithm, the general solution to the original system is a linear combination of the terms $\exp(\alpha_1 \cdot t) = L^{\alpha_1}$ and $\exp(\alpha_2 \cdot t) = L^{\alpha_2}$, i.e., the general solution has the form $g(L) = C_1 \cdot L^{\alpha_1} + C_2 \cdot L^{\alpha_2}$.

- This equation can have a double real-valued root α.

  In this case, according to our algorithm, every solution is a linear combination of the terms $\exp(\alpha \cdot t) = L^{\alpha}$ and $t \cdot \exp(\alpha \cdot t) = \ln(L) \cdot L^{\alpha}$, i.e., the general solution has the form $g(L) = C_1 \cdot L^{\alpha} + C_2 \cdot L^{\alpha} \cdot \ln(L)$.

- This equation can also have two complex conjugate roots $\alpha \pm \beta \cdot i$.

  In this case, according to our algorithm, every solution is a linear combination of the terms $\exp(\alpha \cdot t + i \cdot \beta \cdot t) = L^{\alpha} \cdot \cos(\beta \cdot \ln(L)) + i \cdot L^{\alpha} \cdot \sin(\beta \cdot \ln(L))$ and $\exp(\alpha \cdot t - i \cdot \beta \cdot t) = L^{\alpha} \cdot \cos(\beta \cdot \ln(L)) + -i \cdot L^{\alpha} \cdot \sin(\beta \cdot \ln(L))$ and therefore, the general solution has the form $L^{\alpha} \cdot [C_1 \cdot \cos(\beta \cdot \ln(L)) + C_2 \cdot \sin(\beta \cdot \ln(L))] = C \cdot L^{\alpha} \cdot \sin(\beta \cdot \ln(L) + \varphi)$.

We get exactly the three cases enumerated in Theorem 9.2. Thus, this theorem is proven.

What we will do next. We have just completed the proof of Theorems 9.1 and 9.2, and thus, we have described the best families for all *scale-invariant* optimality criteria. In the remaining part of this Lesson, we will describe families which are the best relative to *shift-invariant* optimality criteria.

Exercises

10.1 Use the general method describe in Section 10.3 to find the general solution to the following system of differential equations:

$$x_1'(t) = x_1(t) + 2 \cdot x_2(t);$$

$$x_2'(t) = 4 \cdot x_1(t) - x_2(t).$$

Describe the particular solution for which $x_1(0) = 2$ and $x_2(0) = 1$.

10.2 Use the general method describe in Section 10.3 to find the general solution to the following system of differential equations:

$$x_1'(t) = x_1(t) + x_2(t);$$

$$x_2'(t) = -x_1(t) + x_2(t).$$

Describe the particular solution for which $x_1(0) = x_2(0) = 1$ (this solution must by in the form that does not use complex numbers).

10.3. Extrapolation families that are the best with respect to shift-invariant optimality criteria

Result. To describe a family, we must choose an integer m, and m functions $f_1(z), \ldots, f_m(z)$. Then, the family consists of all functions of the type $f(z) = C_1 \cdot f_1(z) + \ldots + C_m \cdot f_m(z)$, where $C_1, \ldots, C_m$ are arbitrary real numbers.

Comment. The number m of parameters is often called a *dimension*, due to the following geometric analogy:

- a 1-point set is called 0-dimensional, and one does not need any parameters to specify its only point;

- a line is 1-dimensional, and we need exactly 1 number (1 coordinate) to describe all points on this line;

- a plane is 2-dimensional, and we need exactly 2 numbers (2 coordinates) to describe an arbitrary point on the plane;

- a space is 3-dimensional, and we need three numbers (three coordinates) to describe an arbitrary point in space;

- etc.

Similarly, families that require m parameters $C_1, \ldots, C_m$ are called *m-dimensional families*.

The families that are optimal with respect to some shift-invariant final optimality criteria are described by the following theorem:

Theorem 10.1. *If an m-dimensional family F is optimal in the sense of some optimality criterion that is final and shift-invariant, then every function $f(z)$ from the family F is equal to a linear combination of the functions of the type $z^p \cdot \exp(\alpha \cdot z) \cdot \sin(\beta \cdot z + \phi)$, where p is a non-negative integer, and α, β and ϕ are real numbers.*

These functions are indeed very useful. Before we prove this theorem, let us show that in many practical problems, the functions, that Theorem 10.1 describes as optimal ones, are indeed the best.

Each of these functions is a linear combination of the terms of a certain type. Each term depends on the three coefficients: p, α, and β. In principle, we can have terms for which all these three coefficients are different from 0. However, in practical problems, mainly the *simplest* of these terms are used, namely, the terms for which only one of these three coefficients is different from 0. There are three possibilities:

- $p \neq 0$, $\alpha = \beta = 0$;

- $p = 0$, $\alpha \neq 0$, $\beta = 0$;

- $p = \alpha = 0$, $\beta \neq 0$.

We will show that each of these possibilities leads to an important class of extrapolation techniques.

- If we take $p \neq 0$ and $\alpha = \beta = 0$, we get terms 1, t, t^2, etc. An arbitrary linear combination of these terms, i.e., an arbitrary expression of the type $C_0 + C_1 \cdot t + C_2 \cdot t^2 + \ldots + C_p \cdot t^p$, is a *polynomial*. Polynomial extrapolations are indeed widely used in computing practice.

  Usually, polynomials are used for *local* interpolation and extrapolation so that on different parts of the reconstructed functions, we may use different polynomials. The resulting *piece-wise* polynomial functions are called *splines*. Splines are the major tools in many computer applications:

  - In *Computer-Aided Design*, splines are used to transform approximate drawings into smooth layoffs of modern planes, ships, cars, etc. Splines turned out to lead to the best shapes, with the best aero- and hydrodynamical characteristics.

  - In *Computer Graphics* and *Computer Animation*, splines are used to extrapolate, correspondingly:

    * a picture, outlined by a few points, into a smooth shape;
    * several pictures, that describe two or three intermediate pictures, into a smooth computer-generated transition that provides the viewer with a perfect illusion of a continuous motion.

    Without spline-based algorithms, modern animated cartoons would be all but impossible.

  - In *Computer Typography*. In the old days, workers manually set up a page for printing. Now, a computer does that. How does a computer choose the most elegant shape of a letter? Using a professional term, how does computer design a *font*? The first modern computer system that enabled computers to make high-quality fonts was produced, in the 70s, by Donald Knuth, a famous computer scientist from Stanford University. He came to a conclusion that splines are indeed the best shapes, and his fonts are still used in most of the books. For example, this text was set up using LATEX, a follow-up of Knuth's system TEX, a follow-up that uses Knuth's spline-based algorithm for font design.

- If we take $p = 0$, $\alpha \neq 0$, and $\beta = 0$, we get *exponential* terms $\exp(\alpha \cdot t)$. Such terms (usually, with $\alpha < 0$) describe various fundamental physical processes such as cooling, radioactive decay, etc. It is therefore reasonable to represent an arbitrary function $x(t)$ describing a similar process as a linear combination of functions of this type, i.e., as a linear combination

$x(t) = \sum_p C(p) \cdot \exp(-p \cdot t)$. Such a representation is called *Laplace transform*. It is indeed widely used in engineering to describe the processing of weak signals, and different linear transformations in general.

For example, the radioactivity of an unstable isotope is usually described by a function $x(t) = x(0) \cdot \exp(-\lambda \cdot t)$. Thus, if we have a *mixture* of several radiative isotopes with different decay rates λ_i, then for this mixture, the radiation intensity depends on time as $x(t) = \sum C_i \cdot \exp(-\lambda_i \cdot t)$. So, if we apply Laplace transform to the observation data, we can, hopefully, find the coefficients C_i and λ_i, and thus, determine the contents of the given mixture:

* the values λ_i determine what elements are in this mixture, while
* the values C_i describe how much of each element is in the mixture.

Similar techniques are used to find new elementary particles, etc.

■   Finally, if we take $p = \alpha = 0$ and $\beta \neq 0$, we get *sines* and *cosines*. It is known that an arbitrary continuous function can be represented as a linear combination of sines and cosines. This is a known mathematical fact, first discovered by Fourier in the beginning of 19th century. *Fourier transform* is one of the main engineering tools.

The fact that an arbitrary continuous function can be approximated by a sum of sines and cosines can be easily illustrated by a well-known prism experiment that probably everyone remembers from high school physics.

–   Newton was the first who showed that if we place a prism on the way of the Solar ray, this ray will be transformed into a spectrum of rays that correspond to different colors of the rainbow. If we then place another prism, we can collect these colored rays together and get our white ray back. Since it turned out that a Solar ray "hides" inside itself rays of different color, Newton called Solar rays *polychromatic* (which means "of many colors" in Greek), and he called the resulting rainbow-colored rays (that cannot be further decomposed) *monochromatic* ("of one color").

–   A similar decomposition can be done for an arbitrary light: every light can be decomposed into several monochromatic rays. In physics, this decomposition is called a *spectral decomposition*.

–   In mathematical terms, monochromatic rays are described by sines and cosines $x(t) = A \cdot \sin(\omega \cdot t)$ and $x(t) = A \cdot \cos(\omega \cdot t)$, with the frequency ω corresponding to the color. Therefore, the possibility to represent an arbitrary light as a composition of monochromatic rays means that *an arbitrary continuous function can be represented as a linear combination of sines and cosines.*

Comments.

- By comparing the number of applications of polynomials and of other extrapolation techniques, one can get the impression that polynomials seem to have much more practical applications. This is indeed true, and not only it is *empirically* true, but there is a *theoretical* explanation of this specific role of polynomials. This explanation will be given in the next section.

- We mainly described applications in which only one of the three coefficients is different from 0. However, terms with two or three non-zero terms are also used (although rarer). Let us give two examples:

    - Terms of the type $\exp(-|\alpha| \cdot t) \cdot \cos(\beta \cdot t)$ for which $p = 0$, $\alpha \neq 0$, and $\beta \neq 0$, are used to describe *damped oscillations*.
    - Terms of the type $t \cdot \cos(\beta \cdot t)$, $t^2 \cdot \cos(\beta \cdot t)$, ... (for which $p \neq 0$ and $\beta \neq 0$) and, more generally, terms $t^p \cdot \exp(-|\alpha| \cdot t) \cdot \cos(\beta \cdot t)$ (for which all three coefficients p, α, and β are different from 0) are used to describe *resonance*.

Let us now describe the proof of Theorem 10.1.

The outline of the proof. In proving this result, we will follow the sequence of steps similar to the ones used in the proofs of Theorems 4.2, 5.1, 7.1, and 9.1:

- first, we will find a system of functional equations that describe the functions from the optimal family;

- then, we will show that the functions used in this system are indeed differentiable;

- after that, we will deduce a system of differential equations for the desired functions $f_i(z)$;

- finally, we will solve this system of differential equations and get the desired expression for the optimal functions $f_i(z)$.

First step: deducing a system of functional equations. Similarly to the proof of Proposition 4.1, we come to a conclusion that the optimal family F is shift-invariant. In particular, for every i, the result $f_i(z + s)$ of shifting $f_i(z)$ must belong to the same family, i.e.,

$$f_i(z + s) = C_{i1}(s) \cdot f_1(z) + C_{i2}(s) \cdot f_2(z) + \ldots + C_{im}(s) \cdot f_m(z) \qquad (10.13)$$

for some constants C_{ij} (that may depend on s). This is the desired system of functional equations.

Second step: proving differentiability. Let us prove that the functions $C_{ij}(s)$ are differentiable. Indeed, if we take m different values z_k, $1 \leq k \leq m$, we get m linear equations for $C_{ij}(s)$:

$$f_i(z_k + s) = C_{i1}(s) \cdot f_1(z_k) + C_{i2}(s) \cdot f_2(z_k) + \ldots + C_{im}(s) \cdot f_m(z_k),$$

from which we can determine the values $C_{ij}(s)$ using Cramer's rule. Cramer's rule expresses every unknown as a ratio of two determinants, and these determinants polynomially depend on the coefficients. The coefficients either do not depend on s at all ($f_j(z_k)$) or depend smoothly ($f_i(z_k + s)$) because f_i are smooth functions. Therefore these polynomials are also smooth functions, and so is their ratio $C_{ij}(s)$.

Third step: deducing a system of differential equations. Now, we can follow the path we followed in Lessons 4, 5, and 7:

- differentiate both sides of each functional equation by s;

- take $s = 0$.

When we differentiate both sides of i-th functional equation by s, we get the following system of differential equations:

$$f_i'(z) = c_{i1} \cdot f_1(z) + c_{i2} \cdot f_2(z) + \ldots + c_{im} \cdot f_m(z), \qquad (10.14)$$

where by $c_{ij} = C_{ij}'(0)$ we denoted the derivative of $C_{ij}'(s)$ at $s = 0$.

So the set of functions $f_i(z)$ satisfies a system of linear differential equations with constant coefficients.

Final step: solving the system of differential equations. We already know (from the previous section) the general solution to such systems: it is exactly a linear combination of the desired terms. Thus, the theorem is proven.

Exercise

10.3 Describe all 2-dimensional families that are optimal with respect to shift-invariant optimality criteria. *Hint:* The result should be similar to Theorem 9.2 that describes 2-dimensional families that are optimal with respect to *scale-invariant* optimality criteria.

10.4. What if an optimality criterion is both shift-invariant and scale-invariant

What we already know: either shift-invariance or scale-invariance. In the previous text, we considered the cases when:

- either it is reasonable to assume that the optimality criterion is *shift-invariant*,

- or it is reasonable to assume that the optimality criterion is *scale-invariant*.

Based on each of these assumptions, we described optimal families and optimal multi-dimensional families of functions. In particular, for 1D case, we got the following results:

- If the optimality criterion is *shift-invariant*, then the optimal functions are of the type $x(t) = C \cdot \exp(-\alpha \cdot t)$.

- If the optimality criterion is *scale-invariant*, then the optimal functions are of the type $x(t) = C \cdot x^{\alpha}$.

In 1D case, we cannot assume both shift- and scale-invariance. In some cases, it may seem reasonable to assume that the optimality criterion is *both* shift-invariant and scale-invariant.

For 1D families of functions, we cannot do that, because then, for the optimal function $x(t)$, we would get two inconsistent conclusions:

- First, since the optimality criterion is shift-invariant, this function $x(t)$ is of the type $x(t) = C \cdot \exp(-\alpha \cdot t)$ for some real numbers C and α.

- Second, since the optimality criterion is *scale-invariant*, this function is of the type $x(t) = \tilde{C} \cdot x^{\tilde{\alpha}}$ for some real numbers $\tilde{C}$ and $\tilde{\alpha}$.

An exponential function is different from the power functions and therefore, the equality $x(t) = C \cdot \exp(-\alpha \cdot t) = \tilde{C} \cdot x^{\tilde{\alpha}}$ cannot be true unless the function $x(t)$ is a constant.

In multi-dimensional case, we can assume both shift- and scale-invariance. Luckily, the above negative result does not hold for multi-dimensional families (for which $m > 1$). For such families, we can explicitly describe the optimal families:

Theorem 10.2. *If an m-dimensional family F is optimal in the sense of some optimality criterion that is final, shift- invariant, and scale-invariant, then every function $f(z)$ from this family is a polynomial of order $\leq m - 1$ (i.e., a function of the type $f(z) = a_0 + a_1 \cdot z + a_2 \cdot z^2 + \ldots + a_{m-1} \cdot z^{m-1}$).*

Comments.

- This result explains the above-mentioned fact that polynomials are indeed more widely used than other functions that are optimal with respect to scale-optimal optimality criteria: because polynomials correspond to the reasonable cases when this criterion is also shift-invariant.

- In particular, for $m = 2$ we conclude that functions from the optimal family must be linear functions $f(z) = a_0 + a_1 \cdot z$. This result explains why in genetic algorithms, linear fitness scalings $f(z) = a_0 + a_1 \cdot z$ are often the best to use.

- This theorem also tells what fitness scalings to use if linear scalings work well, but are still not sufficient: in this case, it is reasonable to try quadratic, cubic, and maybe even higher order polynomial scalings.

- The proof of this Theorem is given as an Appendix to this lesson.

Possible future applications of optimal multi-dimensional extrapolation families to computer science. In Lesson 9 and in the previous sections of this lesson, we have described how optimal families can be used in such computer science problems as Internet growth, genetic algorithms, etc. The same families can be used in other problems as well:

- In some practical problems, it is natural to assume that the optimality criterion is *shift-invariant*. We have described several such situations in the previous lessons:

  - a choice of the model for software testing;
  - a choice of the penalty function for continuous optimization;
  - a choice of the scaling function for simulated annealing and for genetic algorithms,
  - etc.

  In all these example, we were looking for a *1-dimensional* family (and found it). If the optimal 1-dimensional family does not lead to satifactory results, a natural idea is to use a *multi-dimensional* family of functions. Since the optimality criterion is assumed to be shift-invariant, the functions from the optimal multi-dimensional family are described by Theorem 10.1.

- Similarly, if it is natural to assume that the optimality criterion is *scale-invariant*, we must use functions described by Theorem 9.1.

- Finally, if it is natural to assume that the optimality criterion is *both* shift-invariant and scale-invariant, we must use polynomials (see Theorem 10.2).

Examples of such applications are given in several Appendices placed at the end of the book.

Appendix: Proof of Theorem 10.2

Part one: proving that all functions from the optimal family are polynomials. Let us first show that all functions $f(z)$ from the optimal family F are polynomials. Since the optimality criterion is both shift- and unit-invariant, both Theorems 9.1 (that describe scale-invariant case) and Theorem 10.1 (which describes shift-invariant case) are applicable here.

- According to Theorem 9.1, every function $f(z)$ from the optimal family is a linear combination of the terms of the type $(\ln(L))^p \cdot L^\alpha$, i.e.,

$$f(z) = \sum_i C_i \cdot (\ln(L))^{p_i} \cdot L^{\alpha_i} \tag{10.15}$$

for some non-negative integers p_i and complex numbers C_i and α_i. (We use expressions with complex numbers simply because these expressions are somewhat easier to write and analyze; a similar proof can be repeated for more complicated real-valued expressions.)

■ According to Theorem 10.1, the same function $f(z)$ is a linear combination of the terms of the type $z^p \cdot \exp(\alpha z)$, i.e.,

$$f(z) = \sum_j D_j \cdot z^{q_j} \cdot \exp(\beta_j \cdot z) \tag{10.16}$$

for some non-negative integers q_j and for some complex numbers D_j and β_j.

From (10.15) and (10.16), we conclude that the two expressions for $f(z)$ must coincide:

$$\sum_i C_i \cdot (\ln(L))^{p_i} \cdot L^{\alpha_i} = \sum_j D_j \cdot z^{q_j} \cdot \exp(\beta_j \cdot z). \tag{10.17}$$

When can these two expressions coincide for all z?

■ If one of the terms in the right-hand of (10.17) has $\beta_j > 0$, then, when $z \to +\infty$, the right-hand side grows exponentially with z, and therefore, it grows faster than any polynomial or than any function from the left-hand side. (The fact that the exponential function grows faster than a polynomial is usually known to computer science students from comparing running times of different algorithms.) Thus, if the right-hand side and the left-hand side are equal, we cannot have $\beta_j > 0$.

■ If one of the terms in the right-hand of (10.17) has $\beta_j < 0$, then, when $z \to -\infty$, the right-hand side, which contains $\exp(|\beta_j| \cdot |z|)$, grows exponentially with $|z|$, and therefore, it also grows faster than any polynomial or than any function from the left-hand side. Thus, we cannot have $\beta_j < 0$ either.

■ If one of the terms in the right-hand of (10.17) has $\mathrm{Re}(\beta_j) > 0$, then, when $z \to +\infty$, the absolute value of the right-hand side, which contains $\exp(|\beta_j| \cdot z)$, grows exponentially with $|z|$, i.e., faster than any polynomial or than any function from the left-hand side. Thus, we cannot have $\mathrm{Re}(\beta_j) > 0$.

■ Similarly, we cannot have $\mathrm{Re}(\beta_j) > 0$. So, the only remaining case is $\mathrm{Re}(\beta_j) = 0$, i.e., when β_j is a purely imaginary number: $\beta_j = i \cdot s_j$. Let us show that the case $s_j \neq 0$ is impossible. Indeed, since the two analytical expressions coincide for all real values z, they must coincide for all complex values z as well. If we take $z = i \cdot t$, then the term $\exp(\beta_j \cdot z)$ turns into the term $\exp(-s_j \cdot t)$, which, depending on the sign of s_j, grows exponentially fast (faster than the left-hand side) either for $t \to +\infty$ or for $t \to -\infty$. Thus, the case $s_j \neq 0$ is also impossible. Hence, $s_j = 0$, and $\beta_j = i \cdot s_j = 0$.

Substituting $\beta_j = 0$ into the formula (10.16), we conclude that

$$f(z) = \sum_j D_j \cdot z^{q_j}$$

for some non-negative integers q_j, i.e., that $f(z)$ is indeed a polynomial. We can re-write this polynomial as $f(z) = a_0 + a_1 \cdot z + a_2 \cdot z^2 + \ldots + a_p \cdot z^p$.

Part 2: proving that the polynomials from the optimal family are of order $\leq m - 1$. To complete the proof of Theorem 10.2, we must prove that the order p of every polynomial $f(z)$ from the optimal family F is not great than $m - 1$. We will do this in a few steps:

- By definition, F is a family of all linear combinations. Therefore, from $f(z) \in F$, we conclude that for an arbitrary real number C, the function $g_p(z) = C \cdot f(z)$ also belongs to the family F. We can use this result to get a simpler function from F than the polynomial $f(z)$: namely, if we take $C = 1/a_p$, we get a function that still belongs to F and in which a coefficient at z^p is equal to 1:

$$g_p(z) = a_0 + a_1 \cdot z + \ldots + a_{p-1} \cdot z^{p-1} + z^p. \tag{10.18}$$

- According to the proof of Theorem 10.1, the optimal family F is *shift-invariant*, i.e., if $f(z) \in F$, then $f(z + s) \in F$ for every real number s. In particular, from $g_p(z) \in F$, we conclude that $g_p(z + 1) \in F$.

- Since F is a family of all linear combinations, the difference between any two functions from F also belongs to F, and hence, the function $h_p(z) = g_p(z + 1) - g_p(z)$ belongs to the optimal family F.

- Substituting (10.18) into this formula, we conclude that

$$h_p(z) = (a_0 - a_0) + a_1 \cdot [(z + 1) - z] + \ldots + a_k \cdot [(z + 1)^k - z^k] +$$

$$\ldots + a_{p-1} \cdot [(z + 1)^{p-1} - z^{p-1}] + [(z + 1)^p - z^p].$$

Using the known expression for the binomial term $(z+1)^k = z^k + k \cdot z^{k-1} + \ldots$, we conclude that each term proportional to a_k is equal to

$$a_k \cdot [(z^k + k \cdot z^{k-1} + \ldots) - z^k] = a_k \cdot [k \cdot z^{k-1} + \ldots].$$

Therefore, the terms corresponding to $k < p$ lead only to terms z^l with $l \le k - 1 < p - 1$. The only term proportional to z^{p-1} comes from the expression $(z+1)^p - z^p$ and is equal to $p \cdot z^{p-1}$. Hence, $h_p(z) = p \cdot z^{p-1} +$ lower order terms. Similarly to the transition from $f(z)$ to $g_p(z)$, we can now divide $h_p(z)$ by the coefficient p at z^{p-1} and thus obtain a new function

$$g_{p-1}(z) = b_0 + b_1 \cdot z + \ldots + b_{p-2} \cdot z^{p-1} + z^{p-1},$$

that also belongs to the optimal family F.

■ From the fact that F contains a polynomial $g_p(z) = z^p + \ldots$ with the highest order term z^p we concluded that F contains a new function $g_{p-1}(z) = z^{p-1} + \ldots$ with the highest order term z^{p-1}. Applying the same arguments to $g_{p-1}(z)$, we conclude that F contains a function $g_{p-2}(z) = z^{p-2} + \ldots$ with z^{p-2} as its highest order term, which, in its turn, implies that F contains functions $g_{p-3}(z) = z^{p-3} + \ldots, g_{p-4}(z) = z^{p-4} + \ldots, \ldots, g_k(z) = z^k + \ldots, \ldots, g_1(z) = z^1 + \ldots, g_0(z) = z^0 = 1$. So, the family F contains $p + 1$ different functions $g_k(z)$.

■ Let us now show that the family F contains the monomials $1, z, z^2, \ldots, z^k, \ldots, z_p$. We will prove this fact by induction over k.

  – *Induction base.* The base for this induction has already been proven: we already know that $g_0(z) = 1 \in F$.

  – *Induction step.* Suppose that we have already proven that $1 \in F, z \in F, \ldots, z^{k-1} \in F$. Let us show that $z^k \in F$.

    Indeed, we know that $g_k(z) \in F$, where $g_k(z) = c_o + c_1 \cdot z + \ldots + c_{k-1} \cdot z^{k-1} + z^k$. We can, therefore, represent z^k as

$$z^k = g_k(z) + (-c_0) \cdot 1 + (-c_1) \cdot z + \ldots + (-c_{k-1}) \cdot z^{k-1}.$$

    The right-hand side of this expression is a linear combination of the functions $g_k(z), 1, z, \ldots, z^{k-1}$ that belong to the family F. Therefore, their linear combination z^k must also belong to the family F. The induction step is proven.

■ So, all the monomial $1, z, \ldots, z^p$ belong to F, and therefore, an arbitrary linear combination of these monomials, i.e., an arbitrary polynomial of order $\le p$, also belongs to F.

■ To describe an arbitrary polynomial of order p, we need $p + 1$ coeffcients. Thus, this family of polynomials is $(p + 1)$-dimensional. Since we assumed that the family F is m-dimensional, we can conclude that $p + 1 \leq m$ and $p \leq m - 1$.

The theorem is proven.

11

NETWORK CONGESTION: AN EXAMPLE OF NON-LINEAR EXTRAPOLATION

As more and more people use computers to communicate, network congestion becomes a more and more serious problem. Ideally, we should be able to prevent congestion by predicting its possibility and undertaking the appropriate actions. For predicting, we need extrapolation. It turns out that for network congestion, the linear extrapolation tools that we have considered so far (i.e., extrapolation tools based on families of functions $\{C_1 \cdot f_1(x) + \ldots + C_m \cdot f_m(x)\}$ that linearly depend on the parameters C_i) do not work very well. In this lesson, we show that continuous mathematics helps to select non-linear extrapolation tools that lead to a better congestion control. Mathematical methods used in the design of these non-linear extrapolation tools will be used in the following lessons to describe neural networks and fuzzy control.

11.1. What is network congestion?

Linear vs. non-linear extrapolation tools.

- In the previous lessons, we described *linear* extrapolation tools, i.e., tools based on families of functions $\{C_1 \cdot f_1(x) + \ldots + C_m \cdot f_m(x)\}$ that depend linearly on the parameters C_i. (In mathematics, such families are called *linear spaces* of functions.) We showed that these tools are very helpful in many areas of computer science. An optimistic reader may even get to a conclusion that these methods are all we need to handle *all* possible extrapolation problems that one can encounter in computer science.

- Alas, there are situation where these methods do not work very well, and one of these situations is *network congestion*.

225

What is it?

What is congestion? A few years ago, after Paul McCartney's concert in Las Cruces, thousands of home-going fans brought the Interstate 10 (where, usually, cars are driving at – and often above – the maximal speed) to a complete stop. Cars were going so slowly that vendors walked by selling additional memorabilia. This is *traffic congestion:*

- everyone wants to hit the road, but

- there is only so much space on the freeway.

Similar problems happen on the Internet (and on the local networks as well):

- everyone wants to communicate, but

- the communication capacity of the network is limited.

The number of messages varies from time to time and inevitably a moment comes when it exceeds the capacity of a router's or gateway's queue. Then some messages are simply dropped from the network.

Network congestion is similar to traffic congestion, but much more difficult to avoid. In some sense, the network congestion problem is much more difficult than the problem of traffic congestion:

- After all, the *traffic* congestion problem can be solved if we build new roads and develop public transportation. Basically, it is the question of investing enough money into the traffic infrastructure.

- With network congestion, no amount of money will help. The more we increase the ability of a network, the more information people will be able to download. A few years ago, people were very happy with sending electronic mail; once in a while, one would download a picture by using a slow tool like ftp or telnet. Now, many people routinely use web browsers that matter-of-factly transfer pictures, and how frustrating it is when the picture does not appear at once! If the new connections are added, we will download animation, routinely transfer each other's videoimages, etc.

From the viewpoint of the *user*, the situation cannot be better: more and more communication ability is added every year, and with this ability, more and more things are available via the world wide web. But from the viewpoint of the *network designer*, it looks like a vicious circle:

- As more and more users join, the network becomes congested.

- To cure this congestion problem, we add new communication channels.

- At first, the congestion goes way.

- However, the the added communication capacity leads to new communication modes that were not possible before.

- Everyone starts using these new modes, and the network rapidly becomes congested again.

In short, even in the distant future, when the communication channels will be thousands times faster, the congestion problem will be as acute as it is now. So, we need to learn how to control network congestion.

To do that, we must first learn the basics of how the networks operate.

11.2. How computer networks operate: the basics

What we are trying to do in this section. The modern world-wide computer network was not designed according to a single plan: it grew, rather chaotically, sometimes inconsistently, often motivated more by the current new ideas than by a long-term planning. Similarly, the way messages are controlled on the network is not the result of a long-term design: there are many features of the current message handling protocols which may be not optimal, but which are there because this is how it started, and it is too expensive to change. When a lay person starts browsing through the literature on networking he, sometimes, gets an impression that *most* of the features are like that. In reality, there is a reason and logic in this seemingly chaotic world of computer networks. This reason and logic is what we will try to describe in this section. So, let us start the logical description of a seemingly chaotic network.

A warning: networking is an extremely complicated area, so all we plan to do within the bounds of one lesson is to briefly introduced a *few* (far from *all*) main terms, and solve a single (and simple) problem. Even this, probably simplest possible, problem will turn out to be so complicated that all mathematical tools that we have described so far will not be sufficient to solve it: we will need to develop these tools even further.

Acks and retransmission timeouts: what are they? The main objective of the computer network is to deliver information from one computer to another speedily and reliably.

- *Speedily* is what we are describing in this lesson.

- *Reliably* is very important: computer networks are fundamental for modern banking, for modern manufacturing, and for modern economy in general.

From the hardware viewpoint, the accidental loss of some messages is unavoidable:

- When a message is transmitted through a wire, a lightning can strike it and erase it.

- When a message resides on an intermediate computer (*server*), this computer can break down, this erasing the message, etc.

With a possibly unreliable communication channel, how can we achieve 100% reliability? Well, computer communication is not the only communication channel that is not 100% reliable: postage service also occasionally loses letters. So, if we want 100% guaranteed delivery, we can use a special service of *return receipts*: To the letter, we attach a special acknowledgment postcard (return receipt):

- If the letter arrives safely, the recipient signs the acknowledge postcard, and the same mail service delivers it back to the sender.

- If after a certain period of time, the return receipt does not arrive, and the message was important, the sender sends a new copy of this same message, etc.

Computer networks achieve reliability in exactly the same way.

- Every message is supplied with the header that describes who sent it, when, etc. When the intended receiving computer receives this message, it sends an acknowledgment message (called an *ack*) back to the sender.

- If, after a certain period of time, the sending computer does not receive the ack for a given message, it automatically re-sends (*re-transmits*) the same message again.

The amount of time during which the node waits for an ack before re-transmitting the message is called a *retransmission timeout*, or simply a *timeout*. It is usually denoted by τ_{out}.

Speed vs. reliability: a trade-off. We started this section by saying that we want to achieve two goals: *speed* and *reliability* of communication. Ideally, we should design the network that is both the fastest and the most reliable. Realistically, as often happens with different goals, we have a *trade-off*: if we make a network more reliable, we, usually, slow it down, and vice versa.

For example, retransmission definitely increases reliability, but at the same time, it can drastically slow the network down. Indeed, once in a while it inevitably happens that the network slows down. Messages that were normally delivered more or less on time (i.e., within the timeout) get slowed down. Some of them may be lost, but mainly, they are simply delayed. As a result, by the deadline (i.e., after the timeout), the senders do not receive the acks and re-send the same messages again. The senders think that their messages are lost, but in reality, the old messages are still there, so this re-sending, basically, doubles the number of message that is circulating on the network, by adding a clone to each original message. This doubling creates an even greater strain on the delivering abilities, and the network becomes even more congested. As a result, after the new timeout period, the second copies are still un-delivered. Therefore, the third copies are sent. The network gets even more congested than before, etc. As a result, we are on the edge of what is called *congestion collapse*, when no messages can go through at all.

This worst-case scenario can happen (and actually happens sometimes) if the actual message delivery time exceed the expected delivery time (that is used to compute the timeout). It is, therefore, extremely important to promptly change τ_{out} every time the network load changes. And this necessity, in turn, brings in the following question: who controls the timeout?

Who controls the timeout? Like the national economy, the network is a big conglomerate of independent highly interactive parts. And, similarly to national economy, there are two basic views on how to control a large computer network:

- Some experts place more emphasis on *centralized control*.

- Other experts think that better results can be achieved with the minimum of centralized control, when most of the decisions are made *locally*.

For timeout, both approaches are represented by widely used protocols:

- In *Transmission Control Protocol* (TCP) the timeout τ_{out} is controlled centrally.

- In *the User Datagram Protocol* (UDP), each node can choose (and change) its own timeout τ_{out}.

Both approaches have their advantages and disadvantages:

- Centralized control is, by definition, well coordinated and, in principle, it can avoid problem caused by the absence of coordination.

   This advantage can be easily illustrated on the example of traffic congestion (computer congestion example is very similar). Suppose that a certain route is congested, and that several alternative routes are available. A natural idea is then to use the least congested of the alternative routes.

   * If the choice is made *locally*, by each driver on his/her own, then, from the viewpoint of each driver, the optimal decision is to choose the least congested of the alternative routes. Therefore, all the drivers will get into this particular route, thus congesting the traffic there too.

   * Centralized planning would allow us to divert part of the traffic to different alternative routes, thus avoiding congestion on each of them.

- On the other hand, local control is more flexible, because it can take into consideration specific aspects of a local situation. For example:

   – If the value of the timeout is chosen *centrally*, then if there is a congestion in California, the timeout increases all over the web, including the communication between Texas and New Mexico. As a result of this (locally unnecessary) increase in timeout, messages that are sent from Texas to New Mexico and that are actually lost will be re-sent much later, thus slowing the average traffic between the authors of the book.

– If the timeout is chosen *locally*, then the California timeout increase will not in any way harm the connection between the authors.

Packet size. Acks and timeout are not the only way to ensure reliability. Timeout is connected with the *time* of the transmission. Another means of ensuring reliability is related to the *size* of transmitted information.

When a message is sent, no matter how large it is, the first natural idea is to deliver this message as is. For *short* messages, it makes perfect sense. However, this is not always the best way to send *long* messages. Indeed, each bit of the message has the same probability that this bit will be corrupted or not delivered at all. As a result, the longer the message, the larger the probability that one of the bits will be corrupted or not delivered at all. For large messages, this probability can be pretty high.

- If we send a message *as a whole*, then, with a high probability, this message will not be delivered safely. Therefore, we will have to re-send the entire long message, maybe again and again. Sending a long message takes time, so this retransmission will drastically slow down the message's delivery.

- It is therefore desirable to send long messages *piece by piece*. Then, even if one of these pieces is lost or corrupted, we do not have to re-send the entire message: we only have to send the lost piece. Re-sending a short piece is definitely faster than re-sending the entire message, so, the whole transmission is much faster. Besides, the probability that this small piece will be mis-delivered is negligible smaller than for re-transmitting the entire big message; therefore, in piece-by-piece transmission, there is a very minor probability that the third re-transmission will be necessary (in contrast to sending the whole message, where this probability is high). As a result, transmission becomes even faster.

Thus, in all transmission protocols, long messages are divided into several pieces (called *packets*) of a certain packet size P.

- We have already described the disadvantage of *large* packet size P, so, P should not be large.

- On the other hand, we should also avoid too small packet sizes. Indeed, with P small:
  - On one hand, in the rare case when a message is lost, we save some time on re-transmissions.
  - On the other hand, in all the other cases, we waste time on the unnecessary disassembling and assembling back the transmitted messages.

It is therefore important to select the appropriate packet size. There are, usually, several packet sizes for the different stages of the communication process.

- First the information is read or written in quanta that are called the *application-level packet size.*
- Then, depending on the transport protocol, these quanta are combined (or subdivided, if necessary) into packets of some other packet size.

Who controls the packet size?

- Usually, the user can control the application-level packet size.
- All other packet sizes are centrally controlled.

Avoiding congestion: interpacket interval and window size. Congestion occurs because too many messages are sent to the network, much more than the network can handle. To avoid congestion, we must, therefore, restrict the number of messages that a node can send. Since a message can be really long and consist of several packets, what we really want to limit is the number of *packets*. There are two possible ways to control this number:

- First, we can directly limit the number of packets per time unit that a network can send. If we decided to limit this number to M messages per minute, this means that during every time interval of length $I = (1/M)$, this node is allowed to send no more than one message. In other words, after sending a message, a node cannot send another message until after time I elapses. This time interval is called *interpacket interval.*

  > This idea is similar to the way traffic is sometimes controlled: if there are repairs on a road, a policeman would waive a car into this road, and let other cars wait for a certain amount of time before letting the next car in.

- For some protocols (e.g., for UDP), interpacket interval is the only way to avoid congestion. Some other protocols (e.g., TCP), in addition to enforcing the interpacket interval, also restrict the total number of the un-acknowledged packets sent by the node. This number is called a *window size* and is usually denoted by W. Protocols that restrict the window size are called *window-based* protocols.

  For example, if $W = 10$, this means that a node can send 10 packets without any problem. After sending these 10 packets, the node cannot send any extra packet before it receives at least one ack. After it has received an ack, the number of un-acknowledged packets drops to 9, and therefore, the node can send a new message, etc.

  Technically, unacknowledged packets are stored in *slots*. There are, totally, W slots. When a packet is acknowledged, the slot is emptied and marked to be empty. A new packet can only be sent if there is at least one empty slot. This packet is then placed into the previously empty slot and stored there until the ack arrives. If, for a certain packet, the ack does not arrive before the timeout, we re-send this packet again.

The window size is usually *centrally* controlled.

How are network parameters tuned now. We have described four parameters that describe the current state of the network: timeout τ_{out}, the packet size P, the interpacket interval I, and the window size W. Each of these parameters helps to avoid congestion by slowing down the communications. Therefore, it is desirable to dynamically vary the values of these parameters:

- When a network becomes clogged, we must choose these parameters so that they help to prevent congestion (even if it means slowing down the network).

- On the other hand, when a network is not congested, we should change the values of these parameters so that the network would not necessarily slow down.

This change of parameters is called *tuning*. How should we tune? If the network is getting clogged (and we know it by the fact that the number of un-delivered or delayed messages increases), we must do the following:

- Increase the timeout τ_{out}.

    When the network gets clogged, the delivery time increases. There-
    fore, if we do not increase the timeout, then after this time τ_{out}, mes-
    sages that are simply slow will be erroneously treated as un-delivered
    and re-sent again, thus adding an extra burden to the network.

- Decrease the packet size P.

    We have already explained that if the network becomes clogged, we
    should subdivide messages into smaller pieces (packets).

- Increase the interpacket interval I.

    Increasing the interval between the packets will decrease the num-
    ber of new packets sent to the network and thus, hopefully, ease the
    congestion.

- For window-based protocols, decrease the window size W.

    Decreasing the allowed number W of un-acknowledged messages also
    limits the number of new messages that each node can send before
    the previous messages (that are currently clogging the network) are
    safely delivered.

We choose tuning so as to satisfy the following two goals:

- First, to avoid congestion.

- Second, to avoid under-utilization and slowing down of the network.

The first goal is much more important, because in modern communication-
centered world, a congestion collapse has very negative (and even sometimes
catastrophic) consequences. So:

- If there is a risk of congestion, we better change the parameters *drastically*,
  to make sure that we avoid the congestion collapse.

- On the other hand, if the risk has gone, we better change these parameters
  back *slowly*, to avoid creating a new risk situation.

What functions should we use for tuning? Our goal is to speed up the communications. If we spend too much time on re-computing the values of the network parameters, we thus slow down the network. Therefore, it is reasonable to use tuning transformations that are as simple to compute as possible. The simplest (and hence, the fastest) to compute are *linear* functions. Because of that, at present, to tune each parameter x, *linear* formulas are mainly used i.e., formulas of the type $x_{new} = a \cdot x_{old} + b$ for some real numbers a and b.

For example, for *interpacket interval*, we need a drastic increase if there is a congestion risk, and a small decrease if the risk is gone. To increase I, we can either multiply $I \to a \cdot I$ (for some $a > 1$) or add $I \to I + b$ (for some $b > 0$), or do both. For large I, the product $a \cdot I$ grows faster than the sum $I \to I + b$, so, if we want a drastic increase, we should use the product. Similarly, if we want a cautious decrease, we should use a difference $I \to I - |b|$ and not a product $I \to a \cdot I$ (with $a < 1$). In view of this, in the current linear tuning method, we choose two real numbers $a > 1$ and $b < 0$, and tune as follows:

- If there is a congestion risk (i.e., if packets start getting dropped at an unusual rate), then we replace the old interpacket interval I_{old} by $I_{new} = a \cdot I_{old}$. This formula is called *multiplicative increase*.

- If there is no congestion risk (i.e., if the packet drop rate is below the pre-determined threshold), we replace the old interpacket interval I_{old} by $I_{new} = I_{old} - |b|$. This formula is called *additive decrease*.

For window size W, similar arguments lead to the following linear tuning method: we choose two real numbers $a \in (0, 1)$ and $b > 0$, and:

- If there is a congestion risk, then we replace the old window size W_{old} to $W_{new} = a \cdot I_{old}$. This formula is called *multiplicative decrease*.

- If there is no congestion risk, we replace the old window size W_{old} by $W_{new} = W_{old} + b$. This formula is called *additive increase*.

Similar multiplicative decrease and additive increase methods are used for tuning the *packet size P*; see, e.g., B. Nowicki, "Transport issues in the network file system" (*Computer Communication Review*, March 1989, pp. 16–20).

The optimal values of a and b are chosen *empirically*: e.g., according to K. K. Ramakrishnan and Raj Jain, "A binary feedback scheme for congestion avoidance in computer networks" (*ACM Transaction on Computer Systems*, 1990, Vol. 8, No. 2, pp. 158–181), the value $a = 0.875$ seems to be the best for window size (this same paper also empirically shows that for window size, an additive decrease is indeed worse than a multiplicative one).

Exercise

11.1 Suppose that we use linear algorithm for tuning the interpacket interval I: multiplicative increase with a coefficient $a = 2$, and additive decrease with a coefficient $b = -10$ msec. Initially, we took $I = 500$ msec. During the first two moments of time, we had congestion, then for 8 more moments of time, we had no congestion. Compute the value of interpacket interval for ten consequent moments of time.

11.3. Non-linear tuning: which tuning should we choose?

Non-linear tuning is necessary. It turns out that even for the best values of the tuning parameters a and b, we still get wild oscillations (typical for linear systems of difference equations), oscillations that under-utilize the network. This fact was noticed ten years ago, in:

- C. Kline, "Supercomputers on the Internet: a case study" (*Proceedings of the SIGCOMM 87 Workshop on Frontiers in Communications Technology*, Stowe, VT, August 1987).

- W. Prue and J. Postel, "Something a host could do with source quench" (ARPANET Working group request for comment, DDN Network Information Center, SRI International, Menlo Park, CA, July 1987, RFC–1016).

- Van Jacobson, "Congestion avoidance and control" (*Proceedings of the SIGCOMM 88 Symposium on Communication Architectures and Protocols*, Stanford CA, 1988, pp. 314–329).

Van Jacobson argued that since linear tuning methods cause this problem, it is reasonable to use *non-linear* tuning methods, i.e., methods for which $x_{new} = f(x_{old})$ for some *non-linear* function $f(x)$.

It is important to choose an appropriate non-linear tuning method.
Van Jacobson has performed the first experiments with non-linear tuning, and
he has shown that:

- on one hand, appropriately chosen non-linear tuning methods indeed drastically improve the network's performance;

- on the other hand, an arbitrarily chosen non-linear tuning method can drastically worsen the network's behavior.

It is therefore very important to *choose the best tuning method.*

This problem is difficult to solve. The problem of choosing the best tuning
method is very difficult to solve because a computer network is a very complicated system. There is no known way to estimate its performance under
different tuning rules other than performing a time-consuming simulation of
the whole network. Therefore we cannot really compare a lot of different tuning functions, and even if we experimentally choose one of them, it will still be
possible that we missed the best tuning algorithm.

In this lesson, we will show how continuous mathematics can help with this
choice.

We must choose a family of tuning functions, not a single function.
Judging by the existing experimental data, the relative quality of a tuning function $f(x)$ essentially depends on the situation (how many packets are sent, how
often, etc). So the only result that we can expect is some *family* F that contains
all tuning functions $f(x)$ that can be optimal in different circumstances. Then,
every time we need to choose an optimal tuning procedure, we'll have to try
only the functions from this optimal family, and find the best of these tuning
functions.

How to describe a family? In the previous lessons, we considered *linear* families, i.e., families of the type $\{C_1 \cdot f_1(x) + \ldots + C_m \cdot f_m(x)\}$, where
$f_1(x), \ldots, f_m(x)$ are fixed functions, and $C_1, \ldots, C_m$ are arbitrary real numbers. As we will see later, for congestion problem, such families are not sufficient, so we will have to consider *general* families $\{f(x, C_1, \ldots, C_m)\}$, in which
the dependence on the coefficients $C_1, \ldots, C_m$ may be non-linear.

Definition 11.1. *Let $m > 0$ be a positive integer.*

■   *By an (m-parametric differentiable) family of functions, we mean a differentiable function $f(x, C_1, \ldots, C_m)$ of $m + 1$ real variables.*

■   *We say that a function $g(x)$ belongs to the family $f(x, C_1, \ldots, C_m)$ if there exists m real numbers $C_1, \ldots, C_m$ for which $g(x) = f(x, C_1, \ldots, C_m)$.*

Comment. In this lesson, we will only consider the tuning transformations $f(x)$ that transform the old value x_{old} of a parameter x into a new value $x_{\text{new}} = f(x_{\text{old}})$. In other words, we only consider tuning in which each of the parameters is tuned independently. It is reasonable, however, to also consider more complicated tunings, in which, say, the new value of the interpacket interval depends not only on the old (previous) value of the interpacket interval, but also on the old values of all other network parameters. In this case, instead of a function $f(x)$ from real numbers to real numbers, we must consider more complicated mappings $\vec{f}(\vec{x})$ from d-dimensional vectors $\vec{x}$ to d-dimensional vectors (for some integer $d > 1$):

■   For window-based protocols, $d = 4$, and $\vec{x} = (\tau_{\text{out}}, P, I, W)$.

■   For other protocols, $d = 3$, and $\vec{x} = (\tau_{\text{out}}, P, I)$, etc.

For this multi-dimensional case, we can also formulate the general notion of a family:

Definition 11.2. *Let $m > 0$ and $d > 0$ be positive integers.*

■   *By an (m-parametric differentiable) family of the transformations of the d-dimensional space, we mean a differentiable function $\vec{f}(\vec{x}, C_1, \ldots, C_m)$ that takes as inputs a d-dimensional vector $\vec{x}$ and m real numbers $C_1, \ldots, C_m$, and returns a d-dimensional vector $\vec{f}$.*

■   *We say that a function $\vec{g}(\vec{x})$ belongs to the family $\vec{f}(\vec{x}, C_1, \ldots, C_m)$ if there exists m real numbers $C_1, \ldots, C_m$ for which $\vec{g}(\vec{x}) = \vec{f}(\vec{x}, C_1, \ldots, C_m)$.*

The optimal family must contain all linear transformations. There are two reasons for this requirement: empirical and theoretical.

- The very fact that the Internet is based on linear tuning algorithms and it performs pretty well, means that in many circumstances, linear tuning methods are indeed good. It is therefore reasonable to assume that the family F of all possibly optimal tuning methods contain all linear functions $x \to a \cdot x + b$ with $a > 0$.

- Another reason is *theoretical*: in many cases, we can use a linear system model as a good approximation to the behavior of a real network, and for many linear systems, linear control algorithms are proved (in control theory) to be indeed the best.

Natural requirements on the family of tuning transformations.

- Let us assume that we have a minor congestion problem. To cure this problem, we applied, to a parameter x (one of the four parameters describing the network), a transformation $f(x)$ from the optimal family F. Thus, we replaced the initial value x_0 of this parameter x by the new value $x_1 = f(x_0)$. If the congestion problem is not repaired by this transformation, we will need to apply a further transformation $g(x)$ from the optimal family F, i.e., replace the value x_1 by the new value $x_2 = g(x_1)$.

  The faster we cure the congestion problem, the better. Thus, if at the moment 0, we knew that the transformation f would not be sufficient, we would apply *both* transformation at the same time, i.e., we would go directly from x_0 to $x_2 = g(x_1) = g(f(x_0))$ without using the intermediate value x_1. Hence:

  > If $f(x)$ and $g(x)$ belong to the class F of reasonable transformations, then their composition $h(x) = f(g(x))$ is also a reasonable transformation, and therefore, should also belong to the class F.

- Nothing in computer networks is 100% reliable, so it is quite reasonable to expect that we may, once in a while, err in deciding whether there is a congestion risk. Supposed that we erred and, based on this erroneous decision, used a transformation $x \to f(x)$ (from the family F) to avoid congestion. If at the next moment of time, we have discovered the error, and thus, discovered that this transformation was unnecessary, we would like to be able to return to the previous value x. The function $g(x)$ that describes this "un-doing" transformation $f(x) \to x$, i.e., the function for which $g(f(x)) = x$, is called an *inverse* function, and denoted by $f^{-1}(x)$. Thus:

> If a function $f(x)$ belongs to the class F of reasonable transformations, then the inverse transformation $g(x) = f^{-1}(x)$ is also a reasonable transformation, and therefore, should also belong to the class F.

In mathematics, there is a special term for families of functions that satisfy these two properties: such families are called *transformation groups*, or simply *groups*:

Definition 11.3. *Let a set X be fixed. A family F of functions from X to X is called a* transformation group *if it satisfies the following two properties:*

- *For every two functions $f, g \in F$, the composition $h = f \circ g$ of these two functions also belongs to F.*

- *Every function $f \in F$ is invertible, and the inverse function f^{-1} also belongs to the family F.*

In these terms, the above requirements can be reformulated as saying that *the family F of all optimal tuning transformations should form a transformation group*.

Examples of groups. The notion of a transformation group (sometimes called *symmetry group*) is very useful in physics and in many other areas. Let us give a few simple examples of groups:

- The set of all shifts $x \to x + s$ is a group: the composition of a shift by s and a shift by s' is a shift by $s + s'$, and an inverse transformation is a shift by $-s$.

- The set of all re-scalings $x \to C \cdot x$ is a group: composition corresponds to the product, and inverse transformation to $1/C$.

- The set of all linear functions is a group.

- For 3D space, the set of all motions (rotations, shifts, etc.) is a group.

Historical comment. The notion of a transformation group was invented in 1820s by a young French mathematician Evariste Galois who used this notion to solve the following problem:

- From antiquity, people know how to solve linear equations: if $a \cdot x + b = 0$, then $x = -b/a$.

- From the days of ancient Egyptians and Babylonians, mathematicians know how to solve quadratic equations: if $a \cdot x^2 + b \cdot x + c = 0$, then

$$x = \frac{-b \pm \sqrt{b^2 - 4 \cdot a \cdot c}}{2 \cdot a}.$$

- The natural next question is: can we write a similar formula for solving *cubic* equations $a \cdot x^3 + b \cdot x^2 + c \cdot x + d = 0$? This general formula was first discovered in the 16th century by an Italian mathematician Tartaglia. The exact year of this discovery is not known, because for several years, Tartaglia did not publish this formula. Instead, he kept it secret, and earned his living by participating (and winning) in numerous mathematical competitions. After a few years of Tartaglia's non-stop success, his rivals guessed that he may know the general algorithm, and succeeded in stealing his manuscript. After that, Tartaglia had to publish this result.

- After Tartaglia's result was published, very soon Cardano found a general formula for solving 4-th order (*quartic*) equations.

- The next question was: how to solve 5th order equations?

This was the question that was open for a several centuries until Galois proved that, unlike equations of 1-st, 2-nd, 3-rd, and 4-th order, the general solution of a 5-th order equation *cannot* be expressed as a composition of arithmetic operations and roots of corresponding (5-th) order.

- To find an algorithm, one does not need any general theory: once found, the algorithm can be easily checked.

- However, to prove a *negative* result, we need to consider *all possible* algorithms (and all possible transformations) and prove that none of them leads to a general solution of 5th order equations. For such a proof, we need a *general* theory, and this theory was developed by Galois.

Galois, a young man and a devoted Republican, was killed in a politically motivated duel with a royalist opponent. Before he died, he only had time to write a brief outlay of the theory of transformation groups (also known as *group theory*). It look a few decades before group theory was fully understood, and a few more decades before it became, in this century, one of the main tools of theoretical physics.

Lie groups. One of the stepping stones in this success story of group theory was a combination of group theory methods and traditional differentiation techniques from calculus. This was done at the end of the 19th century, by a Norwegian Sophus Lie. In his honor, the "differentiable" transformation groups are called *Lie groups*. The most general definition of as Lie group requires some additional math, but for our purposes, it is quite sufficient to consider the following definition:

Definition 11.4. *An m-dimensional (differentiable) family of transformations F that is, at the same time, a transformation group, is called a Lie group.*

In these terms, the problem of describing the optimal tuning transformations for congestion control can be reformulated as follows:

> Find all Lie groups of transformations of a 1D space that contain all linear transformations.

The solution to this problem is known in mathematics:

Theorem 11.1. *If F is Lie group of transformations of 1D space, and F contains all linear transformations, then every transformation $f(x)$ from the family F is a fractionally linear function*

$$f(x) = \frac{a \cdot x + b}{c \cdot x + d}.$$

In the Appendix to this lesson, we will show how one can prove this theorem. Meanwhile, let us describe how exactly we can use this result for congestion control.

The optimal tuning. For congestion control, Theorem 11.1 leads to the following conclusion:

> We should use fractionally linear tuning transformations for congestion control.

What exactly fractionally linear transformations should we use? Let us recall that for each of the four network parameters x, depending on the situation, we were looking either for an *increase* in x, or for a *decrease* in x. So, we arrive at the following definition:

Definition 11.4. *Let F be a class of all fractionally linear functions.*

- *By an increasing tuning transformation, we mean a function $f \in F$ such that $f(x) > x$ for all $x > 0$.*

- *By a decreasing tuning transformation, we mean a function $f \in F$ such that $0 < f(x) < x$ for all $x > 0$.*

Proposition 11.1.

- *Every increasing tuning transformation is a linear function.*

- *Every decreasing tuning transformation is either a linear function, or a function of the type $f(x) = a \cdot x/(x + d)$, where $0 < a \leq d$.*

Proof of Proposition 11.1. Let us first prove the result about *increasing* transformations. For a transformation to be increasing, the inequality $f(x) = (a \cdot x + b)/(c \cdot x + d) > x$ must hold for all $x > 0$. There are two possible cases: $c \neq 0$ and $c = 0$.

- If $c \neq 0$, then, when x tends to $+\infty$, the right-hand side tends to $+\infty$, while the left-hand side tends to $a/c < +\infty$. Thus, the case $c \neq 0$ is impossible.

- Hence, $c = 0$, and $f(x)$ is a linear function.

Let us now describe non-linear *decreasing* transformations, i.e., transformation for which $c \neq 0$.

- From the requirement that this fraction be defined for all positive x we conclude that $d \geq 0$ (else, if $d < 0$, for $x = -d = |d|$ it would be undefined).

- Since $0 < f(x) < x$ for all $x > 0$, we get $f(x) \to 0$ as $x \to 0$. Therefore, $b/d = 0$, and $b = 0$.

- By dividing both the numerator and the denominator of the function by $c \neq 0$, we get a new form in which $c = 1$. Hence, without loss of generality, we can assume that $c = 1$. Thus, $f(x) = a \cdot x/(x + d)$.

- The condition that $f(x) < x$ means that $a \cdot x/(x + d) < x$. Multiplying both sides of this equation by the positive denominator, we conclude that $a \cdot x < x^2 + d \cdot x$. Moving $d \cdot x$ to another side and dividing both sides by x, we conclude that $a - d < x$ for all $x > 0$. For $x \to 0$, we conclude that $a - d \leq 0$, and $a \leq d$. (Vice versa, if $0 < a \leq d$, then it is easy to check that $0 < f(x) < x$.)

The proposition is proven.

Fractionally linear tuning indeed leads to a better congestion control that traditional linear tuning techniques. In his above-cited 1988 paper, Van Jacobson has shown that if, for tuning interpacket interval, we use the multiplicative increase $I_{new} = a \cdot I_{old}$ and fractionally linear decrease $I_{new} = d \cdot I_{old}/(T_{old} + d)$, for some positive constant $d > 0$, then we indeed get a much better congestion control.

Important comment: The optimal family of functions is indeed non-linear. In the beginning of this lesson, we mentioned that for congestion control, we cannot use linear families. Now, we are ready to show that the optimal family F is indeed non-linear:

- In the previous lessons, we considered *linear* families of functions, of the type $\{C_1 \cdot f_1(x) + \ldots + C_m \cdot f_m(x)\}$, where $f_i(x)$ are given functions, and C_i are arbitrary real numbers. Each such linear family had the following easy-to-check property: if $f(x)$ and $g(x)$ are both members of this family, then their sum $f(x) + g(x)$ also belongs to the same family.

- The family of all fractionally linear functions does not have this property: e.g., the sum $x + 1/(x + 1)$ of the two fractionally linear functions $f(x) = x$ and $g(x) = 1/(x + 1)$ is *not* fractionally linear. Thus, the family F of all fractionally linear functions is indeed not a linear family.

Exercise

11.2 Suppose that we use Van Jacobson's fractionally linear algorithm for tuning the interpacket interval I: multiplicative increase with a coefficient

$a = 2$, and fractionally linear decrease with a coefficient $d = 1000$ msec. Initially, we took $I = 500$ msec. During the first two moments of time, we had congestion, then for 8 more moments of time, we had no congestion. Compute the value of interpacket interval for ten consequent moments of time.

Appendix: Proof of Theorem 11.1

1. The transformation group F contains transformations of the type $x \to f(x, C_1, \ldots, C_m)$, where f is a differentiable function of all its $m+1$ arguments and C_i are arbitrary real numbers. Since F is a group, the *identity* transformation $x \to x$ must also belong to this class. In other words, for some values $C_1^{(0)}, \ldots, C_m^{(0)}$, we must have $f(x, C_1^{(0)}, \ldots, C_m^{(0)}) = x$ for all x.

To make our formulas simpler, we can "shift" the parameters C_i to new values $\widetilde{C}_i = C_i - C_i^{(0)}$. In terms of new parameters, $C_i = \widetilde{C}_i + C_i^{(0)}$ and therefore, the transformations take the form $x \to \widetilde{f}(x, \widetilde{C}_1, \ldots, \widetilde{C}_m)$, where we denoted $\widetilde{f}(x, \widetilde{C}_1, \ldots \widetilde{C}_m) = f(x, \widetilde{C}_1 + C_1^{(0)}, \ldots, \widetilde{C}_m + C_m^{(0)})$. This new representation has the advantage that the *identity* transformation corresponds to $\widetilde{C}_i = 0$.

So, without losing generality, we can assume that the identity transformation corresponds to $C_1 = \ldots = C_m = 0$, i.e., that $f(x, 0, \ldots, 0) = x$.

2. After considering the *identity* transformation that corresponds to $C_1 = \ldots = C_m = 0$, the next natural step is to consider transformations that correspond to *small* values of C_i. Here, "small" means that we consider the values C_i that depend on some parameter t (in such a way that $C_i(t) = 0$), and then take $t \approx 0$. If $C_i(t)$ are differentiable functions of t, with $C_i(t) = 0$, then for small t, $C_i(t) = C_i(0) + C_i'(0) \cdot t + o(t) = c_i \cdot t + o(t)$, where by c_i, we denoted the derivative $C_i'(0)$ of the function $C_i(t)$ at $t = 0$. For the transformation itself, we get a similar formula

$$f(x, C_1(t), \ldots, C_m(t)) = f(x, 0, \ldots, 0) + \frac{df}{dt}(x, C_1(t), \ldots, C_m(t))_{|t=0} \cdot t + o(t).$$

The term that does not depend on t is simply equal to x, and we can use the formula for the derivative of the composition to express the second term as $c_1 \cdot f_1(x) + \ldots + c_m \cdot f_m(x)$, where we denoted

$$f_i(x) = \frac{\partial f}{\partial C_i}(x, C_1, \ldots, C_m)_{|C_1 = \ldots = C_m = 0}.$$

In other words, for small C_i, the transformations from F take the form

$$x \to x + (c_1 \cdot f_1(x) + \ldots + c_m \cdot f_m(x)) \cdot t + o(t). \qquad (11.1)$$

If we neglect the $o(t)$ terms, we get an *approximate* formula. The smaller t, the better this formula; this formula is very accurate if t is *extremely* small. In view of this fact, the formula (11.1) is called an *infinitesimal transformation*.

The set of all possible functions of the type $c_1 \cdot f_1(x) + \ldots + c_m \cdot f_m(x)$, that serve as coefficients at t in infinitesimal transformations, is called *Lie algebra*. In these terms, before we describe the desired Lie group, we will first try to describe the corresponding Lie algebra.

3. Since the set F forms a group, any composition of transformations from F also belong to F. In particular, since F contains shifts, for every infinitesimal transformation $x \to x + f(x) \cdot t + o(t)$, F also contains the composition of the following three transformations:

- a shift $x \to x + x_0$;

- the infinitesimal transformation $x \to x + f(x) \cdot t + o(t)$, and

- an inverse shift $x \to x - x_0$.

Let us find the explicit formula for this composition.

- After the first transformation, x turns into $x + x_0$.

- When we apply the infinitesimal transformation to the result $x + x_0$ of the first transformation, we get the following formula

$$x \to x + x_0 + f(x + x_0) \cdot t + o(t).$$

- Finally, when we apply the inverse shift to the result of the first two transformations, we get the following formula:

$$x \to x + f(x + x_0) \cdot t + o(t).$$

The resulting formula describes an infinitesimal transformation and therefore, the corresponding function $f(x + x_0)$ belongs to the Lie algebra. Therefore, we can conclude that:

For every function $f(x)$ from the Lie algebra, its shift $f(x+x_0)$ also belongs to the Lie algebra.

4. Similarly, since F contains a linear function $x \to C \cdot x$, for every infinitesimal transformation $x \to x + f(x) \cdot t + o(t)$, F also contains the composition of the following three transformations:

- a scaling $x \to C \cdot x$;

- the infinitesimal transformation, and

- the inverse scaling $x \to C^{-1} \cdot x$.

Let us deduce the explicit formula for this composition.

- After the first transformation, x turns into $C \cdot x$.

- When we apply the infinitesimal transformation to the result $C \cdot x$ of the first transformation, we get the following formula

$$x \to C \cdot x + f(C \cdot x) \cdot t + o(t).$$

- Finally, when we apply the inverse scaling to the result of the first two transformations, we get the following formula:

$$x \to x + C^{-1} \cdot f(C \cdot x) \cdot t + o(t).$$

The resulting formula describes an infinitesimal transformation and therefore, the corresponding function $\tilde{f}(x) = C^{-1} \cdot f(C \cdot x)$ belongs to the Lie algebra. From the definition of a Lie algebra (as the set of all functions of the type $c_1 \cdot f_1(x) + \ldots + c_m \cdot f_m(x)$), it is easy to conclude that with every function $f(x)$, and for every real number $C > 0$, the Lie algebra also contains an arbitrary function $C \cdot f(x)$. Therefore, the Lie algebra contains the function $C \cdot \tilde{f}(x) = f(C \cdot x)$. Therefore, we can conclude that:

For every function $f(x)$ from the Lie algebra, the "scaled" function $f(C \cdot x)$ also belongs to the Lie algebra.

4. So, the Lie algebra is a set of all functions of the type $c_1 \cdot f_1(x) + \ldots +$ $c_m \cdot f_m(x)$, where $f_1(x), \ldots, f_m(x)$ are given functions and c_i are arbitrary constants. We have already proven that this Lie algebra is invariant with respect to shift and scaling. We already know (from the proof of Theorem 10.2) that if a family of this type is both shift- and scale-invariant, then this family consists of polynomials of order $\leq m-1$, i.e., of polynomials of the type $f(x) = a_0 + a_1 \cdot x + a_2 \cdot x_2 + \ldots + a_{m-1} \cdot x^{m-1}$. So, all elements of the Lie algebra are polynomials.

5. To proceed further, we will consider one more superposition. Namely, if we have two infinitesimal transformations $x \to x + f(x) \cdot t + o(t)$ and $x \to x + g(x) \cdot s + o(s)$, then we can consider the composition of the following four transformations:

■   the first transformation $x \to x + f(x) \cdot t + o(t)$;

■   the second transformation $x \to x + g(x) \cdot s + o(s)$;

■   the transformation $x \to x - f(x) \cdot t + o(t)$ that is inverse to the first one, and

■   the transformation $x \to x - g(x) \cdot s + o(s)$ that is inverse to the second one.

Let us deduce the explicit formula for this composition.

■   After the first transformation, x turns into $x + f(x) \cdot t + o(t)$.

■   When we apply the second transformation to the result of the first one, we get $x + f(x) \cdot t + o(t) + g(x + f(x) \cdot t + o(t)) \cdot s + o(s)$. Here, $g(x + (f(x) \cdot t + o(t))) = g(x) + g'(x) \cdot f(x) \cdot t + o(t)$ and therefore, the composition of the first two transformations takes the form

$$x \to x + f(x) \cdot t + g(x) \cdot s + g'(x) \cdot f(x) \cdot t \cdot s + o(t) + o(s).$$

■   When we apply the first inverse to this result, we get, similarly,

$$x \to x + g(x) \cdot s + (g'(x) \cdot f(x) - f'(x) \cdot g(x)) \cdot t \cdot s + o(t) + o(s).$$

■   Finally, when we apply the second inverse transformation, we get the transformation $x \to (g'(x) \cdot f(x) - f'(x) \cdot g(x)) \cdot (t \cdot s) +$ terms $o(t)$ and $o(s)$.

This transformation is also an infinitesimal transformation, with $t \cdot s$ as a new variable and therefore, the corresponding function $g'(x) \cdot f(x) - f'(x) \cdot g(x)$ must also belong to the Lie algebra.

6. Now, we are ready to show that our Lie algebra can only contain linear or quadratic polynomials. Indeed, if $m - 1 > 2$, then from the fact that the Lie algebra contains the functions $f(x) = x^2$ and $g(x) = x^{m-1}$, we can conclude that the function $g'(x) \cdot f(x) - f'(x) \cdot g(x) = (m-1) \cdot x^{m-2} \cdot x^2 - x^{m-1} \cdot 2 \cdot x = (m-3) \cdot x^m$ also belongs to the Lie algebra. However, we have shown that the Lie algebra is only limited to the polynomials of order $\leq m - 1$. This contradiction shows that $m - 1$ cannot be > 2, so, the Lie algebra can only contains constants, linear functions, and quadratic polynomials.

7. An arbitrary transformation can be represented as a composition of transformations that are close to identity. (Actually, this fact requires some proof, but since this proof is long, and the fact is, more or less, intuitively clear, we skip this part of the proof.) So, to describe what functions form the transformation group, it is sufficient to describe functions that can be obtained as compositions of infinitesimal transformations.

- If we start with *linear* infinitesimal transformations, then, since the composition of two linear functions is also a linear function, we only get *linear* transformations.

- If we add infinitesimal transformations with the *quadratic* function $x \to x + x^2 \cdot t + o(t)$, then, to find out what transformations they lead to, we can consider the compositions $f(x, 0) = x$, $f(x, t) = x + x^2 + o(t)$, $f(x, 2t) = f(f(x, t), t)$, $f(x, 3t) = f(f(x, 2t), t)$, etc. For $C = k \cdot t$, we get the formula

$$f(x, C + t) = f(f(x, C), t) = f(x, C) + f^2(x, C) \cdot t + o(t).$$

This expression relates the value of $f(x, C)$ in a nearby point $(x, C + t)$ with its value at (x, C) and is therefore, looking like a differential equation. To get the actual differential equation, we move the term $f(x, C)$ to the left-hand side, divide both sides by t, and tend to the limit $t \to 0$. As a result, we get the following equation:

$$\frac{\partial f}{\partial C}(x, C) = f^2(x, C).$$

For a fixed x, we get

$$\frac{df}{dC} = f^2.$$

If we move all the terms that contain f into the left-hand side and all the terms that contain C into the right-hand side, we get

$$\frac{df}{f^2} = dC.$$

Integrating both parts of this equation, we conclude that

$$\frac{1}{f} = C + C_0,$$

i.e., that $f = 1/(C + C_0)$.

We obtained this expression by fixing x. The only thing in the right-hand side that can depend on x is C_0. So, we conclude that

$$f(x, C) = \frac{1}{C + C_0(x)}.$$

From the condition that $f(x, 0) = x$, we conclude that $1/C_0(x) = x$, so, $C_0(x) = 1/x$, and

$$f(x, C) = \frac{1}{C + 1/x} = \frac{x}{1 + C \cdot x}.$$

Thus, quadratic terms in the Lie algebra lead to fractional linear transformations.

The composition of linear and fractionally linear transformations is always fractionally linear, so we conclude that the arbitrary transformation from the group F is fractionally linear. The theorem is proven.

12

NEURAL NETWORKS: A GENERAL FORM OF NON-LINEAR EXTRAPOLATION

In many tasks, computers are much faster than humans. However, there are still tasks, such as handwriting recognition, that we humans do much better and much faster than even the fastest modern computers. To help computers solve these tasks, it is natural to simulate the way our brain (which is, in essence, a neural network) solves them. The resulting universal extrapolation tool, called artificial neural network, is described in this lesson. For these networks, continuous mathematics helps to choose the optimal parameters.

12.1. Why neural networks? Why are modern computers not sufficient?

In many tasks, computers are faster than humans. The main objective of the first computers, as one can see from the very word *computer*, was to *compute*. The first computers computed the trajectories of the artillery shells, the distribution of neutrons in a nuclear reactor, etc. Before the computers appeared, these same computations were done by human calculators (before the invention of the computing machines, the human calculators were even called *computers*).

The first computers did the same computational tasks as humans before them, but they did them in a somewhat *different* way than humans. This difference is typical for engineering design in general. The main goal of an engineer is to design *new* things, things that did not exist in nature, that are, from the viewpoint of the design's objective, better than anything in nature. Engineers design cars that run faster than the fastest animals, planes that fly faster than the fastest birds and higher than any bird has even flown, and, of course, they design computers that compute faster than any human being. To an engineer, nature is a *rival*: he may, once in a while, copy some good features of nature's creations, but usually, this is viewed as an intermediate step, after which the artificial creature will suppress the nature. For example, the first primitive designs of the flying machines tried to simulate the way bird use their wings. However, very soon, the wings of the planes became more sophisticated, and the planes started flying faster than the birds.

Similarly, although the first computers were still reasonably slow (so that some very skillful human calculators could still compete with them), very soon computers became enormously faster than human calculators.

This success with which modern computers took over the task that previously required human intelligence – namely, the task of computing – made computer designers and programmers hope that a computer can do other intelligent tasks as well. For many of these tasks, computers and programmers did succeed. Let us give a few examples:

- Several decades ago, skilled editors would spell-check all the texts. Now, a computer program routinely does that (so, if there are still typos in this text, please blame the spell-checking program, not us – just kidding).

- In most universities, computers register students for classes, make sure that the students have necessary pre-requisites, design class lists, etc.

- Computer programs such as Mathematica can find the integral of $\sin^3(x)$ or of any complicated function, find an exact solution of the differential equation, and do many other *symbolic computations*.

These tasks are so numerous that nowadays, computing is no longer the major application of computers: they are used for storing and retrieving data in databases, for communication, for automatic control, etc. In most of these tasks, computers are much better than humans: just like computers compute better than humans, they spell-check better and more accurately than humans (at least than most humans), they can handle student registration much faster than humans would, etc.

In some tasks, human are still faster. In spite of all the successes of modern computers, there are still intellectual tasks in which we humans are much faster.

- Of course, experts are still better than computers in many tasks:

  - Garry Kasparov, the (human) chess champion, still beats the most complicated chess programs (although this particular human superiority may not last for long[1]).

  - Physicists look at the experimental data and design fundamental theories that computers are still unable to do.

  - Mathematicians come up with intelligent proofs of old results that computers are unable to. (Although there have already been a few cases when a computer program found a proof that no human could; the most famous example is the Four Color Theorem, that every map can be colored by four colors so that no two neighboring countries have the same color.)

  - etc.

- But even non-experts, lay people are much faster than computers in many tasks that involve *learning*:

  - After we have met a person a few times, and we see that person again, we immediately recognize that person and say "Hi" (or "Hola", or the corresponding word in whatever language we use) without overstraining our intellectual abilities.

  - When we look at a blurry photo of the office, we immediately recognize the person, the computer, the desk, etc.

  - When we see a handwritten letter, we can usually read and understand it without any effort.

For the computer, any of these learning ("pattern recognition") tasks is extremely difficult. Some of these tasks, even modern super-computers simply cannot do. Some of these tasks computers can do, but humans do them much faster.

Example: mail. As an example, we can take the US Postal Service. In the early 19th century, letters were often slow, because it took a few days or even weeks to drive the letter from one place to another. Nowadays, inside the USA, mail is usually transported by planes. Modern planes take a few hours to fly from any point in the continental USA to any other point. So why are not letters going that fast? Let us give the extreme example: It takes an hour to drive from the first author's home in El Paso to the second author's home in

---

[1] This did happen in May 1997.

Las Cruces. However, a letter sometimes takes a few days. Why? Because most of the time is *not* spent on transportation, it is spent on *handling* the mail. Letters need to be sorted by address, and for that, we need somebody to read the address (or at least the numerical code called *zip* code) and place the letter into the packet going to that particular place. Handling billions of letters by hand is slow and expensive, so the US Postal Service has several computer-based systems in place that try to speed up mail handling by automatically reading the zip code and the address. But, in spite of all the efforts, even the best computer-based systems are 80% to 90% effective, which means that in 10 to 20% of the cases, a postal worker is needed to do the job that computers could not.

It is very surprising that humans are faster.

- There is no mystery is computers being faster in calculations: we know how the computers compute, we program them ourselves, and basically, they do steps that are very similar to what we would have done (slightly similar, because, e.g., computer use binary numbers, while we use decimal ones). The main reasons why computer are faster is not because they use more intelligent and faster algorithms, but because the computer's processing elements are much faster. A modern PC runs 100 million operations per second. We cannot compute that fast.

- So, maybe, the reason why humans are faster in image recognition is because there are some fast physiological processes involved? No, this is not the case. We know in rather great detail how image recognition and other intellectual processes are happening in the brain. All these processes are carried by special cells called *neurons*, and these cells can only operate at a rate of 10 to 100 operations per second.

How come, that with very slow processing elements, the brain can solve the tasks that a modern computer cannot?

The brain is a highly parallel machine. The answer is simple:

- Most modern computers are either *sequential* machines, in which at any other give moment of time, only one operation is being performed, or, at best, have a few processors working in parallel:

  - a PC usually has two parallel processors: the main processor and the math co-processor that takes care of vector operations;

    – a Sun Sparc workstation on which I am typing this text is connected to a server that has 4 to 8 processors running in parallel;

    – a super-computer like Cray has several thousand computers working in parallel.

■ At the same time, the brain has more than *ten billion* neurons, and they all work in parallel.

As a result, during a second, a PC with two processors runs $2 \times 100 = 200$ million operations, while at the same time, the brain runs $10 \times 10 = 100$ *billion* operations.

A natural idea: let us simulate the brain. Since for many tasks, the brain is so much more efficient, the natural idea is: let computers simulate the brain. A computing device that simulates the brain is called an *artificial neural network*, or simply a *neural network*.

Historical comment. The first idea of simulating human brain was due to an interesting coincidence that happened in the 40s. Norbert Wiener, one of the founding fathers of modern data processing (for example, *Wiener filter* is something every electrical and computer engineer knows), was a good friend of a neurophysiologist Arturo Rosenblueth. Once, Rosenblueth noticed Wiener's drawing, and he was surprised to notice that Wiener, who had no prior knowledge in neurophysiology, has actually drawn a part of the vision-processing structure of the human brain. Wiener was even more surprised to learn this because all he did was drew a reasonable electronic design for a vision system.

In other words, it turned out that the 20th century state-of-the-art design of a vision system has already been implemented in the human brain for a few thousand years. This coincidence convinced both of them that a human brain has a much more reasonable design that they may have thought before, and that simulating the human brain is indeed a very reasonable way of designing data processing methods.

12.2. What are neural networks: in brief

How the input information is represented. The human brain processes information that comes from sensing organs such as eyes, ears, etc. Inside these organs, special *sensor* neurons transform this external information into an electric signal. This signal is a series of almost identical pulses (called *spikes*), and the intensity of the external signal is represented by the frequency x of these pulses, i.e., by the number of pulses per second. (This is similar to the way Geiger counters represent the radioactivity level.) This frequency x_i is called the *firing rate*. The signals transmitted inside the brain are of exactly the same type.

A neuron. A typical biological neuron has up to $n = 10,000$ inputs and a single output. The actual biological neurons have a certain degree of *inertia* in the sense that the output signal $y(t)$ at each moment of time depends not only on the values of the input signals $x_1(t), \ldots, x_n(t)$ at this same moment of time but also on the previous values of input and output. This inertia is important if we analyze the human reaction to rapidly changing signals. However, in processes like handwriting recognition, face recognition, etc., the input image does not change much, so, we can neglect the effect of inertia, and assume that the output signal $y(t)$ depends only on the values of $x_1(t), \ldots, x_n(t)$ at this same moment of time. How to describe this dependence?

From the physiological viewpoint, signal transformation inside a neuron can be, roughly, subdivided into three sequential steps:

- First, the pulses coming from different inputs are collected together. Different inputs have different paths. If a path is wide enough, then all x_i pulses per second pass through. If a path is narrow, only some pulses sent towards this particular input reach this neuron. The portion of pulses that reaches a neuron is usually called a *weight* of i-th input and denoted by w_i. As a result, from each input, out of x_i pulses, only $w_i \cdot x_i$ pulses come through, and the total number of pulses that reaches a neuron is equal to the sum of these n numbers: $x = w_1 \cdot x_1 + \ldots + w_n \cdot x_n$.

- In comparison with electronic components of a computer, neurons are not only very slow, they are also very un-reliable and noisy. Once in a while, they emit a random pulse that is not resulting from any input. Hence, in addition to pulses that represent input signals and the result of their processing, there are also many "noise" pulses that circulate through the brain. So, if, e.g., we want a neuron to react to light, we cannot make it react to any small number of input pulses: these pulses can be simply representing the random noise. To avoid reacting to the noise, we must have some *threshold* w_0 for the input pulses, and take the *difference* $x_{\text{in}} = x - w_0$ as the actual input.

■ Finally, there are some chemical reactions occurring in the body of the neuron that transform the input x_{in} into the output y. The function $y = s(x_{in})$ that describes how the output y is related to the input x_{in} of the neuron is called the *activation function*.

In short, the output of the neuron is related to its input by a formula

$$y = s(w_1 \cdot x_1 + \ldots + w_n \cdot x_n - w_0). \tag{12.1}$$

This formula is used to describe artificial (simulated) neurons as well.

A neural network. From the sensory (*input*) neurons, the signals go to some neurons that process the sensory data. The output from these neurons becomes an input for some other neurons, etc., until we reach the neurons whose outputs go outside the network. These final neurons are called *output neurons*. All the neurons in between the input and the output ones are called *hidden* neurons because they are hidden from the outside influence: we cannot send any signal directly to these neurons, and we cannot read any information directly from them.

How many layers in the biological neural network? In principle, the input signal can go through several *layers* of hidden neurons before it reaches the output neuron. In the brain, each neuron takes some time to process. The more layers, the longer the processing. We know the processing time of each neuron, and we know how long it takes to recognize a person, so we can estimate how many layers participate in processing the person's image: at most ten.

How many layers in the artificial neural network? The main reason why we use neural networks in the first place is that they are fast. The fewer layers, the faster the computations. Therefore, we should try to use the smallest possible number of layers.

The biological experience teaches us that we should use no more than ten layers. Shall it be exactly ten? Not necessarily. As we have mentioned, biological neurons are very noisy and un-reliable, so some layers may simply take care of "filtering out" this noise. In electronic implementation, components are usually practically 100% reliable, so we may not need that many layers. How many do we need? To answer this question, we must describe what exactly we want from an artificial neural network.

We want an artificial neural network to be a universal tool: the main idea. In saying that the biological neural network (the brain) is fast, we meant that it is fast in learning *any* image, and, in general, in solving *any* learning problem. Thus, when we design an artificial neural network, we also want it to be a *universal* tool, with the ability to learn *anything*. How can we formulate this requirement in precise (mathematical) terms?

We want an artificial neural network to be a universal tool: towards a mathematical formalization. Ideally, *learning* (or *extrapolation*) means that, given a set of patterns $(x_1^{(p)}, \ldots, x_n^{(p)}, y^{(p)})$, $1 \leq p \leq P$, this tool will be able to construct a function $y = f(x_1, \ldots, x_n)$ that passes exactly through these values, i.e., for which $y^{(p)} = f(x_1^{(p)}, \ldots, x_n^{(p)})$ for all p from 1 to P. In reality, the values from a learning pattern are usually taken from measurements, and therefore, are only approximately known. In this context, it makes sense to require that for an arbitrary accuracy $\varepsilon > 0$, this tool will construct a function $f(x_1, \ldots, x_n)$ that approximates these patterns within a given accuracy, i.e., for which $|y^{(p)} - f(x_1^{(p)}, \ldots, x_n^{(p)})| \leq \varepsilon$ for all p from 1 to P. Moreover, we would like to guarantee that, as we take more and more patterns, the resulting function will closer and closer approximate the actual dependency. If we denote the actual function by $g(x_1, \ldots, x_n)$ and the bound for its domain by Δ, we arrive at the following definition:

Definition 12.1. *We say that a class of function F has a universal approximation property (or is a universal approximator), if for every two real numbers $\varepsilon > 0$ and $\Delta > 0$, and for every continuous function $g(x_1, \ldots, x_n)$ that is defined for all x_i for which $|x_i| \leq \Delta$, there exists a function $f \in F$ for which $|f(x_1, \ldots, x_n) - g(x_1, \ldots, x_n)| \leq \varepsilon$ for all $x_1, \ldots, x_n \in [-\Delta, \Delta]$.*

In these terms, the above requirement means that *we want to choose the smallest number of layers for which the artificial neural network has the universal approximation property.*

Comment. For the reader to be more familiar with the notion of a universal approximator, let us give two examples (simpler than neural networks) of classes of functions with this property. We will not give the proofs of the corresponding results, but we will give intuitive explanations of why these classes have this property:

■   The class F of all *polynomials*.

> *Intuitive explanation:* By definition, an arbitrary *analytical* function is equal to the sum of its (infinite) Taylor series. Hence, if we keep finitely many terms, we get an approximation to the original function;

the more terms we keep, the better the approximation. Thus, an arbitrary analytical function can be approximated by polynomials. To complete the proof, it is sufficient to be able to approximate an arbitrary *continuous* function $g(\vec{x})$ by *analytical* functions. This can be done by different *smoothings*, i.e., by using the function

$$\tilde{g}(\vec{x}) = \int g(\vec{y}) \exp(-\frac{(\vec{x} - \vec{y})^2}{2\sigma^2}) dx_1 \ldots dx_n.$$

The universal approximation property of polynomials was first proven in the 19th century by Cauchy and Weierstrass.

- The class F of all *trigonometric polynomials*, i.e., linear combinations of the functions $\sin(k_1 \cdot x_1 + \ldots + k_n \cdot x_n)$ and $\cos(k_1 \cdot x_1 + \ldots + k_n \cdot x_n)$ for different k_i. (Since $\cos(x) = \sin(\pi/2 - x)$, we can, alternatively, express each polynomial as a linear combination of functions $\sin(k_1 \cdot x_1 + \ldots + k_n \cdot x_n + k_0)$.)

  Intuitive explanation: We have already mentioned, in Lesson 9, that an arbitrary function can be "decomposed" into sinusoids (this decomposition is called *Fourier series*). To get an exact representation, we need infinitely many terms. So, the more terms we get, the better approximation we get.

Two layers are not enough. We need at least one layer of neurons for input, and at least one layer for processing. Hence, we need at least two layers. Let us show that two layers are not sufficient. Indeed, if we use only two layers — input and output, with no hidden layers at all — then we can only compute the functions (12.1) that correspond to a single neuron. Let us show that even the simple non-linear function $g(x_1, x_2) = x_1 \cdot x_2$ cannot be thus represented. Indeed:

- For a function $f(x_1, x_2) = s(w_1 \cdot x_1 + w_2 \cdot x_2 - w_0)$ described by the formula (12.1), for every real number c, the condition $h(x_1, x_2) = c$ is equivalent to $w_1 \cdot x_1 + w_2 \cdot x_2 - w_0 = s^{-1}(c)$. This linear equation describes a straight line, and therefore, the level set $\{(x_1, \ldots, x_n) | f(x_1, \ldots, x_n) = c\}$ is a straight line (or, if there are several different values z for which $s(z) = c$, a union of several straight lines).

- On the other hand, each level set $\{(x_1, x_2) | g(x_1, x_2) = x_1 \cdot x_2 = c\}$ of the product function $g(x_1, x_2)$ is a hyperbola $x_2 = c/x_1$.

Since the level sets of the functions f and g are different, the functions themselves are different. In other words, *two layers are not enough*, and we need at least three.

Three layers are, in principle, sufficient. Let us show that a network with three layers, i.e., with one hidden layer, already has the desired universal approximation property, even if we allow the simplest (linear) neurons in the output layer. Indeed:

- At first, we use the input layer to input the values $x_1, \ldots, x_n$.

- Then, each neuron in the hidden layer transforms these input signals into the value $y_k = s(w_{k1} \cdot x_1 + \ldots + w_{kn} \cdot x_n - w_{k0})$, where $k = 1, 2, \ldots, K$, and K is the total number of neurons in the hidden layer.

- Finally, the linear neuron in the output layer transforms the intermediate values $y_1, \ldots, y_K$ into a linear combination $y = W_1 \cdot y_1 + \ldots + W_K \cdot y_K - W_0$.

If we substitute the expressions for y_k into the formula for y, we get the following formula for y:

$$y = \sum_{k=1}^{K} W_k \cdot s(w_{k1} \cdot x_1 + \ldots + w_{kn} \cdot x_n - w_{k0}) - W_0. \qquad (12.2)$$

For some non-linear function $s(x)$, the corresponding functions indeed have the universal approximation property: e.g., for $s(x) = \sin(x)$, we get exactly all trigonometric polynomials, which are known to have the universal approximation property.

For some non-linear activation functions, three layers are not enough. If we take $s(x) = x^2$, then the expression (12.2) is a quadratic function, and it is known that quadratic functions do not have the universal approximation property. Similarly, if we take a polynomial of an arbitrary order m as $s(x)$, we will get only polynomials of this same order, and thus, we will not get the universal approximation property. (It turns out that polynomials are the only smooth functions $s(x)$ for which three-layer neurons do not have the universal approximation property.)

How to train neural networks? General idea. We start with the patterns $(x_1^{(p)}, \ldots, x_n^{(p)}, y^{(p)})$, $1 \leq p \leq P$, and we try to find the values of the weights W_k and w_{ki} for which errors

$$e^{(p)} = y^{(p)} - f(x_1^{(p)}, \ldots, x_n^{(p)}) \qquad (12.3)$$

are as small as possible. Usually, the least squares method is used, i.e., we try to find the values W_i and w_{ij} for which $J \to \min$, where

$$J = \sum_{p=1}^{P} (e^{(p)})^2. \qquad (12.4)$$

How can we actually find this minimum?

Since the activation function is smooth, the function J also smoothly depends on the weights; so, we face a problem of minimizing a smooth function $J(W_1, \ldots) \to \min$.

In the previous lessons, we usually found a minimum of the smooth function $F(z_1, \ldots, z_n) \to \min$ by equating its partial derivatives to 0: $\partial F / \partial z_i = 0$, $1 \le i \le n$. Unfortunately, in our case, these derivatives are also highly non-linear, and so, by equating them to 0, we reduce the original complex minimization problem to a (not less complex) problem of solving a system of n highly non-linear equations with n unknowns. Since this does not help, we need a new method.

Such a method (called *steepest descent*) is based on the following idea: Suppose that we are standing on the top of the mountain and we want to get down as fast as possible. Where do we go? If we can go in several different directions, then it makes sense to go into the direction in which the descent is the steepest possible, i.e., in which the decrease in the height is the fastest. To find the minimum of a function $F(z_1, \ldots, z_n)$, we will apply this idea to a landscape in which $F(z_1, \ldots, z_n)$ is a height of a point whose horizontal (geographic) coordinates are $z_1, \ldots, z_n$.

Gradient method: towards mathematical formulation. How can we describe this method in mathematical terms? Let us assume that initially, we are at the point with "horizontal" coordinates $z_1, \ldots, z_n$, and, correspondingly, with the height $F(z_1, \ldots, z_n)$. Where do we go next? To describe the direction of our movement, we must describe the velocities $v_i = dz_i/dt$ in each horizontal direction.

In realistic problems, there is a limit v_0 on the total velocity $v = \sqrt{v_1^2 + \ldots + v_n^2}$. Since we want to go down as fast as possible, we will try to go with this very speed; the only question is: in what direction? We will choose the direction in which the decrease in F is the largest. If we go in the direction determined by the velocities $\vec{v} = (v_1, \ldots, v_n)$, then after a small time Δt, we end up in a point with the new coordinates $z_i + \Delta z_i$, where

$\Delta z_i = v_i \cdot \Delta t$. In this new point, the height is equal to $F(z_1 + \Delta z_1, \ldots, z_n + \Delta z_n)$. Since Δt is small, the values Δz_i are also small, and hence, we can use the known fact that $F(z_1 + \Delta z_1, \ldots, z_n + \Delta z_n) \approx F(z_1, \ldots, z_n) + \Delta F$, where $\Delta F = F_{,1} \cdot \Delta x_1 + \ldots + F_{,n} \cdot x_n$, and

$$F_{,i} = \frac{\partial F}{\partial z_i}(z_1, \ldots, z_n).$$

We want to choose v_i for which the change in height ΔF is the smallest possible. Since $\Delta F = (F_{,1} \cdot v_1 + \ldots + F_{,n} \cdot v_n) \cdot \Delta t$, this is equivalent to requiring that $F_{,1} \cdot v_1 + \ldots + F_{,n} \cdot v_n$ is the smallest possible under the condition that $v_1^2 + \ldots + v_n^2 = v_0^2$.

This auxiliary constraint minimization problem can be easily solved by using Lagrange multipliers method that reduces it to the un-constrained optimization problem $F_{,1} \cdot v_1 + \ldots + F_{,n} \cdot v_n + \lambda \cdot (v_1^2 + \ldots + v_n^2 - v_0^2) \to \min$. For this new problem, equating the derivatives with respect to v_i with 0, we conclude that $F_{,i} + 2\lambda v_i = 0$, i.e., that $v_i = -(1/(2 \cdot \lambda)) \cdot F_{,i}$. So, if we started at the point $(z_1, \ldots, z_n)$, the natural next point is $z_i' = z_i + \Delta z_i = z_i + v_i \cdot \Delta t = z_i - (\Delta t/(2 \cdot \lambda)) \cdot F_{,i}$, i.e.,

$$z_i \to z_i - \mu \cdot \frac{\partial F}{\partial z_i},$$

where we denoted $\mu = \Delta t/(2 \cdot \lambda)$. At the new point, we repeat the same calculations, etc. The vector $(F_{,1}, \ldots, F_{,n})$ formed by partial derivatives is called a *gradient* and denoted by ∇F. Because of this, this method $\vec{z} \to \vec{z} - \mu \cdot \nabla F$ is called *gradient descent*.

This simple formula actually defines the method, except for the value μ, because this value was defined in terms of the unknown value λ. In principle, the value μ depends on the local bound v_0 on the "velocity" in a given point. This bound may depend on the point, so, the value μ may change from iteration to iteration. So, the only remaining question is: how to choose the values μ on different steps?

Computationally the easiest way to use the same value μ on each iteration. In this case, the problem is simply to choose a single number $\mu > 0$. How do we choose it? Unfortunately, there are no universal recipes: it is mainly an art; there are many heuristic methods, but few of them are theoretically justified in general situations:

■   If we choose μ too large, i.e., if we make large steps, we may miss the minimum.

- If we choose μ too small, we will get to the minimum, but it will take forever.

So, we can do the following:

- choose an arbitrary value of μ for the first step;

- if with this value of μ, we get a decrease in the value of the objective function, this means that we may be going too slowly; so, we try a larger μ (e.g., a twice larger μ);

- if for some μ, we do not get a decrease in F, this means that our step is too big, so we decrease μ (e.g., to a half of its previous size), etc.

Let us give a simple example of the gradient method.

Gradient method: example. Let us take the simplest possible case when $n = 1$ and $F(z_1) = F(z) = z^2$. Since $dF/dz = 2z$, the gradient method takes the form $z^{(k+1)} = z^{(k)} - 2\mu \cdot z^{(k)}$, i.e., $z^{(k+1)} = (1 - 2 \cdot \mu) \cdot z^{(k)}$. Let us start at a point $z^{(0)} = 1$.

If we use a *large* value of μ, e.g., $\mu = 2$, then we get $z^{(k+1)} = -3z^{(k)}$, i.e., $z^{(1)} = -3$, $z^{(2)} = 9$, etc.; the process diverges. On the other hand, if we use a very *small* value, e.g., $\mu = 0.01$, then we get $z^{(k+1)} = 0.98 \cdot z^{(k)}$. The corresponding geometric progression $z^{(k)} = 0.98^k$ converges to 0, but very slowly: after 20 iterations, we still get ≈ 0.3.

If we adjust μ as we go, we get a much better convergence. For example, if we start with 0.01, we get $z^{(1)} = 0.98$. Since the value of the objective function decreased, we double μ and use $\mu = 0.02$ for the next step; hence, $z^{(2)} = (1 - 2 \cdot 0.02) \cdot z^{(1)} = 0.96 \cdot 0.98 \approx 0.94$. Since we still get a decrease in F, we again double μ, and get $z^{(3)} = (1 - 2 \cdot 0.04) \cdot z^{(2)} \approx 0.87$. Similarly, we get $z^{(4)} = (1 - 2 \cdot 0.08) \cdot z^{(3)} \approx 0.73$, $z^{(5)} = (1 - 2 \cdot 0.16) \cdot z^{(4)} \approx 0.50$, $z^{(6)} = (1 - 2 \cdot 0.32) \cdot z^{(5)} \approx 0.18$, $z^{(7)} = (1 - 2 \cdot 0.64) \cdot z^{(6)} \approx -0.05$. Then, further doubling of μ leads to a new value $(1 - 2 \cdot 1.28) \cdot z^{(7)} \approx 0.08$ for which F increases. Thus, we do not double μ (i.e., we keep μ the same), and take $z^{(8)} = (1 - 2 \cdot 0.64) \cdot z^{(7)} \approx 0.02$. This is clearly much faster than with when we used the same value of μ on each iteration.

Backpropagation: gradient method for neural networks. In 1974, Paul Werbos has shown that for neural networks, the computations of the partial derivatives can be simplified if we store the intermediate results that were used to compute J. Indeed, how do we compute the value J?

- First, for each k and p, we compute the values $z_k^{(p)} = w_{k1} \cdot x_1^{(p)} + \ldots + w_{kn} \cdot x_n^{(p)} - w_{k0}$.

- Second, we compute $y_k^{(p)} = s(z_k^{(p)})$.

- Third, we compute $e^{(p)} = y^{(p)} - W_1 \cdot y_1^{(p)} - \ldots - W_K \cdot y_K^{(p)} + W_0$.

- Finally, we compute $J = (e^{(1)})^2 + \ldots + (e^{(P)})^2$.

So, to compute, e.g., the partial derivative $J_{,0}$ of J with respect to W_0, we can use the formula for the derivative of the composition:

$$J_{,0} = \sum_{p=1}^{P} 2e^{(p)} \cdot \frac{\partial e^{(p)}}{W_0}.$$

Here,

$$\frac{\partial e^{(p)}}{\partial W_0} = 1,$$

so $J_{,0} = 2 \cdot \sum e^{(p)}$.

Similarly,

$$J_{,k0} = \sum_{p=1}^{P} 2e^{(p)} \cdot \frac{\partial e^{(p)}}{w_{k0}},$$

where

$$\frac{\partial e^{(p)}}{\partial w_{k0}} = W_k \cdot \frac{\partial y_K^{(p)}}{\partial w_{k0}},$$

$$\frac{\partial y_K^{(p)}}{\partial w_{k0}} = s'(z_k^{(p)}) \cdot \frac{\partial z_K^{(p)}}{\partial w_{k0}},$$

and

$$\frac{\partial z_K^{(p)}}{\partial w_{k0}} = -1.$$

So, we get the formula $J_{,k0} = 2 \sum e^{(p)} \cdot W_k \cdot s'(z_k^{(p)}) \cdot (-1)$.

In the resulting training method, we start with some (e.g., randomly chosen) values of the weights. Then, we use these weights to simulate, step-by-step, the activity of the neural network; this is called *forward propagation*. After that, we go *back*, from the results to the original weights, and on our way, we use the intermediate results of forward computations to calculate the changes to the weights. Then, we go forward and backward again and again until we find the weights that fit all the patterns (i.e., until the network is *trained*). The *going back* (*back-propagation*) gives the name to the entire training method.

Exercises

12.1 Take any values of weights and of the input signals, and an arbitrary activation function $s(z)$, and use the formula (12.2) to compute the output of the corresponding neural network.

12.2 Use gradient method to find the minimum of a function $F(z) = z^4 - z^3 + z^2 - z$.

12.3 Similarly to the formulas for $J_{,0}$ and $J_{,k0}$, compute the remaining partial derivatives $J_{,k}$ and $J_{,ki}$.

12.4 Use the formulas from the previous exercise to train a neural network with one input on the following four patterns $(x_1^{(1)}, y^{(1)}) = (0,0)$, $(x_1^{(2)}, y^{(2)}) = (1,1)$, $(2,4)$, and $(3,9)$. Start with random weights, and use any non-linear function $s(z)$. *Hint:* For computational simplicity, you can use $s(z) = z^2$.

12.3.   What activation function should we choose?

We need non-linear activation functions: a little bit of history. The first neural networks used *linear* neurons, with a linear function $s(z)$ (e.g., with $s(z) = z$). They were called *perceptrons* because they were very good in automating simple tasks of human perception like recognizing vertically oriented printed letters. The input here is a pixel-by-pixel image of a letter (so that x_i is the brightness of i-th pixel), the output is, e.g., the ordinal number of the letter in alphabetic order, and the patterns are images of different letters at different locations.

For more complicated perception problems, perceptions did not work that nicely. To improve the situation, more and more complicated learning algorithms were tried until, in 1968, M. Minsky (one of the pioneers of Artificial Intelligence) and S. Papert published a book *Perceptrons* (MIT Press, Cambridge, MA). In this book, they use the known fact, that a composition of linear functions is linear, to show that if we only use linear neurons, then we can only compute linear functions.

Surprisingly, this result came as a shock to many neural researchers. Grants were not renewed, papers rejected, conferences cancelled, and even the very word "neuron" became a non-pronounceable "n-word". It took a few years for the neural community to realize that since linear neurons are not sufficient, non-linear activation functions should be used. The next question is: what activation function should we use?

The choice of an activation function is important. We have already mentioned that different activation functions lead to drastically different learning abilities of the network: e.g., networks with a linear function $s(z)$ can only learn linear dependencies, 3-layer networks with quadratic function $s(z)$ can only learn quadratic ones, etc. With non-polynomial activation functions, we get a universal approximation property, but, intuitively, the closer the activation function to a polynomial, the worst it should approximate. It is, therefore, no wonder that different non-polynomial activations functions lead to drastically different learning speeds and learning success rates. Therefore, choosing an *appropriate* activation function is very important. Ideally, we would like to choose the *best* (optimal) activation function. To find this optimal function, let us first formulate the problem of finding the optimal activation function in precise mathematical terms.

We must choose a family of functions, not a single function. The main advantage of neural networks is that they are highly *parallel* and thus, *faster*. At present, neural networks are mainly implemented in *software* on a *sequential* computer, but the main advantage of neural networks will be achieved only if we have *hardware*-implemented neurons working in *parallel*. For hardware neurons, an activation function $s(z)$ means actually transforming the signal z into a signal $y = s(z)$. The numerical representation of the activation function $s(z)$ depends, therefore, on the choice of the units in which we measure y, and on the choice of a starting point of measuring y. We already know that changing a measuring unit changes $s(z)$ into $C_1 \cdot s(z)$, while changing a starting point changes $s(z)$ into $s(z) + C_0$. Hence, if $s(z)$ is a reasonable activation function, then any function $C_1 \cdot s(z) + C_0$ should also be a reasonable activation function. (In terms of the weights, replacing $s(z)$ by $\tilde{s}(z) = s(z) + C_0$ is equivalent to changing the bias W_0, while replacing $s(z)$ by $C_1 \cdot s(z)$ is equivalent to scaling the weights W_k, $k > 0$.)

Since we are talking about non-linear phenomena, it makes sense to assume that, in addition to *linear* re-scalings $y \to C_1 \cdot y + C_0$, some *non-linear* rescaling transformations $y \to g(y)$ may be also applicable. There are many examples of non-linear re-scalings in physics. For examples, we can measure energy of an earthquake in a normal scale, or in a logarithmic (Richter) scale; we can measure noise by its power, or in the logarithmic units (decibels), etc. In other words, we can assume that if $s(z)$ is a reasonable activation function, then $\widetilde{s}(z) = g(s(z))$ is also reasonable, for each reasonable rescaling $g(y)$.

How can we describe these "reasonable re-scalings"? The set G of all reasonable re-scalings must contain all linear functions $g(y) = C_1 \cdot y + C_0$. It must also satisfy the following two properties:

- If a function $y \to \widetilde{y} = g(y)$ is a reasonable re-scaling, then its inverse $\widetilde{y} \to y = g^{-1}(\widetilde{y})$ is also a reasonable re-scaling.

- If $y \to \bar{y} = g_1(y)$ is a reasonable rescaling, and $\bar{y} \to \widetilde{y} = g_2(\bar{y})$ is a reasonable re-scaling, then the direct transformation from y to $\widetilde{y} = g_2(g_1(y))$ is also a reasonable re-scaling.

In short (using a term introduced in Lesson 11), the set G of all reasonable re-scalings must form a *group*. It is reasonable to assume smoothness, so G is a *Lie group that contains all linear transformations*. So, we arrive at the following definition:

Definition 12.1. *Let G be a transformation Lie group that contains all linear transformations $y \to C_1 \cdot y + C_0$, and let a smooth function $s(z)$ be fixed. By a family F, we mean the set of all functions of the type $\widetilde{s}(y) = g(s(z))$, where $g \in G$. The set of all families will be denoted by Φ.*

From all such families, we must choose a family that is the best according to some reasonable optimality criterion.

The optimality criterion must be shift-invariant. In the above text, we discussed the possibility to use different units for the *output* $y = s(z)$ of the hardware neuron. Similarly, we can use different units for its *input* z. In principle, we could both change a unit for measuring z (i.e., *scale* z) and change the starting point (i.e., *shift* z). In real-life implementations of neurons as electronic devices, shifts make much more physical sense. Indeed, most non-linear electronic components have voltage as their input and current as their output:

- *Shifting* all voltages does not change anything. E.g., when a plane is in the air, its potential field with respect to earth is up to a million volts. As a result, this million volts is added to all the voltages, but this does not affect even the most sensitive electronic devices.

- On the other hand, if we *double* all the voltages in a laptop computer, we will probably burn the chip.

Thus, it is reasonable to assume that the optimality criterion is shift-invariant.

Main result. Families optimal with respect to shift-invariant optimality criteria are described by the following theorem:

Theorem 12.1. *If a family F is optimal in the sense of some optimality criterion that is final and shift-invariant, then every function $s(z)$ from this family has one of the following three forms:*

- $s(z) = C_1 \cdot s_0(c_1 \cdot z + c_0) + C_0$, *where* $s_0(z) = 1/(1 + \exp(-z))$, *and* C_0, C_1, c_0, *and* c_1 *are real numbers;*

- $s(z) = C_1 \cdot \exp(c_1 \cdot z) + C_0$ *for some real numbers* C_1 *and* c_1;

- $s(z) = C_1 \cdot z + C_0$.

Comments.

- The function $s_0(z) = 1/(1 + \exp(-z))$ is called a *logistic* function.

- Logistic function is the most popular and the most successful activation functions used in *artificial neural networks*. Exponential and linear functions are also popular and successful; see, e.g., B. Kosko, *Neural networks and fuzzy systems* (Prentice Hall, Englewood Cliffs, NJ, 1992).

- Interestingly, the *biological neurons* are best described by the same logistic function.

    - From the biological viewpoint, this coincidence is not so surprising if we take into consideration that our nervous system is the result of billions years of improving evolution, and therefore, must be close to *optimal*.

- From the computer science viewpoint, the fact that our model of a neuron correctly predicted the actual activation function shows that *our model is a good* approximation to biological neurons.

■ If we did not require that the activation function $s(z)$ is defined for all z, then we would get two more possibilities: $s(z) = \tan(z)$ and $s(z) = 1/z$.

■ In this theorem, we consider "small" families $\{g(s(z))\}$ that originate from a single function $s(z)$. Basically, this means that we consider neural networks in which all neurons are *of the same type*. If for some problems, such networks do not learn well, we can consider networks with neurons *of different type*. For example, we can consider families of the type $\{C_1 \cdot s_1(z) + \ldots + C_m \cdot s_m(z)\}$, where $s_i(z)$ are given functions, and $C_1, \ldots, C_m$ are arbitrary constants. For the cases of shift- and scale-invariance optimal families of this type have been described in Theorems 9.1, 10.1, and 10.2. Neurons with the corresponding activation functions, such as sines and polynomials, have indeed been successfully used in some practical problems: e.g., polynomial activation functions were used in the first commercially successful application of neural networks.

12.4. Proof of Theorem 12.1

The outline of the proof. In this reduction, we will follow the sequence of steps similar to the ones used in the proofs of Theorems 4.2, 5.1, and 7.1:

■ first, we will find a system of functional equations describing the optimal function $s(z)$;

■ then, we will show that the functions used in these equations are indeed differentiable;

■ after that, we will deduce a differential equation for the desired functions $s(z)$;

■ and finally, we will solve this equation.

First step: deducing a functional equation. Similarly to Proposition 4.1, we can conclude that if the optimality criterion is final and shift-invariant, then the optimal family $F = \{g(s(z))\}$ is itself shift-invariant. This means, in particular, that for every shift z_0, the shifted function $s(z + z_0)$ also belongs to the same family, i.e., that $s(z + z_0) = g(s(z))$ for some transformation $g \in G$.

Transformations g can be easily described. Indeed, from Lesson 11, we already know that if a transformation Lie group G contains all linear transformation, that all transformations from G are fractionally linear, i.e., have the form $g(z) = (a \cdot z + b)/(c \cdot z + d)$. If we divide both numerator and denominator of this fraction by d, we conclude that $g(z) = (a \cdot z + b)/(1 + c \cdot z)$. Thus, for every real number z_0, there exist values $a(z_0)$, $b(z_0)$, and $c(z_0)$, for which

$$s(z + z_0) = \frac{a(z_0) \cdot s(z) + b(z_0)}{1 + c(z_0) \cdot s(z)}. \tag{12.5}$$

Second step: proving that the function $a(z_0)$, $b(z_0)$, and $c(z_0)$ are differentiable. If we multiply both sides of the equation (12.5) by the denominator of its right-hand side, and move all terms containing a, b, or c to one side, then we arrive at the following equation:

$$a(z_0) \cdot s(z) + b(z_0) - c(z_0) \cdot s(z) \cdot s(z + z_0) = s(z + z_0).$$

If we write down this same equation for three different values of z, we get a system of three linear equations for three unknowns $a(z_0)$, $b(z_0)$, and $c(z_0)$. Since the solution of a system of linear equations is a differentiable function of its coefficients, and the coefficients are differentiable functions $(1, s(z), s(z+z_0)$, etc.), we conclude that $a(z_0)$, $b(z_0)$, and $c(z_0)$ are differentiable functions.

Third step: reduction to a differential equation. Since all the functions in the equation (12.5) are differentiable, we can differentiate both sides by z_0 and take $z_0 = 0$. For $z_0 = 0$, the transformation $g(y)$ should reduce to $g(y) = y$, i.e., we should have $a(0) = 1$ and $b(0) = c(0) = 0$. As a result, we get the following equation:

$$\frac{ds}{dz} = A \cdot s + B - C \cdot s^2,$$

where $A = a'(0)$, $B = b'(0)$, and $C = c'(0)$.

Final step: solving the differential equation. We have already solved this equation for the case when $C = 0$. In this case, we have either a *linear* function (when $B = 0$), or an *exponential* function (when $B \neq 0$). So, to complete the proof, it is sufficient to consider the case when $C \neq 0$.

This case is solved in exactly the same way as the case $C = 0$: we separate variables s and z by moving all the terms with s to one sides and all the terms with z into another side, and then integrate the corresponding expressions. As a result, we get the following:

$$\int \frac{ds}{A \cdot s + B - C \cdot s^2} = t + \text{const.}$$

Multiplying both sides of this equation by $-C$, we conclude that

$$\int \frac{ds}{s^2 - (A/C)s - (B/C)} = -C \cdot z + \text{const.} \qquad (12.6)$$

The integral in the left-hand side has different values depending on the quadratic polynomial in the denominator:

- If this polynomial has two different real-valued roots $s_1 < s_2$, i.e., if it is equal to $(s - s_1) \cdot (s - s_2)$, then

$$\frac{1}{(s - s_1) \cdot (s - s_2)} = \frac{1}{s_2 - s_1} \cdot \left(\frac{1}{s - s_1} - \frac{1}{s - s_2} \right).$$

  Therefore,

$$\int \frac{ds}{(s - s_1) \cdot (s - s_2)} = \frac{1}{s_2 - s_1} \cdot (\ln(s-s_1) - \ln(s-s_2)) = \frac{1}{s_2 - s_1} \cdot \ln\left(\frac{s - s_1}{s - s_2} \right).$$

  Based on equation (12.6), we can equate this expression with the linear term $C \cdot z + \text{const.}$ Multiplying both sides of the resulting equality by $s_1 - s_1$, we conclude that

$$\ln\left(\frac{s - s_1}{s - s_2} \right) = \tilde{C}_1 \cdot z + \tilde{C}_0.$$

  Applying exp to both sides of this equation, we conclude that

$$\frac{s - s_1}{s - s_2} = \exp(\tilde{C}_1 \cdot z + \tilde{C}_0).$$

  Multiplying both sides of this equation by the denominator, we get a linear equation for $s(z)$, from which we get easily the desired expression related to logistic function. (Alternatively, it can be represented in terms of the hyperbolic tangent function $\tanh(z) = \sinh(z)/\cosh(z)$, where $\sinh(z) = (\exp(z) - \exp(-z))/2$, and $\cosh(z) = (\exp(z) + \exp(-z))/2$.)

- If this polynomial has a double root $s_1 = s_2$, then

$$\int \frac{ds}{(s - s_1)^2} = -\frac{1}{s - s_1}.$$

From equation (12.6), we can then conclude that $s(z)$ is a fractionally linear function.

- Finally, if the polynomial has two complex roots, we get a complex analogue of the tanh solution. It is easy to check (from de Moivre formulas) that $\sinh(i \cdot z) = i \cdot \sin(z)$ and $\cosh(i \cdot z) = \cos(z)$; thus, the complex analogue of $\tanh(z)$ is the tangent $\tan(z) = \sin(z)/\cos(z)$. So, in this case, the activation function is related to $\tan(z)$.

The second and third cases lead to functions that are not defined for all z: tangent $\tan(z)$ is not defined for $z = \pi/2$, and a fractionally linear (non-linear) function $s(z) = (a \cdot z + b)/(1 + c \cdot z)$ with $c \neq 0$ is not defined for $z = -1/c$. The theorem is proven.

Comment. As we can see from the proof, for every function $s(z)$ from the optimal family, its derivative $s'(z)$ is a quadratic function of $s(z)$. In particular, for the logistic function $s_0(z)$, $s_0'(z) = s_0(z)(1 - s_0(z))$. Since computing the values of the derivative is an important part of the backpropagation method, this simplifying formula means that for the *logistic* function (and for other optimal functions) *backpropagation method is the easiest to compute*. This conclusion is in line with the general theorem about the optimality of these functions: if, as a criterion, we choose computational simplicity of backpropagation, then, since this criterion is clearly shift-invariant, we can apply Theorem 12.1, and conclude that the optimal functions are logistic and others.

Historical comments.neural network,history We have already mentioned that the entire area of neural networks started when Wiener noticed that human neural networks are reasonably well designed. It is worth mentioning that the classification of transformation Lie groups (on which our theorem is based) also originated from an observation made by Wiener. Namely, in mid-1940s, Wiener analyzed how we humans recognize an object. According to some physiological studies, there are five clearly distinct levels of recognition:

- When an object is far enough, all we see is a blur. We cannot tell its shape, we cannot tell whether it is a point object or not.

- When we get closer, we can recognize some shape, but we still have trouble telling what shape it is exactly. We may see a circle as an ellipse, a square as a rhombus (diamond).

- As the object gets closer, we can clearly distinguish parallel lines, but we may not yet tell the angles. For example, we are not sure whether what we see is a rectangle or a parallelogram.

- When we get even closer, we can see the shape, but at a large distance, our stereoscopic ability does not work, so we cannot say whether what we see is large or small. For example, we already know this is a square, but we cannot tell whether it is a nearby small square, or a far away large one.

- Finally, when we get really close, we can see both the shape and the size of the object.

In mathematical terms, at each stage (except for the last one), the uncertainty means that we can apply some transformations to the original image without changing the perceived image. So, each stage can be characterized by the set G of all transformations that are, in this sense, possible on this stage.

If each of the two transformations $I \rightarrow g_1(I)$ and $I \rightarrow g_2(I)$ does not change the perceived image, then by applying them one after another, we get a new image $g_1(g_2(I))$ that is also perceived exactly the same as the original image I. Thus, the composition of two transformation from G also belongs to G. Similarly, the inverse transformation g^{-1} also belongs to G. Hence, G is a *group*. The above five stages correspond to the following transformation groups:

- At first, we have the group G of all possible transformations.

- Then, we get down to the group of all *projective* transformations (that describe *projections* from one plane to another).

- Third, we get the group of all *linear* transformations (also called *affine*).

- Fourth, we get the group generated by all of motions (i.e., translations and rotations) and dilations (transformations from this group are called *homotheties*).

- Finally, we get the group of all motions.

From this physiological observation, N. Wiener, in his book *Cybernetics, or Control and Communication in the animal and the machine* (3rd edition MIT Press, Cambridge, MA, 1962; first published in late 40s), made an interesting conclusion: If there was an intermediate group between, e.g., projective and affine transformations, then in an ideal vision system, it would probably be reasonable to use it in situations intermediate between the situations in which we use these two groups. Since a man is a product of billion years of improving evolution, it is therefore reasonable to assume that whatever transformation groups are possible, they are already used by us humans . Since we only use five different groups, he thus concluded that no other transformation groups exist. To be more precise, he conjectured that the only transformation Lie groups that contain the group of all motions are: the group of all homotheties, the group of all affine transformations, and the group of all projective transformations. Mathematicians were at first sceptical about this conjecture, but surprisingly, in mid-60s, two papers appeared that, in effect, proved Wiener's hypothesis:

- V. M. Guillemin and S. Sternberg, "An algebraic model of transitive differential geometry", *Bulletin of American Mathematical Society*, 1964, Vol. 70, No. 1, pp. 16–47.

- I. M. Singer and S. Sternberg, "Infinite groups of Lie and Cartan, Part 1", *Journal d'Analyse Mathematique*, 1965, Vol. XV, pp. 1–113.

For 1-dimensional case, projective transformations are simply fractionally linear, and affine are simply linear; thus, this theorem reduces to our Theorem 11.1.

Hence, interestingly, not only *neural networks themselves* come *from the simulation of human brain*, but even the *mathematical tool* that we used to find the optimal activation function (classification of transformation Lie groups) also *originated from the simulation of brain processes*. In this sense, this proof is not a mathematical trick, unrelated to our computational problem, but a very natural reasoning.

Exercise

12.5 In this section, we described families of activations functions that are optimal with respect to a *shift-invariant* optimality criterion. Show that if we use *scale-invariant* criterion instead, the optimal activation functions will be of the type $s(z) = (A + Bz^{-\alpha})/(C + Dz^{-\alpha})$ for some A, B, C, D and $\alpha > 0$. *Hint:* Describe the functional equation, the differential equation, and then use the same change of variables that we used to reduce differential equations coming from *scale* invariance to an equation originating from *shift* invariance.

Comment. Particular cases of this general formula, such as the Cauchy function $s(z) = 1/(1 + z^2)$, were indeed successfully used.

13

EXPERT SYSTEMS AND THE BASICS OF FUZZY LOGIC

This lesson describes an area from Artificial Intelligence: expert systems. We describe the essentials of fuzzy logic in modeling of expert knowledge. We also touch upon the field of fuzzy control; the general methodology of fuzzy control will be given in the next Lesson 14.

13.1. Expert knowledge and how to represent it in the computer: formulation of the problem

In some extrapolation situations, we have an additional expert knowledge. Many practical problems are naturally formulated as optimization problems. If we have a *complete* information about the object, then optimization becomes a well-defined mathematical problem. However, in many practical situations, we only have a *partial* information about the object. In this case, before we do any optimization, we must first *extrapolate* this partial information. In the previous lessons, we considered the case when this partial information comes from measurements or from experiments, i.e., when it takes the form of *patterns* $(x_1^{(p)}, \ldots, x_n^{(p)}, y^{(p)})$, $1 \leq p \leq P$, from which we can extrapolate the function $y = f(x_1, \ldots, x_n)$ that is consistent with these patterns.

In many practical problems, in addition to the patterns, we also have *experts* whose knowledge about the object we can use. In this lesson, we will describe how we can extract and use this expert knowledge, and how continuous mathematics can help.

275

Why do we need to extract expert knowledge? By definition, an expert is a person who is well knowledgeable about the object. An expert helicopter pilot is a person who can control a helicopter well; an expert doctor is a person who can cure different diseases, etc. If we already have such an expert, what else do we need? The problem is that in every area,

- there are only a *few* top experts, and

- there are *many* problems to be solved.

It is, usually, simply physically impossible to have a top expert for each problem: we cannot have a top surgeon for every surgery, we cannot have a top helicopter pilot for every helicopter in the world. It is therefore desirable to develop a *computer program* that would somehow incorporate the expert knowledge and give advise comparable in quality with the advise of the top experts. In other words, we want to design a computer program that somehow *simulates* the expert. Such programs are called *expert systems*.

Why cannot we extract the function $f(x_1, \ldots, x_n)$ from an expert? At first glance, the problem seems relatively simple: Since the person is a real expert, we simply ask her multiple questions like "suppose that x_1 is equal to 1.2, x_2 is equal to 1.3, ..., what is y?" After asking all these questions, we will get many patterns, from which we will be able to extrapolate the function $f(x_1, \ldots, x_n)$ using one of the known methods. For example, we may ask the medical expert about all possible combinations of symptoms and test results (blood pressure, temperature, etc.), and write down what exactly treatment the expert recommends. However, there are two problems with this idea:

- First, there is a *computational* problem: Since we need to ask a question for each combination of symptoms and test results, we may end up having to ask *too many* questions. This fact makes our idea *difficult* to implement but *not* necessarily *impossible*: Indeed, we only need to do this lengthy questioning once, so we may afford to spend years, if necessary (actually, the design of the first expert systems did take years).

- However, there is another, more serious problem with this idea that makes it, in most problems, *impossible* to implement. Indeed, this idea may sound reasonable until we try applying it to a skill in which practically all adults consider themselves experts: driving a car. If you ask a driver a question like: "you are driving at 55 miles per hour when the car in 30 feet in front of you slows down to 47 miles per hour, for how many seconds do you hit the brakes?", nobody will give a precise number.

What can we do? We might install measuring devices into a car (or into a driving simulator), and simulate this situation, but the resulting braking times may differ from simulation to simulation.

The problem is not that the expert has some precise number (like 1.453 sec) in mind that he cannot express in words; the problem is that one time it will be 1.39, another time it may be 1.51, etc.

An expert usually expresses his knowledge by using words from natural language. An expert *cannot*, usually, express his knowledge in *precise* numerical terms (such as "hit the brakes for 1.43 sec"), but he *can* formulate his knowledge by using *words* from natural language. For example, an expert can say "hit the brakes for a while".

So, the knowledge that we can extract from an expert consists of statements like "if the velocity is a little bit smaller than maximum, hit the breaks for a while".

There may be also another uncertainty involved, e.g., an expert may say something like "if a patient has a temperature of 100 degrees, and a headache, and . . . (several other symptoms), then, most probably, it is a flu". Here, the words "most probably" are the words from natural language that an expert uses to describe his knowledge.

Since our goal is to describe the expert's knowledge inside a computer, we must represent rules formulated in a natural language inside the computer.

Toy example: a thermostat. We will illustrate different methods for representing expert knowledge on another situation in which everyone feels himself an expert: controlling a *thermostat*.

For simplicity, we will consider a *toy*, simplified thermostat in which a thermometer shows the current temperature and a dial allows us to control the temperature:

- turning the dial to the *left* makes it *cooler*;
- turning it to the *right* makes it *warmer*.

The angle on which we turn the dial will be denoted by u.

We will also assume, for simplicity, that we know the comfort temperature T_0 that we try to achieve. Strictly speaking, the input to our control is the temperature T. However, in reality, our control decisions depend not so much on the *absolute* value of T but rather on the *difference* $T - T_0$ between the actual temperature T and the ideal temperature T_0. Since this difference is important, we will use a special notation for it: $x = T - T_0$. Our goal is to describe the appropriate control u for each possible value x, i.e., to describe a function $u = f(x)$.

For such an easy system, we do not need any expert to formulate reasonable rules; we can immediately describe several reasonable control rules:

- If the temperature T is *close* to T_0, i.e., if the difference $x = T - T_0$ between the actual temperature is *negligible*, then no control is needed, i.e., u should also be negligible.

- If the room is slightly overheated, i.e., if x is positive and small, we must cool it a little bit (i.e., u must be negative and small).

- If the temperature is a little lower than we would like it to be, then we need to heat the room a little bit. In other terms, if x is small negative, then u must be small positive.

We can formulate many similar natural rules. For simplicity, in our toy example, we will restrict ourselves to these three. As a result, we get the following three rules:

- if x is negligible, then u must be negligible;

- if x is small positive, then u must be small negative;

- if x is small negative, then u must be small positive.

Toy example re-formulated. Our goal is to make these and similar rules accessible for the computer. Before describing how to do it, let us first reformulate these rules in a way that will somewhat clarify these rules. Namely, in the formulation of these rules, we want to clearly separate the *properties* (like "x is negligible") and *logical connectives* (like "if ... then"). To achieve this separation, let us introduce a shorthand notation for all the properties, and let us use standard mathematical notations for logical connectives:

- $N(x)$ will indicate that x is negligible;

- $SP(x)$ will indicate that x is small positive;

- $SN(x)$ will indicate that x is small negative;

- $A \to B$ is a standard mathematical notation for "if A then B".

In these notations, the above three rules take the form:

$$N(x) \to N(u); \tag{13.1}$$

$$SP(x) \to SN(u); \tag{13.2}$$

$$SN(x) \to SP(u). \tag{13.3}$$

General case. In general, the expert's knowledge about the dependence of y on $x_1, \ldots, x_n$ can be expressed by several rules of the type

If x_1 is A_{r1}, ..., and x_n is A_{rn}, then y is B_r.

Here, $r = 1, \ldots, R$ is the rule number, and A_{ri} and B_r are words from natural language that are used in r-th rule, like "small", "medium", "large", "approximately 1", etc.

- In our toy example, we only had *one* input variable: the temperature T (or, to be precise, the difference x between T and the desired temperature T_0).

- In the general case, we have *several* input variables, so, in addition to the logical connective "if ... then", we need another logical connective "and".

In mathematics, "and" is usually denoted by & (or by $\wedge$). If we use the standard mathematical notation for "if ... then" and "and", we can re-formulate the above rules as follows:

$$A_{r1}(x_1) \& \ldots \& A_{rn}(x_n) \to B_r(y), \tag{13.4}$$

where $r = 1, \ldots, R$. The set of rules is usually called a *rule base*.

Comment. In this text, we will only consider the case when a rules base consists of *straightforward* if–then rules. In reality, experts may have some additional information about the object, which may be formulated in a more complicated form. For example, an expert may formulate several different rules bases, and then formulate "meta-rules" that decide which of the rule bases are applicable.

The problem. The problem is to represent the expert knowledge in a computer and to be able to process this knowledge in a meaningful way.

13.2. Expert system methodology: an outline

Our goal is to represent rule bases. A rule base has a clear structure:

- A rule base consists of rules.
- Each rule, in its turn, is obtained from properties (expressed by words from natural language) by using logical connectives.

In view of this structure, it is reasonable to represent the rule base by first representing the basic elements of the rule base, and then by extending this representation to the rule base as a whole. In other words, it makes sense to follow the following methodology:

- First, we represent the basic properties $A_{ri}(x_i)$ and $B_r(y)$.
- Second, we represent the logical connectives.
- Third, we use the representations of the basic properties and of the logical connectives to get the representations of all the rules.
- Fourth, we combine the representations of different rules into a representation of a rule base.

As a result of these four steps, we get an advising (expert) system. For example, if we apply these four steps to the medical knowledge, we ideally, get a system that, given the patient's symptoms, provides the diagnostic and medical advise. For example, it can say that most probably, the patient has a flu, but it is also possible that he has bronchitis. Such an advice, coming from an expert system, is usually used by a specialist to make a decision.

However, there are situations like helicopter control where there is no time for a human operator to make a decision: we want a system to automatically make a decision based on its own conclusions. For such control situations, we need an *additional*, fifth follow-up step:

- The computer-based system makes a decision.

The resulting five-step methodology is indeed very successful in control. The resulting control (called *fuzzy control*; see Lesson 14) is used in various areas ranging from appliances (camcorders, washing machines, etc.) to automatically controlled subway trains in Japan to cement kilns to regulating temperature within the Space Shuttle.

There are many books that describe fuzzy control. The reader who is interested in learning more about fuzzy methods in general and fuzzy control in particular is advised to start with one of the following textbooks:

- G. Klir and B. Yuan, *Fuzzy sets and fuzzy logic: theory and applications* (Prentice Hall, Upper Saddle River, NJ, 1995); this book gives an *engineering*-oriented introduction;

- H. T. Nguyen and E. A. Walker, *A First Course in Fuzzy Logic* (CRC Press, Boca Raton, Florida, 1997); this textbook is more oriented towards *mathematical* methods.

In the following, we will describe the *basics* all these five steps.

13.3. Representing natural-language properties: main idea

Before we represent a property $P(x)$, let us represent the expert's degree of belief in this property for each x. To represent a property like "x is positive small" ($SP(x)$), we must be able, for every possible value of the temperature difference x, to represent the expert's opinion of whether this particular value x is indeed small. To represent this opinion, we must solve the following problem:

■ All the properties that we have encountered in the previous lessons, such as "x is positive", were *crisp*, in the sense that every real number is either positive, or not.

■ On the other hand, properties like "x is small" (in which we are interested) are *not crisp*. To be more precise:

– Some values x are so small that practically everyone would agree that they are small.

– Some values x are so huge that practically everyone will agree that they are not small.

– However, for many intermediate values x, it is difficult to decide whether x is small or not, and experts may disagree:

  ∗ For a researcher who performs temperature-sensitive experiments in the lab, the difference of $x = 0.5$ degrees may not be small at all.

  ∗ However, for a living room, even the difference of ± 5 degrees is usually not only small, but even negligible.

Such non-crisp properties are called *fuzzy*. This term was introduced by Lotfi Zadeh in 1965.

How can we represent fuzzy properties?

■ If $P(x)$ is a *crisp* property (like "positive"), then for every x, $P(x)$ is either true or false.

  Representing the corresponding truth value in the computer is easy: "true" is usually represented by 1, and "false" by 0.

■ If $P(x)$ is a *fuzzy* property (like "small"), then in general, for a real number x, an expert may not be sure whether x satisfies this property P or not. For example, for "small", the larger the value x, the less confident the expert that x is small. So, for different values x, an expert may have different "degrees of confidence" that x satisfies this property P:

  – For some values x, the expert may be absolutely sure that the statement $P(x)$ is true. In this case, the degree of confidence corresponds to "true".

  – For some other values x, the expert may be absolutely sure that the statement $P(x)$ is false. In this case, the degree of confidence corresponds to "false".

– For yet other values x, the expert is neither sure that $P(x)$ is true, nor he is sure that $P(x)$ is false. In this case, the expert's degree of confidence is *intermediate* between "true" and "false".

How can we represent these intermediate degrees of belief in a computer?

If 0 corresponds to "false", and 1 to "true", then it is natural to represent degrees of confidence that are intermediate between "false" and "true" by numbers from the interval $(0, 1)$.

Thus, to describe *arbitrary* degrees of certainty, we must describe:

- either absolute truth (represented by 1),

- or absolute falsity (represented by 0),

- or intermediate degrees of confidence (represented by numbers from the open interval $(0, 1)$).

Therefore, we arrive at the following conclusion:

We must use real numbers from the interval $[0, 1]$.

Historical comments.

- From the *mathematical* viewpoint, what we are doing is a very natural idea: we extend the traditional *2-valued* logic (in which the truth value of each statement is an element of a 2-element set {"true", "false"} = $\{0, 1\}$), to the *interval* $[0, 1]$. Mathematicians have been developing the corresponding mathematical formalisms (called *multiple-valued logics*) since the pioneer work of K. Lukasiewicz in the early 1920s.

- In *computer science*, the idea of using numbers from the interval $[0, 1]$ to describe different degrees of confidence was also first proposed by Lotfi Zadeh in his pioneer 1965 paper on fuzzy sets.

Comments.

- ■ We have used the term *degree of confidence*; other words are also used, such as *degree of belief, degree of certainty, truth value, subjective probability*, etc. From our viewpoint, all of these terms are good but not perfect:

  - On one hand, each of these terms bring with itself some intuitive meaning.

  - On the other hand, the intuitive meaning brought by these words may be sometimes misleading. For example, the terms "belief" and "subjective probability" may lead to a confusion with the "objective probability", i.e., crudely speaking, with a *frequency* of a certain event (like tossing a coin).

  Since we cannot select one term as the best, we will use these terms interchangingly.

- ■ The interval $[0, 1]$ is probably the most natural to use, but some expert systems use a *different interval* to represent uncertainty. For example, historically the first successful expert system MYCIN, developed at Stanford University for diagnosing rare blood diseases, used values from the interval $[-1, 1]$ to represent uncertainty.

- ■ The main advantage of using an interval (e.g., $[0, 1]$) is that elements from this interval (i.e., real numbers) can be easily represented and easily processed in a computer. However, sometimes, to get a more adequate representation of the expert's degree of confidence, we may need a more sophisticated representation. For example, an expert may not be sure about his degree of confidence. To represent this uncertainty, we may want to describe the expert's degree of confidence not by a *single* number, but by an *interval* of possible numbers:

  - If an expert has no information about the object, he cannot have any definite degree of confidence. To describe this situation, we will use the entire interval $[0, 1]$ as the description of this expert's degree of confidence.

  - If an expert has a certain information, them we may use a narrower interval, or even a number (i.e., a degenerate interval).

  There are several even more sophisticated ways of representing uncertainty. In this text, we will only consider the simplest possible way of representing degrees of confidence: by using real numbers.

We need to elicit a numerical degree of confidence from the expert.
We agreed that for each statement of the type $P(x)$, where:

- P is a fuzzy property (i.e., a property formulated by words from natural language), and

- x is a real number,

a reasonable way to represent the expert's degrees of confidence is to use numbers from the interval $[0, 1]$. The natural next question is: how can we "extract", elicit the value from an expert?

Direct elicitation is impossible. The ideal situation would be if an expert would directly provide us with this number, but this is, unfortunately, not realistic:

- we want numbers from the interval $[0, 1]$, because they are very *natural* for a *computer*; but

- real numbers from the interval $[0, 1]$ are *not natural* for a *human expert*; it is very difficult for most experts to express their degree of confidence in a given statement by a real number.

Since we cannot elicit these numbers *directly*, we have to use *indirect* elicitation techniques.

There are several dozen different elicitation techniques; in this lesson, we will only describe the most frequently used ones.

First elicitation method: selecting on a scale. If we cannot elicit a *real* number from an expert, maybe we can elicit *some* number from him, and then convert the result into a real number from the interval $[0, 1]$. In many cases, this is indeed possible: many polls ask us to estimate our degree of confidence, or our degree of belief, etc., on a certain integer scale, e.g., on a scale from 0 to 5, or on a scale from 0 to 10. So, we can ask an expert to mark his degree of confidence in a given statement $P(x)$ on a given scale 0 to S (0 to 5, 0 to 10, etc). On this scale:

- 0 corresponds to "absolutely confident that $P(x)$ is false";

- S corresponds to "absolutely confident that $P(x)$ is true";

- intermediate marks represent different degrees of uncertainty.

As a result of this procedure, we get an integer from 0 to S. The most natural way to convert these integers $0, 1, 2, \ldots, S$ into real numbers from the interval $[0, 1]$ is to *re-scale* the interval $[0, S]$ into the interval $[0, 1]$ by dividing by S. So, if an expert has chosen the mark s on a scale from 0 to S, we will take s/S as the desired degree of confidence.

For example, if an expert selects, 3 on a scale from 0 to 5, then we take $3/5 = 0.6$ as the desired degree of confidence.

The first elicitation method is not always applicable. The above method seems reasonable and is computationally very simple. The problem with this method is that experts often do not feel comfortable expressing their degree of confidence by numbers on an (integer) scale. Usually, these experts only distinguish between three cases:

- a given statement $P(x)$ is true;

- a given statement $P(x)$ is false;

- we do not know whether a given statement $P(x)$ is true or false.

If we have such experts, how can we elicit their degree of confidence?

Second elicitation method: polling. We have already mentioned that the first elicitation method is taken from the experience of the polls. In addition to the polls that require us to mark a point on a scale, there are other polls in which we only answer "yes", "no", or "undecided". For example, polls conducted before the elections are usually of this type. As a result of each poll of this type, we get a *percentage* of people who answered "yes" ("60% are planning to vote for candidate X"). This percentage represents exactly what we want: a number from the interval $[0, 1]$ (e.g., 60% means 0.6).

So, we arrive at the following *polling* method of eliciting the degree of confidence from the experts: we ask several experts whether $P(x)$ is true or false. Some of these experts may not answer at all, or give "unknown" as an answer; these experts we need not count. If out of N experts who gave a definite answer, M answered "yes" (i.e., "yes, $P(x)$ is true"), then we take the ratio M/N as the desired degree of confidence.

Problems with the second method. For the polling method to give meaningful results, we need many respondents. From the strict mathematical viewpoint, for a pool to give meaningful results (e.g., to give the percentage with a guaranteed 10% accuracy), we must interview at least 1,000 people.

- For an election poll, interviewing 1,000 people is quite possible: millions participate in the elections, and we want to know the opinion of the representative group of people.

- However, in an expert system, we are trying to formalize the knowledge of the top experts. We may not have that many top experts, and even if we do, top experts are usually very busy people, we may not be able to convince all of them to cooperate with this project. Last but not the least, the time of top experts costs money, and we may not have the money to hire 1,000 of them.

Can we somehow ask a single expert and still get a number?

Third elicitation method: using subjective probabilities and bets. Some people feel uncomfortable estimating a probability or marking a value on a scale, but they have a good feel for *bets*. Such people cannot express the probability of their favorite horse winning, but they are absolutely sure that, say, they are willing to bet 4 to 1 but not 5 to 1 that this horse will win.

If we can extract a betting ratio from an expert, then we can easily transform this number into the subjective probability, that will serve as the desired degree of confidence.

For example, if an expert is willing to bet 4 to 1 but not any better that $P(x)$ is true (e.g., that most people will agree that $x = 1$ is a small temperature difference), this means that this experts' subjective probability is $4/(4+1) = 0.8$.

From finitely many degrees of confidence to a membership function: extrapolation is needed. We started this section with a problem of describing a fuzzy property $P(x)$ (of the type "x is small"). Namely, for every possible value x, we would like to know the *degree of confidence* $d(P(x))$ that this value x

satisfies the property P. This degree of confidence is usually denoted by $\mu_P(x)$, and the function that transforms a real number x into a value $\mu_P(x) \in [0, 1]$ is called a *membership function*, or a *fuzzy set*.

For each value x, we can, using one of the above elicitation techniques, find the value $\mu_P(x)$ for this x. However, this is not sufficient:

- On one hand, we want to know the value $\mu_P(x)$ for all possible real numbers x, i.e., for *infinitely many* different values.

- On the other hand, we can only ask an expert *finitely many* questions and therefore, we can only determine the values of the membership function for *finitely many* different values $x^{(1)}, \ldots, x^{(v)}$.

Thus, after all the elicitation is over, we only know the values $\mu_P(x^{(p)})$ of the desired function $\mu_P(x)$ for v different values $x^{(1)}, \ldots, x^{(v)}$. To reconstruct the desired function, we must use *extrapolation*.

Spline extrapolation is most frequently used. We have already argued in the previous lessons that if we have no information about x (and therefore, we have no reason to prefer a certain measuring unit for x), then it is reasonable to choose a class of extrapolation methods that is optimal with respect to a shift- and scale-invariant criterion, which means using a *piece-wise polynomial* (*spline*) extrapolation. This extrapolation is indeed most frequently used in control applications.

Other extrapolation methods and neuro-fuzzy control. Sometimes, more complicated extrapolation techniques such as neural networks lead to better control results. Such control in which neural networks are used together with fuzzy methods is called *neuro-fuzzy*.

Examples of extrapolation. Let us describe a few simple examples of extrapolation. The simplest extrapolation is piece-wise polynomial, and the simplest possible polynomial is a linear function. Thus, the simplest extrapolation is an extrapolation by *piece-wise linear functions*. The simplest case of extrapolation is when we start with the values x for which the expert is either absolutely sure that x is true, or he is absolutely sure that x is false, i.e., with values for which $\mu_P(x) = 0$ or $\mu_P(x) = 1$. Let us give several examples of such situations.

■ Let us first describe the property of the type "x is negligible".

- The only case when we are 100% sure that x is negligible is when $x = 0$. So, we have the value $\mu_P(0) = 1$.

- Usually, we also know the value $\Delta > 0$ after which the difference in temperatures is no longer negligible. For example, for a thermostat that controls the room's temperature, we can take $\Delta = 10$. This means that $\mu_P(x) = 0$ for $x \geq \Delta$ and for $x < -\Delta$.

We know the value of the function $\mu_P(x)$ for $x \leq -\Delta$, for $x = 0$, and for $x \geq \Delta$. We need to use linear extrapolation to find the values of this function for $x \in (-\Delta, 0)$ and for $x \in (0, \Delta)$. In general, the formula for a linear function $f(x)$ that passes through the point $y_1 = f(x_1)$ and $y_2 = f(x_2)$, is

$$f(x) = y_1 + (x - x_1) \cdot \frac{y_1 - y_2}{x_2 - x_1}.$$

For our case, this formula leads to the following membership function:

- To get the values $\mu_P(x)$ for $x \in (-\Delta, x)$, we take $x_1 = -\Delta$, $x_2 = 0$, $y_1 = 0$, and $y_2 = 1$. As a result, we get the expression $\mu_P(x) = (x + \Delta)/\Delta$.

- To get the values $\mu_P(x)$ for $x \in (0, \Delta)$, we take $x_1 = 0$, $x_2 = \Delta$, $y_1 = 1$, and $y_2 = 0$. As a result, we get the expression $\mu_P(x) = 1 - x/\Delta$.

Thus, the function $\mu_P(x)$ takes the following form:

- $\mu_P(x) = 0$ for $x \leq -\Delta$;
- $\mu_P(x) = (x + \Delta)/\Delta$ for $-\Delta \leq x \leq 0$;
- $\mu_P(x) = 1 - x/\Delta$ for $0 \leq x \leq \Delta$; and
- $\mu_P(x) = 0$ for $x > \Delta$.

The graph of this function has the shape of a *triangle* over the x-axis. Therefore, such functions are called *triangular* membership functions.

A similar shape describes properties like "close to a", for some fixed a. In this case, the triangular membership function has the form:

- $\mu_P(x) = 0$ for $x \leq a - \Delta$;
- $\mu_P(x) = (x - (a - \Delta))/\Delta$ for $a - \Delta \leq x \leq a$;
- $\mu_P(x) = 1 - (x - a)/\Delta$ for $a \leq x \leq a + \Delta$; and
- $\mu_P(x) = 0$ for $x > a + \Delta$.

■ For the same property "x is negligible", in some cases, we know the lower bound δ below which the temperature difference x is indeed negligible. In this case, in addition to knowing that $\mu_P(x) = 0$ for $x \leq -\Delta$ and for $x \geq \Delta$, we also know that $\mu_P(x) = 1$ for $-\delta \leq x \leq \delta$. For this function, linear extrapolation to the intervals $(-\Delta, -\delta)$ and (δ, Δ) results in the following membership function:

- $\mu_P(x) = 0$ for $x \leq -\Delta$;
- $\mu_P(x) = (x + \Delta)/(\Delta - \delta)$ for $-\Delta \leq x \leq -\delta$;
- $\mu_P(x) = 1$ for $-\delta \leq x \leq \delta$;
- $\mu_P(x) = 1 - (x - \delta)/(\Delta - \delta)$ for $\delta \leq x \leq \Delta$; and
- $\mu_P(x) = 0$ for $x > \Delta$.

The graph of this function has the shape of a *trapezoid* over the x-axis. Therefore, such functions are called *trapezoidal* membership functions.

■ Another frequent example of piece-wise functions are functions that describe properties like "x is large". For these properties, we usually know that values below a certain δ are definitely not large, and that values about a certain $\Delta \gg \delta$ are definitely large. So, we know that $\mu_P(x) = 0$ for $x \leq \delta$ and $\mu_P(x) = 1$ for $x \geq \Delta$. To get the values of $\mu_P(x)$ for $x \in (\delta, \Delta)$, we use a linear extrapolation. As a result, we get the following function:

- $\mu_P(x) = 0$ for $x \leq \delta$;
- $\mu_P(x) = (x - \delta)/(\Delta - \delta)$ for $\delta \leq x \leq \Delta$;
- $\mu_P(x) = 1$ for $x \geq \Delta$.

We have described the triangular and trapezoid functions because they are *the simplest*. Interestingly, in fuzzy control, they are also, at present, *the most frequently used membership* functions. This fact has two possible explanations:

■ First, fuzzy control is still at its infancy. The first successful application of fuzzy control was produced a little more than 20 years ago (in 1974, by E. Mamdani). Because of that, the potential of simple fuzzy control applications is not yet exhausted. *Maybe, we will need more complicated membership functions* later on, when all simple applications will be used, but so far, we do not seem to need them.

■ Second, we are formalizing *approximate* expert knowledge. If all an expert can say about a control is that it should be *small*, or *medium*, or *large*, it may not be reasonable to try to formalize these notions in a too complicated way. Thus, simple models of expert uncertainty, in particular, simple membership functions, may be more reasonable than complicated ones. From this viewpoint, *we may not need more complicated membership functions at all.*

Probably, the truth lies somewhere in between; but anyway, so far, piecewise-linear functions seem to work just fine.

Exercises

13.1 Choose several values of temperature difference x and, using your friends as experts, apply one of the elicitation methods to describe the degrees of confidence that these values x are negligible.

13.2 Based on your elicitation results, use piecewise-linear extrapolation to form a membership function that corresponds to the word "negligible". Describe an analytical expression for this function, and plot its graph.

13.4. Representing natural-language properties: re-scaling

Different elicitation methods can lead to different results. The main reason why we had to describe *several* different elicitation methods is that each method has its limitations, and so:

■ we need a *second* method to elicit degrees of confidence in the situation when the first method does not work;

■ we need a *third* method to elicit degrees of confidence in the situation when the first two methods do not work;

■ etc.

From this description, it is clear that usually, only one of these methods is applicable. However, there are situations when we can apply several different methods. The reader should be warned that in such situations, different methods can lead to somewhat different results. To be more precise:

- From the *qualitative* viewpoint, the values from different scales are usually consistent: if experts' confidence in a statement A is higher than the experts' confidence in the statement B, then for all reasonmable methods, the degree of belief $d(A)$ in A will be larger than the degree of belief in B: $d(A) > d(B)$.

- However, the actual *numerical, quantitative* values $d(A)$, $d(B)$ of these degrees of belief can differ from method to method.

The reason for this difference is that different methods may result in values corresponding to different *scales* of uncertainty, just like measuring the length in feet or in meters leads to different scales in which the order is preserved but numerical values are different. So, if we have different degrees of belief obtained by different methods, we may need to *re-scale* them to make them comparable.

Comment. As we have mentioned, in most applications of expert systems and fuzzy control, we do not encounter this problem. So, a reader who is only interested in learning about the very basics of fuzzy control, can skip this section.

What re-scaling should we choose? Working intelligent systems use several different procedures for assigning numeric values that describe uncertainty of the experts' statements. The same expert's degree of uncertainty that he expresses, for example, by the expression "for sure", can lead to 0.9 if we apply one procedure, and to 0.8 if another procedure is used. Just like 1 foot and 12 inches describe the same length, but in different scales, we can say that 0.9 and 0.8 represent the same degree of certainty in two different *scales*.

Some scales are different even in the fact that they use an interval different from [0,1] to represent uncertainty. For example, as we have already mentioned, the famous MYCIN system uses the interval $[-1, 1]$.

In some sense all scales are equal, but some are more reasonable than others. From a mathematical viewpoint, one can use *any* scale, but from the practical viewpoint some of them will be more reasonable to use, and some of them less reasonable. We'll consider only *practically reasonable* scales, and we'll try to formalize what that means.

We must describe transformations between the scales. Since we are not restricting ourselves to some specific procedure of assigning a numeric value to uncertainty, we can thus allow values from different scales. If we want to combine them, we must be able to transform them all to one scale. So we must be able to describe the transformations between reasonable scales (*"rescalings"*).

The class $\mathcal{R}$ of reasonable transformations of degrees of uncertainty must satisfy the following properties:

1) If a function $a \to r(a)$ is a reasonable transformation from a scale A to some scale B, and a function $b \to s(b)$ is a reasonable transformation from B into some other scale C, then it is reasonable to demand that the transformation $a \to s(r(a))$ from A to C is also a reasonable transformation. In other words, the class $\mathcal{R}$ of all reasonable transformations must be closed under composition.

2) If $a \to r(a)$ is a reasonable transformation from a scale A to scale B, then the inverse function is a reasonable transformation from B to A.

Thus, the family $\mathcal{R}$ must contain the inverse of every function that belongs to it, and the composition of every two functions from $\mathcal{R}$. In mathematical terms, it means that $\mathcal{R}$ must be a *transformation group*.

3) If the description of a rescaling is too long, it is unnatural to call it reasonable. Therefore, we will assume that the elements of $\mathcal{R}$ can be described by fixing the values of p parameters (for some small p).

In mathematics, the notion of a group whose elements are continuously depending on finitely many parameters is formalized as the notion of a (connected) *Lie group*. So we conclude that reasonable rescalings form a connected transformation Lie group.

4) The last natural demand that we'll use is as follows. Of course, in principle, it is possible that we assign 0.1 in one scale and it corresponds to 0.3 in another scale. It is also possible that we have 0.1 and 0.9 on one scale that comprises only the statements with low degrees of belief, and when we turn to some other scale that takes all possible degrees of belief into consideration, we get small numbers for both. But if in some scale we have

the values 0.5, 0.51 and 0.99, meaning that our degrees of belief in the first two statements almost coincide, then it is difficult to imagine another reasonable scale in which the same three statements have equidistant truth values, say 0.1, 0.5 and 0.9. If this example is not convincing, take 0.501 or 0.5001 for the second value on the initial scale. We'll formulate this idea in the maximally flexible form: *there exist two triples of truth value that cannot be transformed into each other by any natural rescaling.*

Examples of reasonable rescaling transformations. In addition to these general demands, we have some examples of rescalings that are evidently reasonable.

Indeed, one of the natural methods to ascribe the degree of confidence $d(A)$ to a statement A is to take several (N) experts, and ask each of them whether she believes that A is true. If $N(A)$ of them answer "yes", we take $d(A) = N(A)/N$ as the desired certainty value. (This method is a slight modification of the polling method, a modification in which we do not dismiss those whose do not answer anything definite.) If all the experts believe in A, then this value is 1 (=100%), if half of them believe in A, then $t(A) = 0.5$ (50%), etc.

Knowledge engineers want the system to include the knowledge of the entire scientific community, so they ask as many experts as possible. But asking too many experts leads to the following negative phenomenon: when the opinion of the most respected professors, Nobel-prize winners, etc., is known, some less self-confident experts will not be brave enough to express their own opinions, so they will either say nothing or follow the opinion of the majority. How does their presence influence the resulting uncertainty value?

■   Let N denote the initial number of experts, $N(A)$ the number of those of them who believe in A, and M the number of shy experts added. Initially, $d(A) = N(A)/N$. After we add M experts who do not answer anything when asked about A, the number of experts who believe in A is still $N(A)$, but the total number of experts is bigger ($M + N$). So the new value of the uncertainty ratio is

$$d'(A) = \frac{N(A)}{N + M} = c \cdot d(A),$$

where we denoted $c = N/(M + N)$.

■   When we add experts who give the same answers as the majority of N renowned experts, then, for the case when $d(A) > 1/2$, we get $N(A) + M$

experts saying that A is true, so the new uncertainty value is

$$d'(A) = \frac{N(A) + M}{N + M} = \frac{N \cdot d(A) + M}{N + M}.$$

- If we add M "silent" experts and M' "conformists" (who vote as the majority), then we get a transformation

$$d(A) \to \frac{N \cdot d(A) + M'}{N + M + M'}.$$

In all these cases the transformation from an old scale $d(A)$ to a new scale $d'(A)$ is a linear function $d(A) \to a \cdot d(A)b$ for some constants a and b; in the most general case $a = N/(N + M + M')$ and $b = M'/(N + M + M')$. By selecting appropriate values of N, M, and M', we can get arbitrary linear functions with linear coefficients. Thus, we arrive at the following conclusion:

The problem of selecting re-scalings re-formulated in mathematical terms.

The set of possible re-scalings form a Lie transformation group that contains all linear transformations.

This re-formulation leads to a solution.

Description of all possible re-scaling for the case when the property 4) is not required. We already know, from Lesson 11, that each such transformation is fractionally linear. Thus,

Every reasonable rescaling of degrees of confidence can be described by a fractionally linear function.

Description of all possible re-scaling for the case when the property 4) is required. Let us show that if we require property 4), then only linear transformations are possible.

Indeed, to be more precise, the desired Lie group either consists of only linear transformations, or of all fractionally linear ones. In the second case, we get a violation of the property 4): indeed, if we have three degrees $d_1 < d_2 < d_3$, and we want to transform them into three other degrees $d'_1 < d'_2 < d'_3$ by using an appropriate fractionally-linear transformation

$$d \to d' = \frac{a \cdot d + b}{1 + c \cdot d}$$

for appropriate a, b, and c, then we must find these coefficients a, b, and c from the following system of three equations with three unknowns:

$$d'_i = \frac{a \cdot d_i + b}{1 + c \cdot d_i} \quad (i = 1, 2, 3)$$

Multiplying both sides of each equation by its denominator, we get a system of three linear equations with three unknowns:

$$d'_i + c \cdot d_i \cdot d'_i = a \cdot d_i + b, \quad (i = 1, 2, 3)$$

From this linear system, we can easily find a, b, and c. Thus, if we require the property 4), we arrive at the following conclusion:

> *Every reasonable rescaling of degrees of confidence can be described by a linear function.*

Exercise

13.3 Find a fractionally linear transformation $d \to d'$ that transforms the values $d_1 = 0.5$, $d_2 = 0.51$, and $d_3 = 0.99$ into $d'_1 = 0.1$, $d'_2 = 0.5$, and $d'_3 = 0.9$.

14

INTELLIGENT AND FUZZY
CONTROL

In this lesson, we continue the description of mathematical methods for dealing with expert knowledge; namely, we describe, step-by-step, the fuzzy control methodology that transforms the control rules (that expert operators formulate by using words of a natural language) into a precise control strategy.

14.1. Representing logical connectives

How can we represent logical connectives? An ideal solution. Suppose that an expert system contains statements A and B, and we have elicited the degrees of belief $d(A)$ and $d(B)$ from the experts. Suppose now that a user wants to know the degree of belief in a composite statement $A\&B$.

In principle, knowing only the two numbers $d(A)$ and $d(B)$ is not sufficient to describe the expert's degree of belief in $A\&B$: e.g., if $d(A) = d(B)$,

- it could be that the experts perceive A and B as *equivalent*, in which case $d(A\&B) = d(A)$, or

- it could also be that the experts perceive A and B as *independent* statements, in which case the possibility of A *and* B being true is smaller than the possibility that one of them is true: $d(A\&B) < d(A)$.

So, the *ideal* situation would be to elicit, from the experts, not only the degree of belief in the *basic* statements from the knowledge base, but also the degrees of belief in all possible *logical combinations* of these statements.

This ideal solution is not practically possible. The above described ideal solution is practically impossible: If we have N statements $S_1, ..., S_N$ in the knowledge base, then for each of $2^N - 1$ non-empty subsets $\{S_{i_1}, ..., S_{i_k}\}$ of the knowledge base, we need to elicit the degree of belief in the corresponding AND-statement $S_{i_1} \& ... \& S_{i_k}$. For a realistic expert system, N is in hundreds, so asking an expert 2^N questions is impossible.

A practical way to represent logical connectives: logical operations. In view of this practical impossibility, although in *some* cases, we will be able to have the degree of belief $d(A\&B)$ stored in the knowledge base, in *general*, we often have to deal with a following situation:

- we know the degrees of belief $d(A)$ and $d(B)$ in statements A and B;

- we know nothing else about A and B; and

- we are interested in the (estimated) degree of belief of the composite statement $A\&B$.

Since the only information available consists of the values $d(A)$ and $d(B)$, we must compute $d(A\&B)$ based on these values. We must be able to do that for arbitrary values $d(A)$ and $d(B)$. Therefore, we need a *function* that transforms the values $d(A)$ and $d(B)$ into an estimate for $d(A\&B)$. Such a function is called an *AND-operation*. If an AND-operation $f_\& : [0,1] \times [0,1] \to [0,1]$ is fixed, then we take $f_\&(d(A), d(B))$ as an estimate for $d(A\&B)$.

Similarly:

- to estimate the degree of belief in $A \vee B$, we need an *OR-operation* $f_\vee : [0,1] \times [0,1] \to [0,1]$.

- to estimate the degree of belief in the negation $\neg A$, we need a *NOT-operation* $f_\vee : [0,1] \times [0,1]$.

Terminological comment. AND operations are also called *t-norms*, and OR operations are also called *t-conorms*.

Natural properties of logical operations. The logical operations with fuzzy values must satisfy some *natural conditions*.

- For an expert, $A\&B$ and $B\&A$ mean the same. Therefore, the estimates $f_\&(d(A), d(B))$ and $f_\&(d(B), d(A))$ for these two statements should coincide. To achieve that, we must require that $f_\&(a, b) = f_\&(b, a)$ for all a and b; in other words, the operation $f_\&$ must be *commutative*.

- Similarly, from the fact that $A\&(B\&C)$ and $(A\&B)\&C$ mean the same, we can deduce the requirement that $f_\&$ must be *associative*: $f_\&(a, f_\&(b, c)) = f_\&(f_\&(a, b), c)$ for all a, b, and c.

- If A is absolutely false ($d(A) = 0$), then $A\&B$ is also absolutely false, i.e., $f_\&(a, 0) = 0$ for all a.

- If A is absolutely true ($d(A) = 1$), then $A\&B$ is true iff B is true, so, the degree of belief in $A\&B$ must coincide with the degree of belief in B: $f_\&(1, b) = b$ for all b.

- If our belief in A or B increases, then the degree of belief in $A\&B$ must also increase, so $f_\&$ must be *monotonic*.

- If the degree of belief in A changes a little bit, then the degree of belief in $A\&B$ must also change slightly. In other words, $f_\&$ must be *continuous*.

Combining all these requirements, we arrive at the following definition:

Definition 14.1.

- By an *AND-operation*, we mean a commutative, associate, monotonic, continuous operation $f_\& : [0, 1] \times [0, 1] \rightarrow [0, 1]$ with the properties $f_\&(1, a) = a$ and $f_\&(0, a) = 0$.

- By an *OR operation*, we mean a commutative, associate, monotonic, continuous operation $f_\vee : [0, 1] \times [0, 1] \rightarrow [0, 1]$ with the properties $f_\vee(1, a) = 1$ and $f_\vee(0, a) = a$.

How can we determine AND and OR operations? Since our goal is to describe the expert knowledge, we must elicit these operations from the experts. This can be done in the following manner:

- We form several pairs of statements (A_k, B_k), $k = 1, 2, \ldots$

- For each pair from this set, we elicit, from the experts, the degrees of confidence $d(A_k)$, $d(B_k)$, and $d(A_k \& B_k)$.

■   Then, we use an extrapolation procedure to find a function $f_\&(a, b)$ for which $f_\&(d(A_k), d(B_k)) \approx d(A_k \& B_k)$.

Similar procedures enable us to determine OR and NOT operations.

Historical comment. This empirical approach was first implemented and used by the designers of the first successful expert system MYCIN. After spending several years, they came up with AND and OR operations that fit the reasoning of medical experts. Interestingly, it turned out that different medical experts use very similar AND and OR operations. This lead MYCIN designers to a natural hypothesis that these operations are indeed general for human reasoning. Alas, when they tried to apply these same AND and OR operations to another area (geology), the resulting expert system did not lead to good results. It turned out that in different fields, people use different AND and OR operations. This difference is easy to explain:

■   In some applications (e.g., in *medicine*), mistakes can be deadly, so more *cautious* estimates are needed (e.g., $f_\&(a, b) = a \cdot b$).

■   In other applications (e.g., in *geology*), we cannot measure as many parameters as in medicine, so, we have to rely more on expertise, and hence, experts must take risks. In these applications, more brave, more optimistic estimates are needed: e.g., a geologist starts to drill in the uncertainty in which a surgeon is not likely to start an incision. Therefore, for such applications, we need more *optimistic* estimates for $d(A\&B)$ (e.g., $f_\&(a, b) = \min(a, b)$).

Simple AND, OR, and NOT operations: an idea. There are many possible AND and OR operations, many of them very complicated. However, in most applications, we do not need this complexity:

■   Our goal is to apply these operations to the degrees of confidence, that, in their turn, come from the values of membership functions.

■   We have already mentioned that in most applications, it is sufficient to use the *simplest* membership functions, that are obtained by applying the *simplest* spline interpolation methods to crisp values of the corresponding properties.

Since we start with the simplified expressions for degrees of confidence, it makes sense to consider *simple* AND and OR operations for handling these degrees of confidence, i.e., to consider AND and OR operations that are obtained by applying some *simple spline* extrapolation techniques (ideally, linear ones) to the *crisp* values of these operations.

Linear NOT operation. A NOT operation $f_\neg(a)$ must, given the degree of belief $d(A)$ in a statement A, return the degree of belief $f_\neg(d(A))$ in its negation $\neg A$. For crisp values $d(A) = 0, 1$, we have $f_\neg(0) = 1$ and $f_\neg(1) = 0$. One can easily show that for NOT, we can actually have a linear operation, and this linear NOT operation is uniquely defined:

Definition 14.2. *We say that a function $f_\neg(a)$ from $[0,1]$ to $[0,1]$ is a linear NOT operation if $f_\neg(a)$ is a linear function for which $f_\neg(0) = 1$ and $f_\neg(1) = 0$.*

Proposition 14.1. $f_\neg(a) = 1 - a$ *is the only linear NOT operation.*

Comments.

- The proof is straightforward: for every two points (x_1, y_1) and (x_2, y_2), there exists one and only one linear function

$$f(x) = y_1 + (x - x_1) \cdot \frac{y_2 - y_1}{x_2 - x_1}$$

  for which $f(x_1) = y_1$ and $f(x_2) = y_2$. In our case, $x_1 = 0$, $x_2 = 1$, $y_1 = 0$, and $y_2 = 1$.

- The NOT operation $f_\neg(a) = 1 - a$ is indeed most frequently used in fuzzy logic.

- The resulting formula $d(\neg A) = f_\neg(d(A)) = 1 - d(A)$ for the *degree of belief* in $\neg A$ resembles a formula for the *probability* of $\neg A$: If we know the probability $P(A)$ of an event A, then the probability $P(\neg A)$ that this event will not occur is equal to $P(\neg A) = 1 - P(A)$. Thus, the linear NOT operation is consistent with the two probability-like elicitation procedures for degree of belief: namely, with the procedures that are based on polling and betting.

There exist no linear AND and OR operations. For NOT, we have found a linear operation. For AND and OR, the situation is more complicated because there are no linear AND and OR operations:

Definition 14.3.

■ We say that a function $f_\&(a, b)$ from $[0,1] \times [0,1]$ to $[0,1]$ is a linear AND operation if $f_\&(a, b)$ is a linear function for which $f_\&(0,0) = f_\&(0,1) = f_\&(1,0) = 0$ and $f_\&(1,1) = 1$.

■ We say that a function $f_\vee(a, b)$ from $[0,1] \times [0,1]$ to $[0,1]$ is a linear OR operation if $f_\vee(a, b)$ is a linear function for which $f_\vee(0,0) = 0$ and $f_\vee(0,1) = f_\vee(1,0) = f_\&(1,1) = 1$.

Proposition 14.2.

■ There exist no linear AND operations.

■ There exist no linear OR operations.

Comment. In the remaining part of this section, we will only prove the results about AND operations. Corresponding results about OR operations can be proved in a similar manner; these proofs are left for the readers.

Proof. In accordance with what we have just mentioned, we will only prove Proposition 14.2 for AND operations. A general linear function of two variables has the form $f_\&(a, b) = c_0 + c_a \cdot a + c_b \cdot b$ for some coefficients c_0, c_a, and c_b. If we substitute this expression into the four conditions that describe a linear AND operation, we will get a system of four equations for three unknowns: $c_0 = 0$, $c_0 + c_b = 0$, $c_0 + c_a = 0$, and $c_0 + c_a + c_b = 1$. From the first three equations, we can uniquely determine the three coefficients c_0, c_a, and c_b:

■ from the first equation, we conclude that $c_0 = 0$, and then

■ from the second and third equations we conclude that $c_a = c_b = 0$.

Alas, if we we substitute these values into the fourth equation, we get $c_0 + c_a + c_b = 0 \neq 1$. Thus, no linear function satisfies all four equations, and hence, linear AND operations are impossible. Proposition is proven.

Comments.

■ This result is well known to people familiar with the history of *neural networks*: Since neural networks have been proposed as *universal* computing devices, capable of doing both numerical computations and logical reasoning, researchers tried to implement logical operations (such as AND and OR) in neural networks. For quite some time, these attempts were un-successful, until researchers realized that they were trying to use *linear* neurons that can only implement *linear* operations, and with linear operations, we cannot implement AND or OR. Thus, the impossibility of linear AND and OR operations was one of the main motivations for *non-linear* neurons.

- We have decided to use splines (piece-wise polynomial functions) to represent logical operations. Since we cannot use the *simplest* splines (i.e., linear functions), we must take the *next simplest*. A linear function is the simplest case of a spline, in which:

  - there is only *one* piece; and
  - the corresponding polynomial is of the smallest possible degree (*one*).

To get the next simplest case, we must relax one of these two conditions. Thus, we have two options:

  - One option is to still consider splines formed by polynomials of the smallest degree (one), but allow *two* pieces instead of one. In this case, we get *piecewise-linear* functions, that consist of *two* different linear functions on two parts of the square $[0, 1] \times [0, 1]$.

  - Another option is to still consider functions consisting of only one piece, but to allow a higher degree (two). In this case, we get *quadratic* functions.

We will see that in both cases, we get reasonable AND and OR operations.

Piecewise-linear AND and OR operations. We want an AND operation to satisfy four conditions that correspond to $a = 0, 1$ and $b = 0, 1$. Geometrically, if we describe the set of all possible values (a, b) as a square, these four conditions correspond to the *vertices* of this square.

A general linear function of two variables a and b can be described by three coefficients c_0, c_a, and c_b. Initially, we tried to find a linear function that satisfies the four conditions that correspond to all four vertices of the square. As a result, we got an "over-determined" system of four equations with three unknowns. Since we cannot have a linear function that fits all four vertices of the square, we must divide this square into pieces on which a fit is possible.

Each vertex leads to a linear equation for the coefficients c_0, c_a, and c_b. To determine all three coefficients, we must have three equations for each piece. Thus, each of the two pieces must be determined by three vertices. So, from the geometric viewpoint, to find an appropriate piecewise-linear function, we must:

- group the vertices of the square into two groups of three so that the square will be divided into the two triangles, and

- on each of these triangles, determine the corresponding linear function.

There are only two ways to divide the square into two triangles:

- We can divide the square by a diagonal that goes from the lower left vertex $(0, 0)$ to upper right vertex $(1, 1)$. In this case:
    - the first triangle has the vertices $(0, 0)$, $(1, 0)$, and $(1, 1)$;
    - the second triangle has the vertices $(0, 0)$, $(0, 1)$, and $(1, 1)$.

- We can also divide the square by a diagonal that goes from upper left vertex $(0, 1)$ to the lower right vertex $(1, 0)$. In this case:
    - the first triangle has the vertices $(0, 0)$, $(0, 1)$, and $(1, 0)$;
    - the second triangle has the vertices $(0, 1)$, $(1, 0)$, and $(1, 1)$.

In both cases, we can easily determine the coefficients of each of the two linear pieces:

- For the first division, we get $f_\&(a, b) = b$ for the first triangle and $f_\&(a, b) = a$ for the second triangle. These two formulas can be combined into a single expression $f_\&(a, b) = \min(a, b)$.

- For the second division, we get $f_\&(a, b) = 0$ for the first triangle and $f_\&(a, b) = a + b - 1$ for the second triangle. These two formulas can be combined into a single expression $f_\&(a, b) = \max(0, a + b - 1)$.

As a result, we get two possible piece-wise linear AND operations: $f_\&(a, b) = \min(a, b)$ and $f_\&(a, b) = \max(0, a + b - 1)$. Similarly, we get two possible OR operations: $f_\vee(a, b) = \max(a, b)$ and $f_\vee(a, b) = \min(a + b, 1)$.

Comments.

- In knowledge representation, the operations $\min(a, b)$ and $\max(a, b)$ were first proposed in the pioneer 1965 paper of L. Zadeh. Computationally, they are the simplest. The operations $\max(0, a + b - 1)$ and $\min(a + b, 1)$ were used by R. Giles in 1976 under the name of *bold* operations.

- Similarly to the linear NOT operation, these four operations are also consistent with the probability-like interpretation: namely, if we know the probabilities $a = P(A)$ and $b = P(B)$ of two events A and B, then:

  - The probability $P(A\&B)$ that both events A and B will occur can take any real value from the interval $[\max(a + b - 1, 0), \min(a, b)]$.

  - The probability $P(A \vee B)$ that at least one of the events A or B will occur can take any real value from the interval $[\max(a, b), \min(a + b, 1)]$.

Quadratic AND and OR operations. In addition to *piece-wise linear* operations, another option is to use *quadratic* operations. A general quadratic function $f(a, b)$ of two variables has the form

$$f(a, b) = c_0 + c_a \cdot a + c_b \cdot b + c_{aa} \cdot a^2 + c_{ab} \cdot a \cdot b + c_{bb} \cdot b^2$$

with six coefficients c_0, c_a, c_b, c_{aa}, c_{ab}, and c_{bb}.

The four conditions on AND or OR operations, conditions that describe the values of these operations for crisp a and b (i.e., for $a, b \in \{0, 1\}$), lead to only four linear equations. Four linear equations cannot uniquely determine the values of six unknowns. Thus, in contrast to the piecewise-linear case, these four conditions are not sufficient to determine a quadratic function. We will show, however, that if we add a natural requirement of *monotonicity*, we will be able to determine these operations uniquely.

Definition 14.4. *We say that a function $f(a, b)$ is*

- *non-decreasing in a if for every a, a', and b, $a < a'$ implies $f(a, b) \leq f(a', b)$.*

- *non-decreasing in b if for every a, b, and b', $b < b'$ implies $f(a, b) \leq f(a, b')$.*

- *non-decreasing in each argument if it is both non-decreasing in a and non-decreasing in b.*

Definition 14.5.

- We say that a function $f_\&(a, b)$ from $[0, 1] \times [0, 1]$ to $[0, 1]$ is a *quadratic AND operation* if $f_\&(a, b)$ is a quadratic function that is non-decreasing in each argument and for which $f_\&(0, 0) = f_\&(0, 1) = f_\&(1, 0) = 0$ and $f_\&(1, 1) = 1$.

- We say that a function $f_\lor(a, b)$ from $[0, 1] \times [0, 1]$ to $[0, 1]$ is a *quadratic OR operation* if $f_\lor(a, b)$ is a quadratic function that is non-decreasing in each argument and for which $f_\lor(0, 0) = 0$ and $f_\lor(0, 1) = f_\lor(1, 0) = f_\&(1, 1) = 1$.

Proposition 14.3.

- $f_\&(a, b) = a \cdot b$ is the only quadratic AND operation.

- $f_\lor(a, b) = a + b - a \cdot b$ is the only quadratic OR operation.

Proof. Similarly to the previous proposition, we will only prove the result for AND operations; for OR operations, the proof is similar and this proof is left to the reader.

- From the condition $f_\&(0, 0) = 0$, we conclude that $c_0 = 0$.

- Now, from the condition $f_\&(0, 1) = 0$, we conclude that $c_b + c_{bb} = 0$. Therefore, $c_{bb} = -c_b$.

- Similarly, from the condition that $f_\&(1, 0) = 0$, we conclude that $c_{aa} = -c_a$. Hence,

$$f_\&(a, b) = c_a \cdot (a - a^2) + c_b \cdot (b - b^2) + c_{ab} \cdot a \cdot b.$$

- From the condition $f_\&(1, 1) = 1$, we conclude that $c_{ab} = 1$, and hence, $f_\&(a, b) = c_a \cdot (a - a^2) + c_b \cdot (b - b^2) + a \cdot b$.

- The function $f_\&(a, b)$ must take non-negative values for all a and b from the interval $[0, 1]$. In particular, for $a = 0$ and $b = 0.5$, we get $f_\&(0, 0.5) = c_b \cdot 0.25 \geq 0$, hence, $c_b \geq 0$. Similarly, from the condition $f_\&(0.5, 0) \geq 0$, we conclude that $c_a \geq 0$.

- Since the function $f_\&(a, b)$ is non-decreasing in a, we conclude, for $a = 0.5$, $a' = 1$, and $b = 0$, that $f_\&(0.5, 0) = c_b \cdot 0.25 \leq f_\&(1, 0) = 0$, i.e., that $c_a \leq 0$. Since we already know that $c_a \geq 0$, we conclude that $c_a = 0$.

- Similarly, we can conclude that $c_b = 0$ and therefore, that $f_\&(a, b) = a \cdot b$.

The proposition is proven.

Comments.

- In knowledge representation, the operations $a \cdot b$ and $a + b - a \cdot b$ were first proposed in the pioneer 1965 paper of L. Zadeh. They are called *algebraic product* and *algebraic sum*.

- Similarly to the linear NOT operation and to the piece-wise linear AND and OR operations, the two quadratic AND and OR operations are also consistent with the probability-like interpretation: namely, if we know the probabilities $a = P(A)$ and $b = P(B)$ of two events A and B, and we know that the events A and B are *independent*, then:
  - The probability $P(A \& B)$ that both events A and B will occur is equal to $a \cdot b = P(A) \cdot P(B)$.
  - The probability $P(A \vee B)$ that at least one of the events A or B will occur is equal to $a + b - a \cdot b = P(A) + P(B) - P(A) \cdot P(B)$.

More complicated AND and OR operations. The above-described six AND and OR operations are the ones that are most frequently used in expert systems and in fuzzy control. However, these operations are not the only possible ones. Indeed, as we have mentioned in Section 13.4, we can *re-scale* the scale of degrees of belief, i.e., we represent the degree of confidence $a = d(A)$ in the statement A not by a value a, but by a new value $a' = d'(A) = r(a) = r(d(A))$ for some monotonic function $r(a)$.

In Section 13.4, we considered *reasonable* rescalings that correspond to difference ways of assigning the degree of confidence, but we may also have other re-scalings that are motivated by indirect reasons such as computational simplicity.

For example, if we are only interested in applying the AND operation, and we have chosen the quadratic AND operation $f_\&(a, b) = a \cdot b$, then we can simplify computations by using a *logarithmic scale* $(a' = r(a) = \ln(a))$, in which the AND operation is much faster: indeed,

$$d'(A\&B) = \ln(d(A\&B)) = \ln(d(A) \cdot d(B)) =$$

$$\ln(d(A)) + \ln(d(B)) = d'(A) + d'(B),$$

so in the new scale, we need addition instead of multiplication, and on the computers, addition is usually much faster.

This example shows that it makes sense to consider arbitrary monotonic re-scaling functions $r(a)$.

How will an AND operation that has the form $c = f_{\&}(a, b)$ in the old scale look in the new scale $a' = r(a)$? In other words, if we know the degrees of confidence $a' = d'(A) = r(d(A))$ and $b' = d'(B) = r(d(B))$ in the new scale, what will be the degree of belief $c' = d'(A\&B) = r(A\&B))$ in this new scale? To get the expression for c' in terms of a' and b', we must do the following:

- First, we convert the values a' and b' into the old scale by applying the inverse re-scaling r^{-1}: $a = r^{-1}(a')$ and $b = r^{-1}(b')$.

- Then, we apply the AND operation $f_{\&}(a, b)$ to the values a and b expressed in the old scale. As a result, we get the value $c = f_{\&}(a, b) = f_{\&}(r^{-1}(a'), r^{-1}(b'))$.

- Finally, we convert the value c into the new scale by applying the re-scaling $r(a)$: $c' = r(c)$.

As a result of this three-step procedure, we get a new AND operation

$$f'_{\&}(a', b') = r(f_{\&}(r^{-1}(a'), r^{-1}(b'))).$$

Similarly, we get a new OR operation

$$f'_{\vee}(a', b') = r(f_{\vee}(r^{-1}(a'), r^{-1}(b')))$$

and a new NOT operation

$$f'_{\neg}(a') = r(f_{\neg}(r^{-1}(a'))).$$

The new operations are called *isomorphic* to the old ones, because they represent the same operations but on a different scale.

There are special names for operations that are isomorphic to piece-wise linear and quadratic AND and OR operations. (These names may sound somewhat strange for a computer science reader, because they were invented before the computer applications and they describe algebraic properties of the corresponding operations.)

- Operations that are isomorphic to quadratic AND and OR operations, i.e., operations of the type

$$f'_{\&}(a', b') = r(r^{-1}(a') \cdot r^{-1}(b'))$$

and

$$f'_{\vee}(a', b') = r(r^{-1}(a') + r^{-1}(b') - r^{-1}(a') \cdot r^{-1}(b')),$$

for some strictly increasing continuous function $r(a)$, are called *strictly Archimedean* AND and OR operations.

- Operations that are isomorphic to bold AND and OR, i.e., operations of the type

$$f'_{\&}(a', b') = r(\max(r^{-1}(a') + r^{-1}(b') - 1, 0))$$

and

$$f'_{\vee}(a', b') = r(\min(r^{-1}(a') + r^{-1}(b'), 1)),$$

for some strictly increasing continuous function $r(a)$, are called *non-strictly Archimedean* AND and OR operations.

- If we use $f_{\&}(a, b) = \min(a, b)$ and $f_{\vee}(a, b) = \max(a, b)$, then, as one can see, for every strictly monotonic function $r(a)$, the isomorphic operations $f'_{\&}(a', b')$ and $f'_{\vee}(a', b')$ have exactly the same form: $f'_{\&}(a', b') = \min(a', b')$ and $f'_{\vee}(a', b') = \max(a', b')$. These two operations are called *idempotent*.

Most of the AND and OR operations that are actually used belong to one of these three types.

In addition of these three classes of operations, we may consider even more complicated AND and OR operations that are, e.g., isomorphic to an idempotent operation on one subinterval of the interval $[0, 1]$ and to a strictly Archimedean one on another subinterval of this interval $[0, 1]$. It turns out that this combination covers *all* possible AND and OR operations: namely, for an arbitrary AND and OR operation that satisfies several reasonable properties (e.g., the ones described in Definition 14.1), we can sub-divide the interval $[0, 1]$ into sub-intervals on each of which the operation is isomorphic to an operation from one of the three classes. This general classification result was proven by Ling in 1965; see the textbooks by Klir–Yuan and Nguyen–Walker (mentioned in Lesson 13) for details.

Exercise

14.1 Prove the above results about OR operations.

14.2. Representing rules

Problems with implication. We started the description of the expert knowledge by mentioning that the expert knowledge is usually represented by "if-then" rules. Since we want to formalize these rules, it seems reasonable to formalize the statements of the type "if A then B", i.e., using a logical term for such statements, to formalize *implication* in the same way as we formalized "and", "or", and "not" operations. There are, however, two problems with this idea:

- First, a *minor* problem: unlike "and", "or", and "not", implication is usually *not directly implemented* in the computers. Our goal is to describe the expert knowledge for a computer. Most computer languages have built-in logical operations "and", "or", and "not", but usually, not implication. Therefore, even if all the statements are crisp, when we formalize these statements for the computer, we will still need to first *reformulate* implication in terms of other logical operations.

- Second, a *major* problem: implication is somewhat *counter-intuitive*. Namely, researchers working in mathematical logic are using a formalization of implication in which implication is defined by a truth table in which $A \rightarrow B$ is true in all cases except when A is true and B is false. So, if A is false, then $A \rightarrow B$ is true for an arbitrary statement B. For example, we can conclude that "if $2 + 2 = 5$, then witches have six wings". Such statement may be mathematically correct, but they are absolutely *counter-intuitive*, because our intuitive understanding of "if-then" assumes some relation, while in the witches example, there is no relation whatsoever between the assumption (that $2 + 2 = 5$) and the conclusion (that witches have six wings).

Due to these problems, most expert systems and intelligent control systems do not *directly* formalize implication but instead, try first to *reformulate* if-then rules in terms of "and", "or", and "not". How can we do that?

Reformulating if-then rules in terms of "and", "or", and "not": example. Let's first consider our toy thermostat example, with three rules (13.1)–(13.3). If we know the difference x between the actual and the desired temperature, what control u should we apply? We have three rules that describe when a control is reasonable. Therefore, u is a reasonable control if one of the the three rules is applicable, i.e., when either:

- the first rule is applicable (i.e., x is negligible) and u is negligible; or

- the second rule is applicable (i.e., x is small positive), and u is small negative; or

- the third rule is applicable (i.e., x is small negative), and u is small positive.

Summarizing, we can say that u is an appropriate choice for a control if and only if either (x is negligible *and* u is negligible), *or* (x is small positive *and* u is small negative), etc.

Let us describe this statement in more succinct terms. Let us use the following notations:

- $R_k(x, u)$ will indicate that k-th rule is applicable for a given x, and that this rule recommends to use the control value u;

- $C(x, u)$ will indicate that u is a reasonable control for a given input x.

Then, in the above example, we get

$$C(x, u) \equiv R_1(x, u) \vee R_2(x, u) \vee R_2(x, u),$$

where

$$R_1(x, u) \equiv N(x) \& N(u); \quad R_2(x, u) \equiv SP(x) \& SN(u);$$
$$R_3(x, u) \equiv SN(x) \& SP(u).$$

We already know how to formalize the properties (as membership functions, or fuzzy sets), and we know how to formalize the "and" and "or" operations.

Thus, for every input x, we can define the degree of belief $d_r(x, u)$ that r-th rule will be fired:

$$d_1(x, u) = f_\&(\mu_N(x), \mu_N(u)); \quad d_2(x, u) = f_\&(\mu_{SP}(x), \mu_{SN}(u));$$

$$d_3(x, u) = f_\&(\mu_{SN}(x), \mu_{SP}(u)).$$

Reformulating if-then rules in terms of "and", "or", and "not": general case. For a general rule base with R rules of the type with

$$A_{r1}(x_1) \& \ldots \& A_{rn}(x_n) \to B_r(y), \tag{13.4}$$

we get

$$C(x_1, \ldots, x_n, y) \equiv R_1(x_1, \ldots, x_n, y) \lor \ldots \lor R_R(x_1, \ldots, x_n, y), \tag{14.1}$$

where

$$R_r(x_1, \ldots, x_n, y) \equiv A_{r1}(x_1) \& \ldots \& A_{rn}(x_n) \& B_r(y). \tag{14.2}$$

Therefore, for each rule, the "firing degree" is equal to:

$$d_r(x_1, \ldots, x_n, y) = f_\&(\mu_{r1}(x_1), \ldots, \mu_{rn}(x_n), \mu_r(y)), \tag{14.3}$$

where:

- $\mu_{ri}(x_i)$ is the membership function corresponding to the property A_{ri};

- $\mu_r(y)$ is the membership function corresponding to the property $B_r(u)$;

- $f_\&(a, b, c)$ stands for $f_\&(f_\&(a, b), c)$, and, in general, $f_\&(a_1, \ldots, a_n, a_{n+1}) = f_\&(f_\&(a_1, \ldots, a_n), a_{n+1})$.

14.3. Representing rule bases

General idea. In the previous section, we have shown how to transform the formula (14.2) into an algorithm that describes the degree of belief $d_r(x_1, \ldots, x_n, y)$ that each rule is applicable. After we have computed the firing degree of each rule, we can similarly formalize the formula (14.1) and get a numerical value that describes to what extent each possible control value y is reasonable for a given input $x_1, \ldots, x_n$:

$$\mu_C(x_1, \ldots, x_n, y) = f_\lor(d_1(x_1, \ldots, x_n, y), \ldots, d_R(x_1, \ldots, x_n, y)), \tag{14.4}$$

where $f_\vee(a, b, c)$ stands for $f_\vee(f_\vee(a, b), c)$, and, in general, $f_\vee(a_1, \ldots, a_n, a_{n+1}) = f_\vee(f_\vee(a_1, \ldots, a_n), a_{n+1})$.

Toy example. In particular, in our toy example,

$$\mu_C(x, u) = f_\vee(d_1(x, u), d_2(x, u), d_3(x, u)),$$

i.e.,

$$\mu_C(x, u) = f_\vee(f_\&(\mu_N(x), \mu_N(u)), f_\&(\mu_{SP}(x), \mu_{SN}(u)), f_\&(\mu_{SN}(x), \mu_{SP}(u))).$$

Numerical example. Let us assume that all three membership functions are piece-wise linear, and that they are described by the following graph:

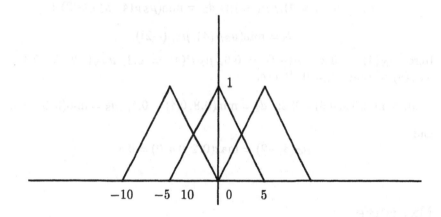

What is the degree of confidence $\mu_C(4, -2)$ that for $x = 4°$ the control $u = -2°$ is reasonable? According to our formulas, let us first compute the values of the membership functions. From the above general formula for linear extrapolation, we can find the analytical formulas for these membership functions:

■ The term "negligible" is described by the following formulas:

- $\mu_N(x) = 1 + x/5$ for $-5 \leq x \leq 0$;
- $\mu_N(x) = 1 - x/5$ for $0 \leq x \leq 5$;
- $\mu_N(x) = 0$ for all other x.

■ The term "small positive" is described by the following formulas:

  • $\mu_{SP}(x) = x/5$ for $0 \le x \le 5$;
  • $\mu_{SP}(x) = 2 - x/5$ for $5 \le x \le 10$;
  • $\mu_{SP}(x) = 0$ for all other x.

■ The term "small negative" is described by the following formulas:

  • $\mu_{SN}(x) = 2 + x/5$ for $-10 \le x \le -5$;
  • $\mu_{SN}(x) = -x/5$ for $-5 \le x \le 0$;
  • $\mu_{SN}(x) = 0$ for all other x.

If we use $f_{\&}(a,b) = \min(a,b)$ and $f_{\vee}(a,b) = \max(a,b)$, then we get $\mu_C(4,-2) = \max(d_1, d_2, d_3)$, where

$$d_1 = \min(\mu_N(4), \mu_N(-2)); \quad d_2 = \min(\mu_{SP}(4), \mu_{SN}(-2));$$

$$d_3 = \min(\mu_{SN}(4), \mu_{SP}(-2)).$$

Here, $\mu_N(4) = 0.2$, $\mu_N(-2) = 0.6$, $\mu_S P(4) = 0.8$, $\mu_{SN}(-2) = 0.4$, and $\mu_{SN}(4) = \mu_{SP}(-2) = 0$. Hence,

$$d_1 = \min(0.2, 0.6) = 0.2; \quad d_2 = \min(0.8, 0.4) = 0.4; \quad d_3 = \min(0,0) = 0,$$

and

$$\mu_C(4,-2) = \max(0.2, 0.4, 0) = 0.4.$$

Exercise

14.2 Compute $\mu_C(3,-3)$ assuming that the properties "negligible", "small positive", and "small negative" are described by the same membership functions, and that we use algebraic product and algebraic sum as AND and OR operations.

14.4. Decision making (defuzzification)

The problem. As a result of applying the previous steps, we get a *fuzzy set* that describes, for each possible control value u, how reasonable it is to use this particular value. In other words, for every possible control value u, we get a degree of confidence $\mu(u)$ that describes to what extent this value u is reasonable to use. In automatic control applications, we want to transform this *fuzzy* information into a *single* value $\bar{u}$ of the control that will actually be applied.

This transformation from a *fuzzy* set to a (*non-fuzzy*) number is called a *de-fuzzification*. What defuzzification should we apply?

Main idea. In order to find out what defuzzification is the best, let us recall the meaning of the values $\mu(u)$. We have obtained these values *indirectly*, by processing the expert rules formulated for the general inputs. However, we could, in principle, have obtained them *directly*, by:

- asking experts about the possible controls for this very situation, and then

- by applying known elicitation techniques to get the degrees of confidence $\mu(u)$.

In Section 13.3, we described two major elicitation methods: estimating on a *scale* and *polling* (we also described a third, less frequently used, method based on betting).

- Estimation on a *scale* method does not explain why this or that number is chosen by an expert, so if we assume that $\mu(u)$ is obtained by this method, this does not help much in figuring out how we can defuzzify this information.

- The second, *polling* method, as we will see, turns out to be much more helpful.

According to this second elicitation method, to get a value $\mu(u)$, we poll several (N) experts and then define $\mu(u)$ as the ratio $M(u)/N$, where $M(u)$ is the total number of experts who believe that for this particular situation, u is a reasonable control value.

The function $\mu(u)$ is usually different from 0 for *infinitely many* different values u. However, in reality, we can only ask experts about *finitely many* different values. So, to use this interpretation, let us assume that there are only finitely many possible control values.

In this case, as a result of the elicitation, we have several (finitely many) control values u that different experts deemed reasonable. We can order these values into a sequence $u_1, \ldots, u_n$. In this sequence, each value u is proposed by $M(u)$ experts and is, therefore, repeated $M(u)$ times.

If we select a value $\bar{u}$, and the actual best control is u, then we get a control error $e = \bar{u} - u$. Thus, if it turns out that u_i is the best control, we get an error $e_i = \bar{u} - u_i$. We do not know which of the control values is the best, therefore, we would like all these errors e_i to be as small as possible. We cannot get all of these errors e_i equal to 0, so we would like to combine the errors $e_1, \ldots, e_n$ into a reasonable combination $J(e_1, \ldots, e_n)$, and then select $\bar{u}$ for which this combination takes the smallest possible value.

Implementation of the main idea: case of finitely many possible control values. We have already encountered a similar problem in Lesson 3, and we provided arguments in favor of the combination function $J(e_1, \ldots, e_n) = (e_1)^2 + \ldots + (e_n)^2$. This method is called the *least squares* method. If we apply the least squares method to our problem, we will choose the value $\bar{u}$ from the condition that $(\bar{u} - u_1)^2 + \ldots + (\bar{u} - u_n)^2 \to \min$.

This formula can be simplified if we group together the terms $(\bar{u} - u_i)^2$ that correspond to equal values u_i. Since each value u is repeated $M(u)$ times, the minimized sum can be re-written as $J = \sum M(u) \cdot (\bar{u} - u)^2$, where the sum is taken over all possible control values u, and the minimization problem takes the form

$$J = \sum_u M(u) \cdot (\bar{u} - u)^2 \to \min_{\bar{u}}.$$

We cannot directly apply this formula, because we do not know the values $M(u)$. Instead, we only know the values $\mu(u) = M(u)/N$. To use this knowledge, we can divide the minimized quantity J by N.

We can do that, since an arbitrary function J attains its minimum if and only if J/N attains its minimum, where N is an arbitrary constant.

The problem $J/N \to \min$ can be, thus, re-written as

$$\sum_u \mu(u) \cdot (\bar{u} - u)^2 \to \min_{\bar{u}}.$$

Now, we have formulated the problem of finding the defuzzification $\bar{u}$ as a precise mathematical problem, and we can solve this problem.

One of the main reasons for choosing the least squares method in Lesson 3 was that for this method, equating the derivative with 0 (a standard techniques for solving optimization problems) leads to linear equations. To utilize this advantage, let us differentiate the left-hand side of the minimized quantity with respect to the unknown $\bar{u}$ and equate the result to 0. As a result, we get the equation

$$\sum_{u} 2 \cdot (\bar{u} - u) = 0.$$

To simplify this equation, we can divide both sides by 2, and then move all the terms that do not contain the unknown $\mu(u)$ into the right-hand side. As a result, we get the equation

$$\bar{u} \cdot \left(\sum_{u} \mu(u) \right) = \sum_{u} u \cdot \mu(u),$$

and

$$\bar{u} = \frac{\sum u \cdot \mu(u)}{\sum \mu(u)}. \tag{14.5}$$

This formula is called *centroid defuzzification*, because it resembles a formula from mechanics that describes the *center of mass* $\vec{r}$ of a system of several points with masses m_i at locations $\vec{r}_i$ as

$$\vec{r} = \frac{\sum m_i \cdot \vec{r}_i}{\sum m_i}.$$

Here,

- u is the analog of a location, and

- $\mu(u)$ is the analog of the mass.

Defuzzification for the realistic case of infinitely many possible values of possible control. To get the formula (14.5), we made a simplifying assumption that only *finitely many* different control values u are reasonable (i.e., that only for finitely many values u, we have a positive degree of confidence $\mu(u)$ that u is reasonable). In reality, there are *infinitely many* possible values u. To get a formula for this realistic case, we must take more and more points and then tend this number of points to infinity.

When we take denser and denser values of u, with a step $\Delta u \to 0$, then the sum $\sum \mu(u) \cdot \Delta u$ tends into an *integral* of $\int \mu(u)du$ of the function $\mu(u)$. Thus, for small Δu, the sum $\sum \mu(u)$ (the denominator of the expression (14.5)) is approximately equal to $(\int \mu(u)du)/\Delta u$. Similarly, the sum $\sum u \cdot \mu(u)$ (the numerator of the expression (14.5)) is approximately equal to the ratio

$$(\int u \cdot \mu(u)du)/\Delta u.$$

Thus, when $\Delta u \to 0$, the fraction (14.5) is approximately equal to

$$\bar{u} \approx \frac{(\int u \cdot \mu(u)du)/\Delta u}{(\int \mu(u)du)/\Delta u}.$$

The fraction will not change if we multiply both the numerator and the denominator by Δu. Thus, we get a simplified expression

$$\bar{u} \approx \frac{\int u \cdot \mu(u)du}{\int \mu(u)du}.$$

The smaller Δu, the closer $\bar{u}$ to the right-hand side. Thus, in the limit $\Delta u \to 0$, $\bar{u}$ will be exactly equal to the right-hand side. Hence, we can select

$$\bar{u} = \frac{\int u \cdot \mu(u)du}{\int \mu(u)du}. \tag{14.6}$$

This formula is also called *centroid defuzzification*. It is actually successfully used in fuzzy control.

Warning: centroid defuzzification does not always work. In most real-life situations, centroid defuzzification leads to a meaningful control, but in some cases, it may lead to an unreasonable control.

Let us give a simple example. Suppose that we are designing an automatic controller for a car. If the car is traveling on an empty wide road, and there is an obstacle straight ahead (e.g., a box that fell from a truck), then a reasonable idea is to *swerve* to avoid this obstacle. Since the road is empty, there are two possibilities:

- we can swerve to the right; and

- we can swerve to the left.

For swerving, the control variable u is the angle to which we steer the wheel. Based on the distance to the obstacle and on the speed of the car, an experienced driver can describe a reasonable amount of steering u_0.

In reality, u_0 will probably be a fuzzy value, but for simplicity, we can assume that u_0 is precisely known.

Thus, as a result of formalizing expert knowledge, we conclude that there are two possible control values:

■ the value u_0 with degree of confidence $\mu(u_0) = 1$, and

■ the value $-u_0$, with degree of confidence $\mu(-u_0) = 1$.

If we apply the defuzzification formula (14.5) to this situation, we get

$$\bar{u} = \frac{u_0 \cdot \mu(u_0) + (-u_0) \cdot \mu(-u_0)}{\mu(u_0) + \mu(-u_0)} = \frac{u_0 \cdot 1 + (-u_0) \cdot 1}{1 + 1} = 0.$$

So, the recommended control means that *no swerving* will be applied at all, and the car will run straight into the box.

To avoid such situations, we must *modify* the centroid defuzzification. We could have screened out the value $\bar{u} = 0$ because for this value, $\mu(\bar{u}) = 0$. Thus, instead of using a simply centroid defuzzification, we can do the following:

■ First, we apply defuzzification to the original membership function $\mu(u)$.

■ Then, we check whether the resulting control value $\bar{u}$ is reasonable, i.e., whether the degree of confidence $\mu(\bar{u})$ is big enough (e.g., larger than some pre-defined value μ_0).

    – If $\mu(\bar{u}) \geq \mu_0$, then we apply the control $\bar{u}$.

    – If $\mu(\bar{u}) < \mu_0$, this means that there are several areas of reasonable control separated by a gap, and $\bar{u}$ happens to be in this gap. In this case, instead of applying the the centroid defuzzification to the entire *membership* function $\mu(u)$, we do the following:

        * we select *one of the areas*, and then

        * we apply centroid defuzzification only to value u from this area.

This idea was first proposed and successfully implemented by John Yen.

14.5. Summary: fuzzy control

Let us summarize what we have learned in these two lessons:

■ We start with the if-then expert rules of the type "If x is small, and ...,
   then y is small". In general, each of these rules can be represented by a
   formula

$$A_{r1}(x_1)\& \ldots \&A_{rn}(x_n) \to B_r(y), \qquad (13.4)$$

where $x_1, \ldots, x_n$ are inputs, and A_{ri} and B_r are words that describe properties of inputs and output.

■ For each words w used in these rules, we pick several values $x^{(1)}, \ldots, x^{(k)}$,
   and use one of the elicitation techniques described in Lesson 13 to determine the degrees of confidence $\mu_w(x^{(1)}), \ldots, \mu_w(x^{(n)})$ that these values
   satisfy the property w.

■ Then, we use some extrapolation technique to determine the membership
   functions $\mu_w(x)$ (that describe the degrees of confidence that different values of x satisfy the property w).

■ We choose AND and OR operations $f_\&(a, b)$ and $f_\vee(a, b)$.

■ For each rule r, and for each possible values of input and output, we
   compute the *firing degrees* $d_r(x_1, \ldots, x_n, y)$ using the formula

$$d_r(x_1, \ldots, x_n, y) = f_\&(\mu_{r1}(x_1), \ldots, \mu_{rn}(x_n), \mu_r(y)), \qquad (14.3)$$

and then we compute the membership function for control as

$$\mu_C(x_1, \ldots, x_n, y) = f_\vee(d_1(x_1, \ldots, x_n, y), \ldots, d_R(x_1, \ldots, x_n, y)). \quad (14.4)$$

■ For every input $x_1, \ldots, x_n$, we get a function $\mu_C(y)$ that describes the
   degree of confidence that this very y is a reasonable control. To get a
   single recommended control value $\bar{y}$, we use a formula

$$\bar{y} = \frac{\int y \cdot \mu_C(y)dy}{\int \mu_C(y)dy} \qquad (14.6a)$$

or a more complicated defuzzification method described in the previous
section.

Comment. It is quite possible that the resulting control is sometimes inadequate. There are two reasons for this:

- First, when an expert formulates the rules, he usually remembers *specific* rules but sometimes forgets to explicitly mention *common sense* rules that are absolutely evident to any human, but that need to be explicitly spelled out for the computer.

- Second, all knowledge elicitation methods are *approximate*, and if we add distortions caused by this approximate character on each step of fuzzy control methodology, we may end up with a rather distorted representation of expert's control.

In view of this possibility, before implementing this control, we must first *test* it. If it turns out that in some situations, the resulting control is inadequate, we have two options:

- If the inadequacy is *huge*, this probably means that we are missing or misinterpreting some of the rules. In this case, we need to confront the experts with these results. Since the rules that the experts have formulated lead to not adequate results, the experts will be able to either *modify* these rules, or *add* new rules that cover these situations.

- If the inadequacy is *small*, then probably, the experts' rules were adequate, and the inadequacy is caused by the approximate character of the expert system methodology. In this case, to make a control better, we can *tune* the parameters of the resulting control.

For further reading on the fundamentals of fuzzy control, see:

- D. Driankov, H. Hellendoorn, and M. Reinfrank, *An introduction to fuzzy control* (Springer-Verlag, Berlin, 1993).

- R. Palm, D. Driankov, and H. Hellendoorn, *Model based fuzzy control* (Springer-Verlag, Berlin, Heidelberg, 1997).

Exercise

14.3 For the toy thermostat example, write a program that will compute the optimal control value $\bar{u}$ for every given x. To compute the integrals, use either an analytical expression or some simple numerical method.

14.4 Apply fuzzy control methodology to design an extrapolation algorithm. Namely, we start with the patterns $(x^{(i)}, y^{(i)})$, and for each pattern, we formulate the rule "if x is close to $x^{(i)}$, then y is close to $y^{(i)}$". To describe a notion "x is close to x'", use a simple membership function $\mu(\Delta x)$, where $\Delta x = x - x'$ (e.g., a triangular function $\mu(\Delta x)$).

15

RANDOMNESS, CHAOS, AND FRACTALS

In the previous lessons, we considered the extrapolation problems in which the result was, in principle, predictable, in the sense the more patterns we know, the better our predictions can be. In many practical problems, however, the analyzed process has a random component that is impossible to predict even if we have many patterns. For example, we may be able to predict how the Internet grows with time, but it is difficult to predict the delay of a single message sent through the Internet. In such situations, when we cannot determine the exact value, we may be able to determine the probabilities of different possible values. In this lesson, we show how continuous mathematics can help in in computing these probabilities, and how fractal theory can help to interpret the results in visual geometric form.

15.1. Situations when exact predictions are impossible: randomness, chaos, and how to deal with them

Traditional extrapolation problem: a reminder. In the previous lessons, we considered the *interpolation* and *extrapolation* problems, in which we have a *partial* information about the system, and we want to make reasonable predictions based on this partial information. In general, the partial information may contain the recorded reaction of the system to different inputs, expert knowledge about the system, etc.

For example, we may have only a partial information on how a helicopter reacts to different control actions.

As a result of the extrapolation procedure, we get a function $f(x_1, \ldots, x_n)$ that describes how the output y of the system depends on its inputs: $y = f(x_1, \ldots, x_n)$. If the system has *several* outputs $y_1, \ldots, y_m$, we need *several* such functions $f_1(x_1, \ldots, x_n), \ldots, f_m(x_1, \ldots, x_n)$, one for each of these outputs.

In the helicopter example, we get the functions $f_1(x_1, \ldots, x_n)$, $f_2(x_1, \ldots, x_n), \ldots$ that describe how the vertical acceleration y_1, the horizontal acceleration y_2, and other parameters of the helicopter's trajectory depend on the control parameters $x_1, \ldots, x_n$.

In the previous lessons, we implicitly assumed that each output y is uniquely determined by the inputs, i.e., that there is some *actual* (unknown) function $y = f_{\text{actual}}(x_1, \ldots, x_n)$, and the main goal of the extrapolation is to come up with a function $f(x_1, \ldots, x_n)$ that is as close to $f_{\text{actual}}(x_1, \ldots, x_n)$ as possible.

Large systems are often unpredictable. The above-described implicit assumption (that the output is uniquely determined by the inputs) is not always true. Let us give two simple and related examples from computer science:

- If we plot the empirical data on how the *size* of the Internet changes with time t, we see that there is a clearly visible trend. Moreover, we see that the resulting curve is *smooth*. In this case, the main problem is to describe this curve.

- Alternatively, we may describe the same Internet not by its *size*, but by its *efficiency*, e.g., by the time τ it takes for a message to travel between two fixed sites. To analyze how this time delay τ changes with time, we may repeatedly send messages at different moments of time t, measure τ, and plot τ as a function of t. In this case, we will not see any trend, we will see a "random", "chaotic", seemingly unpredictable behavior.

In large systems, such "unpredictable" situations are very typical. There are, usually, two reasons for this unpredictability:

First reason for unpredictability: randomness. First, usually, very *many* different factors influence the value y of the desired quantity; many of these factors cannot be measured.

For example, a small (unpredictable) turbulence can modify the helicopter's position. It is not possible to trace all airflows.

In other words, the desired quantity y depends not only on the characteristics $x_1, \ldots, x_n$ that we measure, but also on some other characteristics whose values we do not measure. Since we do not know the values of these additional characteristics, we cannot uniquely predict y based on $x_1, \ldots, x_n$. At best, if we know $x_1, \ldots, x_n$, we can determine:

- the *set* Y of possible values of y; and

- the *probabilities* of different values $y \in Y$.

Since probability is involved, such situations are called *random*.

Second reason for unpredictability: chaos. Randomness explains that if we *do not know* the values of all the characteristics that describe a system, we cannot uniquely predict the *values* of the desired quantities, and we have to resort to predicting *probabilities* of different possible values. However, often, even if we *know* the values of all the characteristics $x_1, \ldots, x_n$ that influence y, we still are not able to predict y uniquely. The reason is that in many complicated systems, a *small* change in the initial state can, after a certain time, lead to *drastic* changes.

For example, in the Internet routing, we usually have several alternative routes to go from one site to another. The message is routed to the path that is, e.g., currently the freest (and thus, seems the most promising). If the routing algorithm is fair, the loads on all the paths are more or less equal to each other, and whichever path is currently freer changes rather unpredictably. Thus, a message that was actually routed to path 1 could have been, if it arrived 1 millisecond later, routed to route 2. The actual time delay τ depends on which of these routes the message was routed to. Thus, a *minor* (1 msec) change in the message sending time can lead to a *drastic* change in the actual time delay of this message. In other words, *small changes in input data can be drastically amplified.*

One can find even more impressive examples of this amplification. For example, if a communication network has been designed properly, then it should normally operate close to its own capacity limits: Otherwise, if the system is always operating at, say, 50% of its potential, then we could have spent half the money and still carry the same communication load. By operating a network at the edge of its capacity, we save on its design, but we risk congestion. So, we need to build-in some anti-congestion control. Thus, even when we send a very small message, the network is usually so close to be overloaded, that it is highly probably that this particular message will actually overload the network, and thus, trigger the anti-congestion control that will drastically slow down all the messages. The actual load fluctuates, so whether this control will be triggered or not depend on the exact timing of this message. Again, a difference in a few milliseconds can drastically change the delay time, but this time, not only the delay time of *this* very message, but of *all* messages that are currently in the net.

In short, if we know the timings and other characteristics of all the messages, we can, in principle, predict the behavior of the network at any given moment of time. But in reality, this would require knowing all the times of all the messages with a millisecond (or better) accuracy. This is unrealistic. If we know these times with a slightly worse accuracy, we cannot predict the resulting time delays.

In other words, there are situations is which very small changes in the input parameters can lead to drastic changes in the output.

- In *regular* extrapolation situations described in the previous lessons, the output continuously (and even smoothly) depends on the inputs $x_1, \ldots, x_n$. Thus, if we start at two slightly different states $x = (x_1, \ldots, x_n)$ and $\tilde{x} = (\tilde{x}_1, \ldots, \tilde{x}_n)$ ($x_i \approx \tilde{x}_i$ for all i), we get slightly similar values $y \approx \tilde{y}$. In other words, in such situations, there is an *order*:

  - if we know x_i precisely, we can get y precisely;

  - if we know x_i approximately, we can get y approximately.

- In the situations when the input differences are amplified, seemingly equal (but actually slightly different) values $x = (x_1, \ldots, x_n)$ and $\tilde{x} = (\tilde{x}_1, \ldots, \tilde{x}_n)$ lead to drastically different values y and $\tilde{y}$. So, the approximate knowledge of x_i no longer enables us to predict anything reasonable about y. In other words, we do not have order, we have *chaos*.

Chaos is actually the "official" scientific term for such situations. Another name for the same phenomenon is *stochastization*:

- Strictly speaking, we have a *deterministic* situation, in which the knowledge of x_i should uniquely determine y.

- However, in reality, even when we measure x_i with a reasonable accuracy, we cannot exactly predict y. In other words, the situation is actually as unpredictable as if y depended on some *random, stochastic* factors.

When a system is unpredictable, we can predict probabilities. Due to randomness and/or chaos, in many large and complex systems, we cannot predict the *values* y of the desired quantity. Instead, we can only predict the *probabilities* of different values.

Computing all probabilities is often computationally impossible. If we want to compute *all* probabilities, we immediately run into a serious computational problem. Indeed:

- For regular systems, we only need to predict a *single* number y. This number takes a *small* computer *space* to store, and usually, a *small time* to compute.

- However, if we aim at determining the probabilities of all possible values of y, we run into a computational problem: Indeed, the set Y of possible values of y is, usually, an interval, and in principle, contains infinitely many real numbers. Even if we take into consideration that only finitely many real numbers can be represented in a computer, we still get a large amount of numbers. For example, if the computer allows 6 decimal digits, then on the interval $[0, 1]$, we have 1,000,000 different real numbers: 0.000,000, 0.000,001, 0.000,002, ..., 0.999,999, 1.000,000. For each of these million numbers, we need to compute and store the corresponding probability. The resulting million-times increase in computation time and computer storage makes computing all probabilities completely un-realistic.

The idea of solving this computational problem: let us predict the average and the deviation from the average. It would be nice to know all the probabilities, but if we cannot get all of them, which characteristics is then reasonable to compute?

This is a typical situation that happens in many real-life situations. For example, we may want to know how tall people in the state of Texas are. Ideally, we should know the heights $h^{(1)}, \ldots, h^{(N)}$ of all the people in Texas, but realistically, a person cannot deal with this huge amount of data. Instead, we can ask the two basic questions:

- "What is the *average* height of a Texan?" and

- "What is the *average* deviation from this average height"?

How do we formalize these notions?

The average. The average height can be naturally defined as

$$h_{av} = \frac{h^{(1)} + \ldots + h^{(N)}}{N}, \tag{15.1}$$

where N is the total number of people in the State of Texas. From the theoretical viewpoint, this is a reasonable and natural formula. However, if we try to use this formula, we run into two computational problems.

- The first problem is that there are millions of people in the State of Texas, and even getting the height of each them may take quite some time, and make this formula *computationally un-feasible*. To avoid this problem, researchers, usually, take a *sample* of the population and estimate an average based on this sample. If $h^{(1)}, \ldots, h^{(n)}$ are the heights of the people from this sample, then we can estimate the average height as

$$\bar{h} = \frac{h^{(1)} + \ldots + h^{(n)}}{n}. \tag{15.2}$$

- The second problem is that both formulas (15.1) and (15.2) are *computationally inefficient*. For example, when we use formula (15.1), we add the height of each person from the State of Texas. In Texas, we may have several thousand people of height 170 cm; so, according to the formula (15.1), we must add 170 several thousand times. It is, of course, inefficient: instead of computing $170 + \ldots + 170$, it is much faster to *multiply* 170 by the total number $N(170)$ of people whose height is equal to 170. If we thus group all the people by their heights, then, instead of the formula (15.1), we will get an equivalent formula (the new formula may look slightly more complicated, but it is, in reality, much easier to compute):

$$h_{av} = \frac{\ldots + N(150) \cdot 150 + \ldots + N(170) \cdot 170 + \ldots}{N}.$$

In other words,

$$h_{av} = \ldots + \frac{N(150)}{N} \cdot 150 + \ldots + \frac{N(170)}{N} \cdot 170 + \ldots$$

Similarly, if we use the formula (15.2), we can group people of the same height together, and get a computationally simpler formula

$$\bar{h} = \ldots + \frac{n(150)}{n} \cdot 150 + \ldots + \frac{n(170)}{n} \cdot 170 + \ldots$$

The ratio $N(170)/N$ is the *probability* $p(170)$ that a Texan has the height of 170 cm, while the ratio $n(170)/n$ is the (empirical) *frequency* of Texans of height 170 cm. Thus, the formula for h_{av} and for $\bar{h}$ can be rewritten as

$$h_{\mathrm{av}} = \ldots + p(150) \cdot 150 + \ldots + p(170) \cdot 170 + \ldots,$$

$$\bar{h} = \ldots + f(150) \cdot 150 + \ldots + f(170) \cdot 170 + \ldots$$

In our transformations, we did not, in any way, use the fact that h is a height. So, the same transformations are applicable in the very general case, when we have several different values of y, and we know the probability $p(y)$ of each of these possible values. In such case, a variable y is *randomly* defined, so it is called a *random variable*. For an arbitrary random variable, the "true average" can be described by a formula

$$E[y] = \sum p(y) \cdot y, \qquad (15.3)$$

where the sum goes over all possible values of y, and $p(y)$ is the probability of a value y. This "true average" is called an expected value, or a *mean*, and it is, therefore, denoted by $E[y]$ or by $M[y]$.

If, instead of the true probabilities $p(y)$ of different values y, we only know *frequencies* $f(y)$ of different values, we get a formula for the (empirical) average:

$$\bar{y} = \sum f(y) \cdot y, \qquad (15.4)$$

where the sum goes over all possible values of y.

By definition, a probability of a certain event is a limit of its frequency when we repeat the experiment more and more times. For example, the fact that a probability of the coin falling on its head, means that as we toss the coin more and times, the frequency of having heads will get closer and closer to 1/2. So, when we have sufficiently many samples, the frequencies $f(y)$ of different values tend to the probabilities, and the empirical average computed by the formula formula (15.4) tends to the true average described by the formula (15.3).

The average deviation. If we have n values $h^{(1)}, \ldots, h^{(n)}$, then we can describe the empirical average $\bar{h}$ by using one of the formulas (15.2) and (15.4).

How can we describe the average deviation? For each of n values, we can compute the deviation $e^{(i)} = h^{(i)} - \bar{h}$, $1 \leq i \leq n$. We are considering the situation when keeping all n values $e^{(1)}, \ldots, e^{(n)}$ would require too much space and too much computations time. Thus, we need a single numerical characteristic $J(e^{(1)}, \ldots, e^{(n)})$ that describe all these values, so that this "combined" characteristic J will be equal to 0 if all error are 0, and large when at least some of these errors are large.

In Lesson 3, we have already analyzed reasonable characteristics of this type, and we came to a conclusion that the best characteristic to use is the sum of squares, or, to be more precise, $J = c_1 \cdot ((e^{(1)})^2 + \ldots + (e^{(n)})^2) + c_0$ for some real numbers $c_1 > 0$ and c_0.

■   In Lesson 3, we were *not interested in the exact value* of J, only in finding the values of the model's parameters for which this J is the smallest possible. Thus, the choice of the parameters c_1 and c_0 did not matter.

■   In our case, we *want the actual value* of J, so we want to choose c_0 and c_1 appropriately.

   −   Since we want J to be equal to 0 when all the errors are equal to 0 (i.e., when $e^{(1)} = \ldots = e^{(n)} = 0$), we must have $c_0 = 0$.

   −   In Lesson 3, we took, for simplicity of computations, $c_0 = 0$ and $c_1 = 1$. The value $c_0 = 0$ is appropriate here as well, but the value $c_1 = 1$ is not, because as a result, the characteristic J depends not only on the random variable itself, but on the number N of observations: If all errors are approximately of the same size $|e^{(i)}| \approx e$, we get

$$((e^{(1)})^2 + \ldots + (e^{(n)})^2) \approx n \cdot e^2.$$

   Thus, if we want the value J to characterize the random variable itself, and not the number of observations, we should divide this sum by N, and get an *average* square rather than the sum of squares:

$$\bar{V} = \frac{(e^{(1)})^2 + \ldots + (e^{(n)})^2}{n} = \frac{(h^{(1)} - \bar{h})^2 + \ldots + (h^{(n)} - \bar{h})^2}{n}.$$

This average square deviation is called the *population variance*. Similarly to the formula for the average, we can combine terms that correspond to equal values of $h^{(i)}$, and thus get, for an arbitrary random variable y, the following formula

$$\bar{V} = \sum f(y) \cdot (y - \bar{y})^2.$$

In the limit, when we have many observations, frequencies tend to probabilities, the (empirical) average $\bar{y}$ tends to the mean $E[y]$, and we get the formula

$$V[y] = \sum p(y) \cdot (y - E[y])^2. \tag{15.5}$$

The expression $V[y]$ is called the *variance* of a random variable y. Comparing (15.5) with (15.3), we can see that (15.5) is actually the mean of the square $(y - E[y])^2$. Thus, the formula (15.5) can be rewritten as

$$V[y] = E[(y - E[y])^2].$$

In the above example of heights, when $y = h$ was measured in centimeters, the value $V[y]$ is measured in square centimeters. To get the estimate comparable with the the values themselves, we must, therefore, take the square root of $V[y]$. This square root $\sigma[y] = \sqrt{V[y]}$ is called a *standard deviation* of the random variable y.

For background on probability and statistics see, e.g., H. T. Nguyen and G. S. Rogers, *Fundamentals of mathematical statistics*, Vol. I and II (Springer-Verlag, Berlin, 1989).

What characteristics should we choose if we have several outputs? We have just shown that if we have a *single* random variable y (that characterizes a *single* output y), then we can described this variable by two numbers: its mean $E[y]$ and its standard deviation $\sigma[y]$. (For some reasonable classes of distributions, called *Gaussian*, these two numbers uniquely determine all the probabilities; see Appendices for details.)

In many practical extrapolation problems, we need to find *several* variable $y_1, \ldots, y_m$. In this case, instead of a random *variable*, we have a random *vector* $\vec{y} = (y_1, \ldots, y_m)$, i.e.:

- a *set* $Y \subset R^m$ of all possible values of $\vec{y}$, and

- *probabilities* of different values $\vec{y} \in Y$.

Even for the case of one variable, we concluded that it is unrealistic to store *all* the probabilities. It is even more unrealistic for several variables. So, which characteristics should we compute and store instead?

First of all, we can store the *mean* $E[y_j]$ and the *standard deviation* $\sigma[y_j]$ of each of the variables $y_1, \ldots, y_n$. However, as we will show, this is not sufficient. For example, the quantities y_1 and y_2 may represent spatial *coordinates* of a *robot*. The actual values of the coordinates depend on the choice of the coordinate axes (and on the choice of the origin). If we choose different axes, then we will have different numerical values $\tilde{y}_1$ and $\tilde{y}_2$ that are linear combinations of the original values: $\tilde{y}_1 = \cos(\theta) \cdot y_1 + \sin(\theta) \cdot y_2$ and $\tilde{y}_2 = -\sin(\theta) \cdot y_1 + \cos(\theta) \cdot y_2$.

- If we know the *exact* values y_1 and y_2, then we can apply this formula and find the exact values of $\tilde{y}_1$ and $\tilde{y}_2$.

- Similarly, we would like to be able to transform the chosen characteristics of y_j into the mean and standard deviation of the new coordinates $\tilde{y}_1$ and $\tilde{y}_2$.

In general, there are many other practical situations in which, in addition to the original physically meaningful variables $y_1, \ldots, y_m$, some linear combinations

$$y = \alpha_1 \cdot y_1 + \ldots + \alpha_m \cdot y_m \tag{15.6}$$

of these variables also have a direct physical sense. We would, therefore, like to choose the characteristics of the random vector $\vec{y}$ in such a way that we would be able, based on these characteristics, to compute the mean $E[y]$ and the standard deviation $\sigma[y] = \sqrt{V[y]}$ of each linear combination y.

For mean, the situation is simple: From the formula (15.3), one can easily see that:

- first, for every random variable y and for every real number α, we have $E[\alpha \cdot y] = \alpha \cdot E[y]$, and

- second, the mean of the sum of two random variables is equal to the sum of the means: $E[y_1 + y_2] = E[y_1] + E[y_2]$.

Thus, for every linear combination of the type (15.6), we can conclude that $E[y] = \alpha_1 \cdot E[y_1] + \ldots + \alpha_m \cdot E[y_m]$. Hence, if we know the means $E[y_j]$ of all m variables y_j, we can compute the mean of an arbitrary linear combination of these variables.

For the standard deviation, the situation is slightly more complicated. Namely, the difference $y - E[y]$ can still be represented as a linear combination of the differences corresponding to y_j:

$$y - E[y] = \alpha_1 \cdot (y_1 - E[y_1]) + \ldots + \alpha_m \cdot (y_m - E[y_m]).$$

However, when we square both sides of this formula, we conclude that

$$(y - E[y])^2 = \sum_{j=1}^{m} \sum_{k=1}^{m} \alpha_j \cdot \alpha_k \cdot (y_j - E[y_j]) \cdot (y_k - E[y_k])),$$

i.e., that $(y - E[y])^2$ is a linear combination of the products

$$(y_j - E[y_j]) \cdot (y_k - E[y_k]).$$

Thus, the variance $V[y] = E[(y - E[y])^2]$ is equal to the linear combination of the expected values of the product terms:

$$V[y] = \sum_{j=1}^{m} \sum_{k=1}^{m} \alpha_j \cdot \alpha_k \cdot + C_{jk}, \qquad (15.7)$$

where

$$C_{jk} = E[(y_j - E[y_j]) \cdot (y_k \cdot E[y_k])]. \qquad (15.8)$$

The $m \times m$ matrix formed by the values C_{jk} are called a *covariance matrix*. In particular, for $j = k$, we have $C_{jj} = V[y_j]$.

Thus, to be able to determine the mean and standard deviation of arbitrary linear combinations of the variables $y_1, \ldots, y_m$, we must know:

- the means $E[y_j]$ of all m variables, and
- the covariance matrix $C_{jk} = E[(y_j - E[y_j]) \cdot (y_k \cdot E[y_k])]$, $1 \leq j, k \leq m$.

Back to the Internet example: stochastic processes. So far, we have considered the situation in which we are interested in *finitely many* different output quantities $y_1, \ldots, y_m$. In this case, instead of describing all possible probabilities, we restricted ourselves to mean and covariance.

In our Internet example, we were actually interested in predicting the time delay $\tau(t)$ for all possible real values t; hence, here we have *infinitely many* different random variables $\tau(t)$ that correspond to infinitely many different possible moments of time. A situation in which we have a random value $y(t)$ for each moment of time t (i.e., in which we have a random *function*) is called a *stochastic process*. For a stochastic process $y(t)$, even if we restrict ourselves to mean and covariance, we still need infinitely many numbers. Namely:

■   for every real number t, we must know the mean values $M(t) = E[y(t)]$;
    and

■   for every two real numbers t and t', we must know the covariance function
    $C(t, t') = E[(y(t) - E[y(t)]) \cdot (y(t') - E[y(t')])]$.

From the purely *statistical* viewpoint, we therefore need to describe these two
function $E(t)$ and $C(t, t')$. However, as we will show, from the *computational*
viewpoint, it is often more reasonable to compute slightly different functions
(from which $E(t)$ and $C(t, t')$ can be easily reconstructed):

■   In many cases, we are only interested in the values of $y(t)$ for different t,
    and *not* in their linear combinations. To cover these cases, it is sufficient
    to know the mean values $E[y(t)]$ and the variance

$$V(t) = C(t, t) = E[(y(t) - E[y(t)])^2].$$

■   Probably the most frequent case when we *need* a linear combination of the
    values of the output $y(t)$ at different moments of time is when we estimate
    the *change* $\Delta y(t, t') = y(t) - y(t')$ that occured between the moments of
    time t and t'. To characterize this change, we must know the mean and
    the variance of this change.

    –   The mean $E[\Delta y(t, t')]$ is equal to the difference of the means $E[y(t)] -$
        $E[y(t')] = M(t) - M(t')$, so, to find this mean, it is sufficient to know
        the function $M(t)$.

    –   The variance

$$V[\Delta y(t, t')] = E[(\Delta y(t, t') - E[\Delta y(t, t')])^2] =$$

$$E[(y(t) - y(t') - (E[y(t)] - E[y(t')]))^2].$$

    needs to be determined. This variance of "delta y" is called *delta
    variance*. We will denote it by $D(t, t')$.

From the definition of $D(t, t')$ and from the linearity of the mean, we conclude
that

$$D(t,t') = E[((y(t) - E[y(t)]) - (y(t') - E[y(t')]))^2] =$$
$$E[(y(t) - E[y(t)])^2] + E[(y(t') - E[y(t')])^2] -$$
$$2 \cdot E[(y(t) - E[y(t)]) \cdot (y(t') - E[y(t')])] =$$
$$V(t) + V(t') - 2 \cdot C(t,t').$$

Hence, if we know the variance $V(t)$ and delta-variance $D(t,t')$, we can reconstruct $C(t,t')$ as $C(t,t') = (1/2) \cdot (V(t) + V(t') - D(t,t'))$.

The mean values $M(t) = E[y(t)]$ form a *trend*, which is exactly what we have been extrapolating in the previous lessons. So, we know what models we can use, and how to do the extrapolation.

The new problem is determining the *variance* $V(t)$ and the *delta variance* $D(t,t')$ that describe random deviations from this (smooth and predictable) trend $M(t)$. The problem of finding these functions is a typical extrapolation problem:

- From the experimental data, we may be able to find the estimates for $V(t)$ and $D(t,t')$ for *some* values t and t'.

- Based on these estimates, we must find the values $V(t)$ and $D(t,t')$ for *all* real numbers t and t'.

If we have *a large amount* of experimental data, that cover many densely located values t and t', then, however we extrapolate, we will probably get a pretty good result. However, in many practical situations, we only have *a small amount* of data. In such cases, the quality of the extrapolation result drastically depends on the choice of the extrapolation method. Therefore, a question arises: *how to choose the best extrapolation technique?*

In this lesson, we will only consider the simplest case of this problem: when the process is *stationary*. Before we start solving this problem, let us explain what "stationary" means.

Stationary stochastic processes. In many practical situations, the properties of a stochastic process do not change if we start measuring time at a different starting point, i.e., if we use a new variable $\tilde{t} = t + t_0$ instead of t. (This fact is definitely not surprising to us, because we have used the corresponding shift-invariance in many computer science problems.) In physical terms, this means that the properties of a stochastic process do not change with time, i.e., that this process is *stationary*.

For background on stochastic processes, see, e.g., D. Bosq and H. T. Nguyen, *A course in stochastic processes: stochastic models and statistical inference* (Kluwer Academic, Dordrecht, 1996).

All characteristics of a stochastic process, in particular, the expressions for the mean, variance, and delta variance, should not change if we simply change the starting point for time, i.e., if we replace t by $t+t_0$. In mathematical terms, this means that $M(t+t_0) = M(t)$, $V(t+t_0) = V(t)$, and $D(t,t') = D(t+t_0,t'+t_0)$ for all real numbers t, t', and t_0. These equalities enables us to simplify the description of a covariance function: namely,

- if we take $t_0 = -t$, we conclude that $M(t) = M(0)$ and $V(t) = V(0)$, i.e., that both the mean and the variance do not depend on t at all;

- if we we take $t_0 = -t'$, we conclude that $D(t,t') = D(t-t',0)$.

Thus, in order to describe a function $D(t,t')$ of *two* variables t and t', it is sufficient to know the function of a single variable $D(t,0)$. In the following text, we will denote $D(t,0)$ simply by $D(t)$, and call it the *delta variance function* of a stationary stochastic process.

The problem is: *how to extrapolate $D(t)$?*

Exercises

15.1 A random variable y takes only two possible values: 0 and 1. Let $y^{(1)} = 0$, $y^{(2)} = 1$, and $y^{(3)} = 0$. Compute the empirical average using both formulas (15.2) and (15.4). Compute the standard deviation $\sigma[y]$.

15.2 Let y_1 and y_2 be two random variables with mean values $E[y_1] = E[y_2] = 1$ and covariance matrix $C_{11} = 0.5$, $C_{12} = C_{21} = 0.1$, and $C_{22} = 0.6$. Compute the mean and the standard deviation of $y = y_1 - 2 \cdot y_2$.

15.2. Which stochastic processes are the most adequate? Optimization in case of high uncertainty

Towards the mathematical formulation of the problem. To extrapolate, we must choose a family of possible functions $D(t)$, and then choose a functions from this family based on the experimental data. As in the previous lessons, we will use the possibility of choosing different units and different starting points to describe the possible families and to choose the optimal family.

Our ultimate goal is to describe the dependence of y on t. Thus, we have two physical quantities: y and time t, for each of which we may:

- choose different units and,

- (possibly) choose different starting points.

Let us first show that we cannot use the change in starting points:

- We have already restricted ourselves to stationary processes, and thus, we have already taken into consideration the possibility of changing a starting point for time.

- If we change a starting point for y, i.e., if we take $\tilde{y} = y + y_0$ instead of the original values y, then the mean will also increase by y_0: $E[\tilde{y}] = E[y+y_0] = E[y]+E[y_0] = E[y]+y_0$. Hence, the difference $y(t)-E[y(t)]$ will not change: $\tilde{y}(t) - E[\tilde{y}(t)] = (y(t) + y_0) - (E[y(t)] + y_0) = y(t) - E[y(t)]$. Hence, the delta variance function that is defined in terms of these differences, will not change: $\tilde{D}(t) = D(t)$. In other words, changing the starting point for y does not affect the delta variance function $D(t)$ at all.

Thus, the only symmetries left are the changes in the measuring unit.

- First, we can choose a different unit for measuring y. If we replace the original unit by a new unit that is λ times smaller, then every old numerical value y is replaced by a new numerical values $\tilde{y} = \lambda \cdot y$. In the new units, $\Delta y(t,0) = y(t) - y(0)$ is changed into $\Delta \tilde{y}(t,0) = \tilde{y}(t) - \tilde{y}(0) = \lambda \cdot \Delta y(t,0)$, which is exactly λ times larger than the old value of $\Delta y(t,0)$.

If we multiply an arbitrary variable z by a number λ, then its variance is multiplied by λ^2: Indeed, since $E[\tilde{z}] = E[\lambda \cdot z] = \lambda \cdot E[z]$, we have $\tilde{z} - E[\tilde{z}] = \lambda \cdot z - \lambda \cdot E[z] = \lambda \cdot (z - E[z])$, and therefore,

$$V[\tilde{z}] = V[\lambda \cdot z] = E[(\tilde{z} - E[\tilde{z}])^2] = E[(\lambda \cdot (z - E[z]))^2] =$$

$$\lambda^2 \cdot E[(z - E[z])^2] = \lambda^2 \cdot V[z].$$

Therefore, when we choose a new unit that is λ times smaller, the delta variance function $D(t) = V[\Delta y(t, 0)]$ gets multiplied by λ^2:

$$\tilde{D}(t) = \lambda^2 \cdot D(t).$$

Thus, if $D(t)$ is a reasonable delta variance function, then, for every λ, the function $\tilde{D}(t) = \lambda^2 \cdot D(t)$ which describes the same process in different units, must also be a reasonable delta variance function. Since λ^2 can be an arbitrary positive real number, we this conclude that the set of all possible reasonable delta variance functions must contain, with every function $D(t)$, the entire family $\{C_1 \cdot D(t)\}$ of functions that correspond to different values $C_1 > 0$. In this lesson, we will consider the simplest case, if which the families are the "smallest" possible (i.e., have the fewest possible parameters). Therefore, we will consider 1-parametric families of the type $\{C_1 \cdot D(t)\}$, where $D(t)$ is a given (smooth) function, and C_1 runs over arbitrary real numbers.

■ Changing the unit for measuring time hould not change the relative quality of different families. In mathematical terms, changing the unit for measuring time means *scaling*, i.e., replacing t by $\tilde{t} = C \cdot t$. Thus, the optimality criterion must be *scale-invariant*.

So, the problem of choosing the optimal delta variance function $D(t)$ is naturally re-formulated as the following mathematical problem:

On the set of all families of the type $\{C_1 \cdot D(t)\}$, a scale-invariant final optimality criterion is given. We must find the family that is optimal with respect to this criterion.

Choosing the best delta covariance function: solution. We have already solved this mathematical problem in Lesson 8. According to Theorem 8.1, every element of the optimal family is of the type $D(t) = C_1 \cdot t^\alpha$ for some real numbers C_1 and α. Thus, if we have no *a priori* information about the process,

The optimal extrapolation method is to use the power-law delta covariance function $D(t) = C_1 \cdot t^\alpha$.

Such processes have indeed been successfully used; the two most used examples are:

- For $\alpha = 1$, we get the *Brownian motion* that describes the random movement of tiny particles in a liquid or a gas, and the *Wiener process* that describes several types of noise in electrical engineering.

- For $\alpha = 0$, we get a *white noise*, a process in which the noises at different moments of time are not related at all: $C(t, t') = \sqrt{V(t) \cdot V(t')}$.

Most examples of real-life processes are indeed well described by this formula with $\alpha \in [1, 2)$. Many examples of this type are given by Benoit B. Mandelbrot in his monograph *The fractal geometry of Nature* (Freeman, San Francisco, 1982).

How to actually extrapolate? The idea. We already know that the optimal approximation for the delta variance function $D(t)$ is $C_1 \cdot t^\alpha$. The natural next question is: how to determine C_1 and α from the experimental data?

In this lesson, we will only describe the *main idea* of determining C_1 and α. We will illustrate this main idea on a frequent case when we know the values $y(t_1), \ldots, y(t_n)$ in equally distributed moments of time t_1, $t_2 = t_1 + \Delta t$, $t_3 = t_1 + 2 \cdot \Delta t$, $\ldots$, $t_i = t_1 + (i - 1) \cdot \Delta t$, $\ldots$, $t_n = t_1 + (n - 1) \cdot \Delta t$. In this case, we can do the following:

- Since we are assuming that the process is stationary, the mean $M(t)$ is the same for all t. Thus, to estimate this mean, we can simply take an average of all observed values:

$$\bar{M} = \frac{y(t_1) + \ldots + y(t_n)}{n}. \tag{15.9}$$

- The variance $V(t)$ also does not depend on t, so, we can estimate it as an average of the squares of the differences $(y(t_i) - E[y(t_i)])^2$. We cannot do this directly, because we do not know the exact values of $E[y(t_i)] = M(t_i)$, but we do know an estimate $\bar{M}$ for $M(t_i)$. Therefore, we can estimate V as the average of the squares $(y(t_i) - \bar{M})^2$:

$$\bar{V} = \frac{(y(t_1) - \bar{M})^2 + \ldots + (y(t_n) - \bar{M})^2}{n}. \qquad (15.10)$$

■ By definition, $D(t)$ is the variance of $\Delta y(t', t' + t) = (y(t') - y(t + t')) - (E[y(t') - E[y(t+t')]])$. Since $y(t)$ is a stationary process, we have $E[y(t')] = E[y(t + t')]$ and therefore, $\Delta y(t', t + t') = y(t') - y(t + t')$. So, $D(t)$ is the mean of the squares $(y(t')) - y(t + t'))^2$. We can, therefore, estimate the values of $D(t)$ by taking an average of the squares $(y(t') - y(t''))^2$ for all pairs (t', t'') for which we know both values $y(t')$ and $y(t'')$ and for which $t'' - t' = t$. We only know the values $y(t')$ and $y(t'')$ for t' and t'' of the type $t' = t_1 + i' \cdot \Delta t$ and $t'' = t_1 + i'' \cdot \Delta t$ for some integers i' and i''. For such t' and t'', $t'' - t' = (i'' - i') \cdot \Delta t$. Hence, we can only estimate $D(t)$ for $t = i \cdot \Delta t$. For each such t, corresponding pairs (t', t'') range from (t_1, t_{i+1}) to (t_{n-i}, t_n). Thus, we have $n - i$ such pairs, and the estimating average for $D(i \cdot \Delta t)$ is

$$\bar{D}_i = \frac{(y(t_1) - y(t_{i+1}))^2 + (y(t_2) - y(t_{i+2}))^2 + \ldots + (y(t_{n-1}) - y(t_n))^2}{n - i}.$$
$$\qquad (15.11)$$

■ After the previous estimate, we have the values $\bar{D}_i$ that approximate $D(i \cdot \Delta t) = C_1 \cdot (i \cdot \Delta t)^\alpha$:

$$\bar{D}_i \approx C_1 \cdot (i \cdot \Delta t)^\alpha. \qquad (15.12)$$

To determine C_1 and α, we can use, e.g., the *least squares* method. We have already used this method in Lesson 3. The only problem here is that the dependence on α is highly non-linear and therefore, the least squares method will lead to a difficult-to-solve system of non-linear equations. To simplify the situation, we can turn to *logarithms*. Taking logarithms of both sides of (15.12), we get an equivalent *linear* system

$$\ln(\bar{D}_i) \approx \alpha \cdot \ln(i \cdot \Delta t) + c, \qquad (15.13)$$

where we denoted $c = \ln(C_1)$. We can solve this system using the least squares method, and then find C_1 as $\exp(c)$.

So, we arrive at the following algorithm.

How to actually extrapolate? The algorithm. Suppose that we have the values $y(t_1), \ldots, y(t_n)$ of a process $y(t)$ for equally spaced moments of time $t_i = t_1 + (i - 1) \cdot \Delta t$. Then, we do the following:

- to estimate the mean, we use the formula (15.9);

- to estimate the variance, we use the formula (15.10);

- to estimate the delta variance function, we:

  - compute the values $\bar{D}_i$ by using the formula (15.11),

  - apply the least squares method to the system (15.13), and find α and c;

  - compute $C_1 = \exp(c)$;

then, we estimate $D(t)$ as $C_1 \cdot t^\alpha$.

Exercise

15.3 Apply the above algorithm to the following simplified version of the Internet time delay data: $t_1 = 0$, $\Delta t = 0.1$ sec, $y(t_1) = 3.0$, $y(t_2) = 1.6$, $y(t_3) = 2.7$, $y(t_4) = 5.0$, $y(t_5) = 0.9$.

15.3. Fractals: a geometric interpretation of those stochastic processes which are the most adequate for the situations of high uncertainty

A theoretical description of a stochastic process is not sufficient: we need computer simulations. From the *theoretical* viewpoint, we are almost done: we have described a class of stochastic processes that describes the process $y(t)$, and we learned how to determine the parameters of the process from the experimental data. As a result, we may get, e.g., a description of the stochastic process that characterizes the fluctuations of time delays on the Internet.

However, from the *practical* viewpoint, we often want more than simply probabilities. For example, for Internet, sometimes, we are interested in knowing the probabilities of different delays, but more frequently, we want to develop tools that would decrease these delays (or, at least, somehow decrease their influence). In general, the information about a system is interesting not so much by itself, but as a means to develop methods for *controlling* this system.

In some cases, there exist *analytical* methods that transform the probabilities of different events into the explicit control formulas. Alas, such methods are only known for the simplest systems. For more complicated situations, we have to rely on *simulations*; namely:

■ we design a *random function* generator that simulates a function with the given probabilities of different values;

■ we run this generator several times, and get simulated functions; and then

■ we test different control algorithms on these simulations, and select an algorithm whose performance on these simulation is the best.

What is "simulating a function"? What does it mean to "simulate a random function"?

■ A real number is one of the standard computer-represented data types. Thus, when we simulate a random *number*, we simply generate one real number, and we design a random number generator in such a way that it generates each possible real number with a desired probability. (Of course, computers have limited precision, so, we have to truncate the real number to whatever decimal digits a given computer can represent, but this precision is usually, pretty good, and therefore, for most practical problems, this truncation is un-noticeable.)

■ On the other hand, a *function* cannot be easily represented in a computer. To be precise, most programming languages have a special data type "function", but it means an *algorithm* that transforms real numbers into real numbers, and in our simulations, we want to represent a *random*, non-algorithmic function.

　　– From the *mathematical* viewpoint, a function $y(t)$ is defined as a set of pairs $(t, y(t))$ that correspond to all possible real numbers t.

　　– In *practice*, we cannot represent infinitely many pairs, so, we must represent a function by storing a *finite* number of pairs $(t_i, y(t_i))$ for different values $t_1 < \ldots < t_i < \ldots < t_n$. (Then, to estimate $y(t)$ for the values of t that are different from all these t_i, we can find the interval $[t_i, t_{i+1}]$ to which t belongs and use, e.g., linear extrapolation.)

Computational complexity of simulating a random function. To simulate a random function, we need to store several pairs $(t_i, y(t_i))$. The more pairs we need to store, the more computer memory it takes, and the more computer time it takes to process all these values. So, we would like to store as few values as possible.

If we want to reconstruct the value $y(t)$, and we store only a few values $y(t_i)$, then we will have to interpolate the value $y(t)$ based on the values of the function $y(t)$ in points t_i and t_{i+1} that are, in general, rather distant from t. As a result, this interpolation may be reasonable inaccurate. The more points we store, the closer these points t_i to each other and therefore, the closer to each desired value t. The more accuracy we want, the more points we have to use. It is, therefore, natural to describe the "complexity" of a function $y(t)$ by the number of points $(t_i, y(t_i))$ that we need to store for every possible accuracy $\varepsilon > 0$. In order to formalize this idea, let us reformulate it in geometric terms.

Complexity reformulated in geometric terms: epsilon-entropy. In geometric terms, a function $y(t)$ is represented by its *graph*, i.e., by the set F of points $X(t) = (t, y(t))$ whose x-coordinate is the time t, and whose y-coordinate is the value $y(t)$ of the represented function at this particular point t.

We know the desired representation accuracy $\varepsilon > 0$. Ideally, we want to represent this graph F by its finite subset $\{X_1, \ldots, X_n\}$ in such a way that every other point X from this graph is ε-close to one of these points $X_1, \ldots, X_n$. Here, we say that two points $X = (t, y)$ and $X' = (t', y')$ are *ε-close* if both their coordinates are ε-close, i.e., if $|t - t'| \leq \varepsilon$ and $|y - y'| \leq \varepsilon$. These two inequalities can be combined into a single inequality $\rho(X, X') \leq \varepsilon$, where we denoted $\rho((t, y), (t', y')) = \max(|t - t'|, |y - y'|)$.

For a function of several variables, we can use a slightly more complicated expression $\rho(X, X')$. The resulting notion has a special name in approximation theory: it is called an *ε-net*. The precise definition is as follows:

Definition 15.1. *Let a set F be given, and let a function $\rho(X, X')$ be defined that transforms every two points $X, X' \in F$ into a non-negative real number so that $\rho(X, X') = 0$ if and only if $X = X'$, and $\rho(X, X') = \rho(X', X)$. Let $\varepsilon > 0$ be a real number.*

- *We say that a finite set $\{X_1, \ldots, X_n\}$ is an $\varepsilon-net$ for F iff for every $X \in F$, there exists an i for which $\rho(X, X_i) \leq \varepsilon$.*

- *The smallest possible number of elements in an ε-net is denoted by $N_\varepsilon(F)$.*

- *The logarithm $H_\varepsilon(F) = \ln(N_\varepsilon(F))$ is called the ε-entropy of the set F.*

In these terms, the question is: to estimate $N_\varepsilon(F)$ (or, equivalently, the ε-entropy) of the graph F of a function $y(t)$ that corresponds to a given stochastic process with the power law delta variance $D(t) = C_1 \cdot t^\alpha$.

ε-entropy of the simplest possible function graph. Before we start answering this question, let us first describe how many points we need to represent the graph of the simplest possible function: a constant $y(t) = 0$. For a constant function, the values y are always equal, so the main problem is to choose the values $t_1, \ldots, t_n$ in such a way that every other point t is ε-close to one of these points. On every interval $[0, T]$, how can we choose the smallest possible number of points with this property?

■ Let $t_1, \ldots, t_n$ be the desired ε-net. This means that each point t from the interval $[0, T]$ is ε-close to one of the points t_i. The fact that t and t_i are ε-close means that $t \in [t_i - \varepsilon, t_i + \varepsilon]$. Hence, every point from the interval $[0, T]$ belongs to one of the intervals $[t_i - \varepsilon, t_i + \varepsilon]$. So, the interval $[0, T]$ is *covered* by these n intervals, and therefore, the total width T of the interval $[0, T]$ does not exceed the sum of the widths of these small intervals. Each of the small intervals is of width $2 \cdot \varepsilon$, and there are n of them. So, we can conclude that $T \leq 2n \cdot \varepsilon$, and therefore, that $n \geq T/(2 \cdot \varepsilon)$. In other words, an arbitrary ε-net cannot have fewer than $T/(2 \cdot \varepsilon)$ points. In particular, the ε-net with the smallest possible number of elements must also have not fewer than $T/(2 \cdot \varepsilon)$ points, i.e., $N_\varepsilon(F) \geq T/(2 \cdot \varepsilon)$.

■ Let us show that this *lower bound* for $N_\varepsilon(F)$ is at the same time almost equal to the upper bound. Indeed, let us place the points t_i so as to minimize their total number.

  – The first point t_1 must be ε-close to $t = 0$. The further away we place this first point, the more values will be covered by it, so it is desirable to place this point t_1 at the farthest possible location that is still ε-close to 0, i.e., to take $t_1 = \varepsilon$. This point covers all t from 0 to $2 \cdot \varepsilon$.

  – Points that are not covered by t_1 start with $2 \cdot \varepsilon$. We want these points to be covered by the point t_2. For that, we want this new point t_2 to be ε-close to $2 \cdot \varepsilon$. Again, to minimize the total number of points, we will choose t_2 as the farthest possible point that is still ε-close to $2 \cdot \varepsilon$, i.e., as $t_2 = 2 \cdot \varepsilon + \varepsilon = 3 \cdot \varepsilon$.

Similarly, we can take $t_3 = 5 \cdot \varepsilon, \ldots, t_i = (2i - 1) \cdot \varepsilon$, and we select n as the index of the first point t_n that covers the right endpoint T of the interval

$[0, T]$, i.e., as the first integer n for which $T \leq t_n + \varepsilon = 2n \cdot \varepsilon$. In other words, we can take, as n, the first integer that is greater than $T/(2 \cdot \varepsilon)$.

Thus, we can conclude that asymptotically,

$$N_\varepsilon([0, T]) \sim \frac{0.5 \cdot T}{\varepsilon}, \tag{15.14}$$

where $f \sim g$ means that $f/g \to 1$ as $\varepsilon \to 0$.

ε-entropy of a random process with a power law delta variance: an idea. In this lesson, we will give some reasonable arguments in favor of the estimate. (We will not prove that this estimate is actually correct, but the reader may be assured that such a proof exists.)

We know the delta variance of a process, and we must reformulate this knowledge in terms of ε-entropy. The delta variance $D(t)$ is an "average" square deviation of the difference $y(t') - y(t + t')$ from its average value. Since we are dealing with stationary processes, for which $E[y(t)] = M(t) = $ const, we can conclude that the average value $E[y(t') - y(t + t')]$ of the difference $y(t') - y(t + t')$ is equal to $M(t') - M(t + t') = 0$. Hence, the delta variance describes the average square of $y(t') - y(t + t')$. In other words, on average, $(y(t') - y(t+t'))^2 \approx D(t) = C_1 \cdot t^\alpha$, and so, $|y(t') - y(t+t')| \approx \sqrt{D(t)} = c \cdot t^{\alpha/2}$, where we denoted $c = \sqrt{C_1}$.

We must choose the values $t_1 < \ldots < t_n$ in such a way that for every point $X = (t, y(t))$ from the graph F, there exists a point $X_i = (t_i, y(t_i))$ for which the "distance" $\rho(X, X_i)$ between the points X and X_i is $\leq \varepsilon$. By definition of the "distance" $\rho(X, X')$, this means that

$$|t - t_i| \leq \varepsilon \tag{15.15}$$

and that $|y(t) - y(t_i)| \leq \varepsilon$. Since we know that $|y(t') - y(t + t')| \approx c \cdot t^{\alpha/2}$, the second inequality can be reformulated as $c \cdot |t - t_i|^{\alpha/2} \leq \varepsilon$, i.e., as $|t - t_i|^{\alpha/2} \leq \varepsilon/c$ and

$$|t - t_i| \leq C_2 \cdot \varepsilon^{2/\alpha}, \tag{15.16}$$

where we denoted $C_2 = c^{-(2/\alpha)}$.

We have already mentioned that for real-life processes, the value α is usually from the interval $[1, 2)$. In this case, $2/\alpha > 1$. Hence, for small ε, we have $\varepsilon^{2/\alpha} \ll \varepsilon$, and hence, the condition (15.16) follows from the condition (15.15). Thus, when we look for the ε-net with the smallest number of elements, we must only consider the inequality (15.16). In other words, we must look for an arrangement in which for every point t, there exists i for which $|t - t_i| \leq \delta$, where we denoted $\delta = C_2 \cdot \varepsilon^{2/\alpha}$. We already know that for that, we need $\sim (0.5 \cdot T)/\delta$ points. Thus,

$$N_\varepsilon(F) \sim \frac{C_3}{\varepsilon^{2/\alpha}}, \qquad (15.17)$$

where we denoted $C_3 = (0.5 \cdot T)/C_2$.

Geometric interpretation of this formula: the relation between epsilon-entropy and dimension. In this section, we consider the problem of estimating the computational complexity of the random process simulation.

We have started with a *geometric* reformulation of this problem, and we ended up with a purely *analytical* formula (15.17). It is desirable to reformulate this solution in *geometric* terms.

In order to give a geometric interpretation to the expression (15.17) for $N_\varepsilon(F)$, let us recall the expressions for $N_\varepsilon(F)$ for simple geometric sets F in a 3-dimensional space with Euclidean metric ρ:

■   For a *0-dimensional* set F, e.g., for a set F consisting of a single *point X*, this very point X is its own ε-net. If the set F has several points, then, for small enough ε (to be more precise, for ε smaller than the smallest of the distances between these points), we need all these points in an ε-net. Therefore, for such sets F, asymptotically, $N_\varepsilon(F) = \text{const}$, i.e.,

$$N_\varepsilon(F) \sim \frac{C}{\varepsilon^0} \qquad (15.18)$$

for some constant C.

■   We already know that for the *1-dimensional* interval $F = [0, T]$, the function $N_\varepsilon(F)$ has an asymptotic

$$N_\varepsilon(F) \sim \frac{C}{\varepsilon^1} \qquad (15.19)$$

for some constant $C > 0$. One can show that a similar asymptotic formula holds for an arbitrary smooth curve in a 3-dimensional space (for such curves, C is equal to one half of the curve's length).

■   Let us now consider *2-dimensional* sets F. The simplest 2-dimensional set is a *square F*. For squares, we can repeat the same argument that we used for intervals:

- On one hand, every point X_i serves as a close to all the points from a *circle* of radius ε. Thus, if $\{X_1, \ldots, X_n\}$ is an ε-net, the entire set F is covered by n circles of radius ε centered in the points X_i. Hence, the area $A(F)$ of the set F cannot exceed the sum of the areas of these circles, i.e., $A(F) \leq n \cdot \pi \cdot \varepsilon^2$. Therefore, for every ε-net, we have $n \geq c_1/\varepsilon^2$, where we denoted $c_1 = A(F)/\pi$. In particular, we can conclude that the number of points in the smallest ε-net cannot be smaller than this number, i.e., $N_\varepsilon(F) \geq c_1/\varepsilon^2$.

- On the other hand, if we place the points $X_1, \ldots, X_n$ on a rectangular grid with a step $\varepsilon/\sqrt{2}$, we get an ε-net consisting of $\approx A(F)/(\varepsilon/\sqrt{2})^2 = (2 \cdot A(F))/\varepsilon^2$ elements. Hence, the smallest possible number of elements in an ε-net cannot be larger than this number. Hence, $N_\varepsilon \leq c_2/\varepsilon^2$, where we denoted $c_2 = 2 \cdot A(F)$.

So, we have inequalities $c_2/\varepsilon^2 \leq N_\varepsilon(F) \leq c_1/\varepsilon^2$. One can show that asymptotically,

$$N_\varepsilon(F) \sim \frac{C}{\varepsilon^2} \qquad (15.20)$$

for some constant C. It is also possible to show that a similar asymptotic formula holds not only for a square, but also for an arbitrary smooth surface (i.e., for an arbitrary smooth 2-dimensional set F).

■ Similarly, if F is a *3-dimensional* set, e.g., an interior of a body, then we can get an estimate $N_\varepsilon(F) \leq c_1/\varepsilon^3$ from the fact that F is covered by balls of radius ε with centers in X_i, an estimate $c_2/\varepsilon^3 \leq N_\varepsilon(F)$ from the possibility of using a grid, and the asymptotics

$$N_\varepsilon(F) \sim \frac{C}{\varepsilon^3}. \qquad (15.21)$$

By comparing the formulas (15.19), (15.20), and (15.21), we conclude that for a regular (smooth) geometric set of dimension d, we have

$$N_\varepsilon(F) \sim \frac{C}{\varepsilon^d} \qquad (15.22)$$

for some constant C. For smooth sets, this formula determines the *dimension* of the set in terms of its *computational complexity* $N_\varepsilon(F)$. This very formula has been used by Hausdorff in the beginning of this century to *generalize* the notion of dimension to non-smooth sets (such as the graphs of the random functions):

Definition 15.2. *If a set F satisfies the formula (15.21) for some real number $d > 0$, then this value d is called the dimension of the set F.*

From the formula (15.17), we can conclude that the graph F of a random function $y(t)$, that corresponds to the stochastic process with $D(t) = C_1 \cdot \varepsilon^\alpha$ ($1 \leq \alpha < 2$), is a set of dimension $d = 2/\alpha$. Except for $\alpha = 1$, this value $2/\alpha$ is *not* an integer, it is a fraction. Sets of *fractional* (non-integer) dimension are called *fractals*.

For $\alpha = 1$, we get dimension $d = 2/\alpha = 2$. This dimension is an integer, but it is not the integer that we would expect from a graph of function, i.e., from a *curve*. The integer dimension that we "expect" from a set (1 for a curve, 2 for a surface, etc.) is called a *topological dimension* as opposed to the above-described *metric* dimension (defined in terms of metric $\rho(X, X')$). Sets for which the metric dimension is an integer but an integer different from the (precisely defined) topological dimension are as irregular as set of fractional dimension, and they are, therefore, also called *fractals*. Thus, a more general definition of a fractal is:

> A set F is called a *fractal* if its metric dimension is different from its topological dimension.

Fractals are actively used in different areas. In particular, in computer science, in addition to computer communications, fractals are used:

- in *computer graphics* to simulate terrain, coastal lines, and other elements of geographic environment (actually, in geographic simulations, we need a fractal *field*, i.e., a random function $y(x_1, x_2)$ of *two* spatial variables x_1 and x_2);

- in *image processing*, to describe the noise; see, e.g., C. V. Stewart, B. Moghaddam, K. J. Hintz, and L. M. Novak, "Fractional Brownian motion models for synthetic aperture radar imagery scene segmentation" (*Proceedings of the IEEE*, 1993, Vol. 81, No. 10, pp. 1511–1521); etc.

Comment. Sometimes, we need more complicated stochastic processes and more complicated sets F, for which the asymptotic formula for $N_\varepsilon(F)$ is also more complicated. There exist generalizations of Definition 15.2 that cover such cases.

Random walk: a simple example of a fractal. To give a reader some hands-on experience with fractals, we will describe the simplest possible fractal: a *random walk*. It is a description of a path of a drunkard along the street. He starts at moment 0, at a point $x(0) = 0$. If at a moment t, he is at a location $x(t)$, then, independently on his previous path, with probability 0.5, he makes a step to the right, and with probability 0.5, he makes a step to the left. In other words, $x(t+1) = x(t) + \Delta x(t)$, where $\Delta x(t) = 1$ with probability 0.5 and $\Delta x(t) = -1$ with probability 0.5. What is the mean and delta variance of this process?

First of all, let us describe $x(t)$ in terms of $\Delta x(t)$. We have $x(0) = 0$, $x(1) = x(0) + \Delta x(0) = \Delta x(0)$, $x(2) = x(1) + \Delta x(1) = \Delta x(0) + \Delta x(1)$, and, in general, $x(t) = \Delta x(0) + \Delta x(1) + \ldots + \Delta x(t-1)$. For each i, the mean value of $\Delta x(i)$ is equal to $0.5 \cdot 1 + 0.5 \cdot (-1) = 0$. Therefore, the mean value $M(t) = E[x(t)]$ of the sum $x(t)$ of t such terms is equal to sum of the mean values $E[\Delta x(i)]$, i.e., to 0.

Let us now compute the delta variance $D(t)$. Since the mean of the difference $x(t + t') - x(t')$ is equal to 0, the delta variance is equal to the mean of the square of this difference: $D(t) = E[(x(t + t') - x(t'))^2]$. This difference can be also represented as a sum of the terms $\Delta x(i)$:

$$x(t + t') - x(t') = \Delta x(t') + \ldots + \Delta x(t' + t - 1).$$

If we take the square of both sides of this equation, we conclude that

$$(x(t+t') - x(t'))^2 = (\Delta x(t'))^2 + \ldots + (\Delta x(t'+t-1))^2 + 2 \cdot \Delta x(t') \cdot \Delta x(t'+1) + \ldots$$

Since $\Delta x(i)$ is equal to $+1$ or to -1, the square of $\Delta x(i)$ is always equal to 1. Hence, the sum of t such squares is equal to t, and

$$(x(t + t') - x(t'))^2 = t + 2 \cdot \Delta x(t') \cdot \Delta x(t' + 1) + \ldots$$

The mean $D(t)$ of the square is, therefore, equal to the sum of t and of the means $2 \cdot E[\Delta x(i) \cdot \Delta x(j)]$ for $i \neq j$. We assumed that the values $\Delta x(i)$, that correspond to different i, are independent of each other. Therefore, $E[\Delta x(i) \cdot \Delta x(j)] = E[\Delta x(i)] \cdot E[\Delta x(j)] = 0$, and $D(t) = t$.

Thus, a random walk is indeed a discrete analogue of the fractal: namely, of the fractal that corresponds to the Brownian motion ($\alpha = 1$).

For a background on fractals and on fractal simulations, see, e.g., H. M. Hastings and G. Sugihara, *Fractals: A user's guide for the natural sciences* (Oxford University Press, Oxford, New York, 1993).

Exercises

15.4 Toss a coin several times and simulate a random walk; plot how the position x changes with time.

15.5 Program the random walk and plot the result. What is the metric dimension of the resulting graph?

A

SIMULATED ANNEALING REVISITED

Simulated annealing: brief reminder. In Lesson 7, we described the *simulated annealing* method for solving discrete optimization problems $J(x) \to$ max. This is an iterative method, in which:

- We *start* the computation process with a (randomly chosen) point x.

- At every iteration, we start with a point x, and exploit all the points y in the neighborhood of x.

  - If in this neighborhood, there exists a point y for which $J(y) > J(x)$, we take the neighbor y with the largest possible value of $J(y)$ as the next choice of x.

  - If for all neighboring points y, we have $J(x) \geq J(y)$ (i.e., if x is the *local maximum*), then we move to one of the neighboring points y with the probability $p(y)$ that is proportional to $f(J(y))$ for some function $f(z)$. (We consider each point x to be its own neighbor, so, it is possible that we will choose $y = x$ and thus, stay in x.)

- We *stop* the computation process if we get the same point x on *two* iterations in a row (we may also want to be more cautious and require, e.g., that the same point x appears on *three* iterations in a row).

In Lesson 7, we used the fact that the problem $J(x) \to$ max is equivalent to $J(x) + s \to$ max and concluded that the optimal choice of the function $f(z)$ is $f(z) = \exp(\beta \cdot z)$ for some constant $\beta > 0$.

The higher β, the greater the probability that we stay in the local maximum, and the smaller the probability that we jump out.

How can we improve simulated annealing? In Lesson 7, we assumed that we are using *the same* function $f(z)$ (i.e., the same value β) on all iterations. However, intuitively, it seems reasonable to use *different* functions $f(z)$ (i.e., different values β) on different iterations. Indeed:

- *In the beginning*, when we have just started the search, the fact that we have reached a local maximum probably means exactly this, that this maximum is only local. It is highly unprobable that after the first few search steps, we have already reached a global maximum. In this situation, we would like to jump out of this local maximum with a reasonably high probability. In other words, during the *first* iterations, we would like to have a *small* value of β.

- *After many iterations*, it becomes quite probable that the local maximum that we are currently visiting is actually the global maximum. If we jump out of the global maximum, and then spend some time getting back, we will waste time. To avoid this time wasting, we would like to *decrease* the probability of jumping out, i.e., *increase* β.

In short, it makes sense, instead of using the *same* value β for all iterations, to *increase* the value β from iteration to iteration, i.e., to use different values $\beta_1 < \beta_2 < \ldots < \beta_k < \ldots$ on each iteration.

How can we change β? Since we are talking about an *iterative* method, in which the next value of x is determined by the value of x on the previous iteration, it is reasonable to change β in the same iterative fashion, i.e., to choose $\beta_{k+1} = g(\beta_k)$ for some function $g(z)$.

We must choose a function $g(z)$. Numerical experiments show that the success of this idea depends on the choice of the function $g(z)$. So, the question is: *which function $g(z)$ is the best?*

Optimality criterion must be scale-invariant. In Lesson 7, we have used *shift-invariance* of the optimization problem, i.e., its invariance with respect to going from $J(x)$ to $J(x) + s$. We have also mentioned in Lesson 7 that the optimal values x_{opt} do not change if we use a different unit for measuring the values of the objective function, i.e., if we replace the original objective function $J(x)$ with a *re-scaled* function $\tilde{J}(x) = \lambda \cdot J(x)$, for an arbitrary $\lambda > 0$.

Since the "new" problem $\tilde{J}(x) \to \max$ is, in essence the same optimization problem as $J(x) \to \max$, it makes sense to re-scale β accordingly to make sure that the simulated annealing algorithm follows the same sequence of values x for the "new" objective function $\widetilde{J}(x)$ and for the "old" objective function $J(x)$. To guarantee this, we must select the new value $\widetilde{\beta}$ in such a way that the new and the old probabilities of jumping to any value y will coincide. In other words, we must choose $\widetilde{\beta}$ in such a way that $\exp(\widetilde{\beta}\cdot\widetilde{J}(x)) = \exp(\widetilde{\beta}\cdot\lambda\cdot J(x)) = \exp(\beta\cdot J(x))$. It is easy to see that for this, we must choose $\widetilde{\beta} = \lambda^{-1}\cdot\beta$.

If, instead of using the same value β on all iterations, we use different values β_k, then the corresponding re-scaled values $\widetilde{\beta}_k$ are determined by a similar formula $\widetilde{\beta}_k = \lambda^{-1}\cdot\beta_k$.

How will the function $g(z)$ look in the new scale of β? Suppose that we given a value $\widetilde{\beta}_k$ that is used on k-th iteration. Then, to find the value $\widetilde{\beta}_{k+1} = \widetilde{g}(\widetilde{\beta}_k)$ to use on the next iteration, we can do the following:

- First, we express the original value of β in the old scale: $\beta_k = \lambda\cdot\widetilde{\beta}_k$.

- Second, we apply the function $g(z)$ to the value β expressed in the old scale: $\beta_{k+1} = g(\beta_k) = g(\lambda\cdot\widetilde{\beta}_k)$.

- Finally, we express the next value of β_{k+1} in the new scale:

$$\widetilde{\beta}_{k+1} = \lambda^{-1}\cdot\beta_{k+1} = \lambda^{-1}\cdot g(\lambda\cdot\widetilde{\beta}_k).$$

In other words, the transition from $\widetilde{\beta}_k$ to $\widetilde{\beta}_{k+1}$ uses the function

$$\widetilde{g}(z) = \lambda^{-1}\cdot g(\lambda\cdot z).$$

Since $g(z)$ and $\widetilde{g}(z)$ represent *the same* algorithm (but in different scales), it is natural to require that if a function $g_1(z)$ is better than the function $g_2(z)$, then the re-scaled function $\widetilde{g}_1(z) = \lambda^{-1}\cdot g_1(\lambda\cdot z)$ should be better than the re-scaled function $\widetilde{g}_2(z) = \lambda^{-1}\cdot g_2(\lambda\cdot z)$. In other words, we want the optimality criterion to be *scale-invariant*. Let us formulate this condition in precise mathematical terms.

How to choose $g(z)$: mathematical formulation of the problem and the result.

Definition A.1. *Let $\lambda > 0$ be a real number. By a λ-scaling of a function $g(z)$ we mean a function $\widetilde{g}(z) = \lambda^{-1}\cdot g(\lambda\cdot z)$. This function $\widetilde{g}(z)$ will be denoted by $S_\lambda(g)$.*

Definition A.2. *We say that an optimality criterion on the set of all possible functions $g(z)$ is scale-invariant if for every real number $\lambda > 0$, the following two conditions are true:*

- *If g_1 is better than g_2 in the sense of this criterion (i.e., if $g_1 \succ g_2$), then $S_\lambda(g_1) \succ S_\lambda(g_2)$.*

- *If g_1 is equivalent to g_2 in the sense of this criterion (i.e., if $g_1 \sim g_2$), then $S_\lambda(g_1) \sim S_\lambda(g_2)$.*

Proposition A.1. *If a function $g(z)$ is optimal in the sense of some optimality criterion that is final and scale-invariant, then $g(z) = c \cdot z$ for some real number c.*

Proof. Similarly to the proofs of Proposition 4.1 and 4.2, we can conclude that the optimal function $g(z)$ must be *scale-invariant*, i.e., that $S_\lambda(g) = g$ for all $\lambda > 0$. By definition of S_λ, this means that $g(z) = \lambda^{-1} \cdot g(\lambda \cdot z)$ for every λ and z. Multiplying both sides of this equality by λ, we conclude that $g(\lambda \cdot z) = \lambda \cdot g(z)$. In particular, for $z = 1$, we conclude that $g(\lambda) = \lambda \cdot g(1)$, i.e., $g(\lambda) = c \cdot \lambda$ for $c = g(1)$. The proposition is proven.

Comments.

- The optimal values β_k satisfy the equation $\beta_{k+1} = c \cdot \beta_k$. Thus, these values form a *geometric progression* $\beta_k = \beta_0 \cdot c^k$.

- Experiments show that the geometric progression is indeed the best way of changing β in simulated annealing.

B

SOFTWARE COST ESTIMATION

The problem. When we start a large software project, it is desirable to estimate the required amount of time T and the required development effort E (measured, e.g., in man-months). If we know the effort E, then the personnel cost can be determined by multiplying the estimated effort E (in man-months) by an average monthly salary of a software developer.

Usually, we more or less know what procedures we will need, so we can pretty well estimate the total length L of the resulting code (measured, e.g., in lines of code). The question is: *Based on this estimate L, how can we estimate T and E?*

Formulating the problem in precise (mathematical) terms. Let us first find a good model $g(L)$ for the dependence $T = g(L)$. Since time can be measured in different units, and changing to a different time unit means that all numerical values of time are multiplied by a constant $(g(L) \rightarrow C \cdot g(L))$, the optimal family F must contain a function $C \cdot g(L)$ together with each function $g(L)$. The simplest case is when we consider 1-dimensional families $\{C \cdot g(L)\}$.

The additional problem is that the length of a program is not a well-defined notion: For example, if we measure the number of lines of code, then this number of lines depend on the programmer's style:

- Some programmers want to place as much code on one page as possible, so they prefer longer lines. For example, these programmers will place the following simple (self-explanatory) Pascal if-then statement on a single line:

```
IF (x>=0) THEN y:=x ELSE y:=-x;
```

■ Other programmers prefer to use short lines, with lots of indentation. This style makes the whole program somewhat longer, but it provides a clear visual understanding of the logic of each part of the program. For example, the above if-then statement will look something like this:

```
IF
   (x>=0)
THEN
   y:=x
ELSE
   y:=-x;
```

It is reasonable to assume that the optimality criterion is such that the relative quality of any two software cost models should not depend on the programmer's style. If we denote by λ the average decrease in the program's length caused by a change in style, then after changing the style, the length of the same code changes from L to $L_{new} = L/\lambda$, and, correspondingly, the dependence of T on L changes from $g(L)$ to $g_{new}(L_{new}) = g(\lambda \cdot L_{new})$ (i.e., from $g(L)$ to $g(\lambda \cdot L)$). The above assumption about the optimality criterion means that if, according to this criterion, $f(L)$ is better than $g(L)$, then $f(\lambda \cdot L)$ should be better than $g(\lambda \cdot L)$. In other words, the optimality criterion must be *scale-invariant*.

Similarly, the model that describes the dependence of the effort E on length L should also be optimal with respect to some scale-invariant criterion.

Results.

■ For *1-dimensional families* $\{C \cdot g(L)\}$, we can, therefore, conclude (from Theorem 8.1) that all functions from the optimal family (that describe the dependence $T(L)$ or $E(L)$) are of the type $g(L) = C \cdot L^\alpha$ for some real numbers C and α.

Such models have indeed been successfully used since the early 80s, when B. W. Boehm introduced them in his book *Software Engineering Economics* (Prentice-Hall, Englewood Cliffs, NJ, 1981) under the name of COCOMO (COnstructive COst MOdel).

Boehm not only introduced the formulas $T = C \cdot L^\alpha$ and $E = D \cdot L^\beta$, he also described the connection between the parameters C, α, D, and β, and numerical characteristics of the desired software and of the software company.

- For the cases when the original COCOMO formulas do not work sufficiently well, Theorem 9.1 provides more general formulas that are worth trying.

Exercises

B.1 A software company has finished programming two software projects with 5,000 and 20,000 lines of code. These projects required, correspondingly, 20 and 40 man-months. The company is starting a new project (of the same type as the previous two) that would require delivering 80,000 lines of code. Use the COCOMO model to predict the effort required for the new project. *Hint:* for a power-law model $C \cdot L^{\alpha}$, an extrapolation algorithm was described in Lesson 9.

C

ELECTRONIC ENGINEERING: HOW TO DESCRIBE *PN*-JUNCTIONS

Problem. Computers consist of electronic components. There are two basic types of electronic devices:

- Some components, e.g., resistors, connecting wires, etc., are made of *conductors*. In these components, current is transmitted by free electrons (negatively charge particles). The equations that describe these devices are usually linear, simple, and well known.

- More complicated electronic components such as *transistors*, that are used for amplification and complicated logical operations, are made of *semiconductors*. A semiconductor device usually has several regions in which current is transmitted differently:

  - *n* (negative) regions, where electric current is conducted by moving *electrons* (as in the metals), and

  - *p* (positive) regions, where the current is transmitted by moving an electron from one immobile ion to the neighboring one. What actually moves in *p* regions is not free electrons, but *holes* in the electron structure of the atoms (that are "frozen" to their places in the crystal order).

Inside each region, electrical properties are described by simple linear laws similar to the ones known for metals. The *interaction* between *p* and *n* regions is what makes these devices non-linear and interesting. Therefore, to describe an arbitrarily complicated electronic device, we must be able to describe what happens in the border between the two regions; this border is called a *pn junction*. We would like to describe the junction's *volt-ampere characteristic*

$I = g(V)$, i.e., the dependency of the electric current I on the voltage V applied across the junction.

Formulating the problem in mathematical terms. Since we can measure current in different units, with each function $g(V)$, a function $C \cdot g(V)$ is also a reasonable description of a junction. Thus, in the first approximation, we should consider families of the type $\{C \cdot g(V)\}$.

With respect to voltage, changing the unit is no longer a physically reasonable option, because there is, usually, a physically meaningful unit of voltage, for which, e.g., the junction is "fried" (damaged). However, there is a different symmetry here. Namely, from the physical viewpoint,

■ the current I describes the *number* of moving electrons (or other charges), while

■ the voltage V describes the *energy* of each electron.

What really matters is not the *absolute* energy, but the *difference* between the energies. For example:

■ When a plane is flying, usually, there is about a 1,000,000 Volt difference in potential between people flying the plane and the Earth. However, since there is no direct contact, neither people not sensitive electronic devices on board feel this difference.

■ On the other hand, if a person accidentally touches two 127 Volt household wires, he experiences a painful shock.

Since what matters is the *difference* in potentials, not the absolute values of them, the relative quality of different approximation models should not change if we simply add a constant to all the values of voltage, i.e., if we replace V by $V + s$ and, correspondingly, $g(V)$ by $g(V + s)$. Thus, the optimality criterion should be *shift-invariant*.

Results.

■ For *1-dimensional* families $\{C \cdot g(V)\}$, we can conclude, from Theorem 7.1, that $g(L) = C \cdot \exp(\alpha \cdot V)$.

This model indeed describes a *pn* junction pretty well; see, e.g., E. M. Rips, *Discrete and integrated electronics* (Prentice Hall, Englewood Cliffs, NJ, 1986).

■ For *multi-dimensional* families $\{C_1 \cdot g_1(V) + \ldots + C_m \cdot g_m(V)\}$, Theorem 10.1 provides more general functions that be used for a better description.

Some of these functions indeed lead to a better description, e.g., functions of the type $g(L) = C_1 + C_2 \cdot \exp(\alpha \cdot V)$ (see Rips's book, Section 1-1).

D

LOG-NORMAL DISTRIBUTION JUSTIFIED: AN APPLICATION TO COMPUTATIONAL STATISTICS

Traditional engineering statistical methods are mostly based on the use of Gaussian (normal) distribution. In some practical problems, however, we must process quantities that take only positive values. For such quantities, Gaussian-based methods are not applicable because for each Gaussian random variable, there is a positive probability that this variable takes a negative value. To process such quantities, we must pre-transform positive numbers into arbitrary real numbers. We show that continuous mathematics helps to choose the optimal pre-transformation, and that this pre-transformation is indeed helpful in many practical problems.

Standard methods of engineering statistics: brief reminder

For the benefit of the readers who are not well familiar with standard engineering statistical methods (or who would simply like a brief reminder), we provide the reminder in this section. Readers who already know these standard methods can skip this reminder and start with the next section.

Random variables are needed. In many real-life situations, the measured (or observed) value of the same physical quantity changes in an unpredictable (*random*) way from measurement to measurement. There can be two reasons for that:

363

- In some situations, the actual value changes.

  For example, due to high temperature, the actual temperature and density in a combustion engine can fluctuate, and therefore, the values measured at different moments of time can differ.

- In some situations, the *actual* value does not change (or, at least, does not change much), but, due to unpredictable (random) measurement errors, subsequent measurements will lead, in general, to different measurement results.

In such situations, we cannot predict the precise measurement result, we can only predict how frequently different values can occur as the result of the measurement. The more measurements we make, the more accurately we can determine these frequencies. The limit values of the frequencies are called *probabilities*.

A quantity x about which we only know probabilities is called a *random variable*, or a *random number*.

Comment. To avoid misunderstanding, we should mention that although in this appendix, we consider only probabilities that stem from frequencies, in general, probability theory has many other applications where such a frequentist interpretation is not possible (or, at least, not directly possible). For example, in Artificial Intelligence, we often have to consider *subjective* probabilities that describe expert's degree of belief rather than frequencies of actual values.

How to describe a random variable: a brief and informal reminder. To describe a random variable x, we must describe, for an arbitrary interval $[a, b]$, the *probability* $P(x \in [a, b))$ that the (unknown) value x will belong to this interval.

This probability has a convenient *additivity* property that relates, for every three values $a < b < c$, the probabilities $P(x \in [a, b))$, $P(x \in [b, c))$, and $P(x \in [a, c))$. Indeed, in order to determine these three probabilities, we must measure x many times, and determine how many times x was in each of the three intervals. If we denote the total number of measurements by N, and the number of measurements for which x belongs to each interval $[a, b)$, as $N([a, b))$, etc., then the corresponding frequencies $f_N(x \in [a, b)), \ldots$, are determined by the formulas

$$f_N(x \in [a, b)) = \frac{N([a, b))}{N}; \quad f_N(x \in [b, c)) = \frac{N([b, c))}{N};$$

$$f_N(x \in [a, c)) = \frac{N([a, c))}{N}.$$

Due to our choice of the three numbers a, b, and c, a value x gets into the interval $[a, c)$ if and only if it either gets into $[a, b)$, or it gets into $[b, c)$. Thus, the number of cases in which $x \in [a, c)$ is equal to the sum of the number of values for which $x \in [a, b)$ and of values for which $x \in [b, c)$:

$$N([a, c)) = N([a, b)) + N([b, c)).$$

Dividing both sides of this equality by N, we conclude that

$$f_N(x \in [a, c)) = f_N(x \in [a, b)) + f_N(x \in [b, c)).$$

When N increases ($N \to \infty$), the frequencies tend to the corresponding probabilities. Thus, in the limit, we conclude that

$$P(x \in [a, c)) = P(x \in [a, b)) + P(x \in [b, c)).$$

This equality is called *additivity*.

Due to additivity, if we take a twice longer interval, then the probability, crudely speaking, increases twice, and, in general, probability $P(x \in [a, b))$ of belonging to an interval $[a, b)$ is roughly proportional to the width $b - a$ of this interval. It is therefore reasonable to characterize the behavior of the probabilities near a point a by the coefficient of this proportionality.

The narrower the interval, the more accurate is this proportionality property. Thus, it makes sense to consider the *limit* of these proportionality coefficients when the width ε of an interval tends to 0:

$$\rho(a) = \lim_{\varepsilon \to 0} \frac{P(x \in [a, a + \varepsilon))}{\varepsilon}.$$

This limit is called the *probability density function* (PDF, for short).

If we know the probability density, how can we reconstruct the probabilities $P(x \in [a, b))$?

- For *narrow* intervals, when the width $\varepsilon = b - a$ of the desired interval is small, the ratio $P(x \in [a, a + \varepsilon))/\varepsilon$ is close to the limit $\rho(a)$. From $P(x \in [a, a + \varepsilon))/\varepsilon \approx \rho(a)$, we conclude that $P(x \in [a, a + \varepsilon)) \approx \rho(a) \cdot \varepsilon$. The narrower the interval, the more accurate this formula.

■ To get the probability for *wider* intervals, we can use the additivity property:

$$P(x \in [a, b)) = P(x \in [a, a + \varepsilon)) + P(x \in [a + \varepsilon, a + 2 \cdot \varepsilon)) + \ldots +$$

$$P(x \in [a + k \cdot \varepsilon, b)),$$

where $a + k \cdot \varepsilon \leq b \leq a + (k + 1) \cdot \varepsilon$. Since we already know the approximate formula for small probabilities, we conclude that

$$P(x \in [a, b)) \approx \rho(a) \cdot \varepsilon + \rho(a + \varepsilon) \cdot \varepsilon + \ldots + \rho(a + k \cdot \varepsilon) \cdot (b - (a + k \cdot \varepsilon)).$$

The smaller ε, the more accurate this formula. Thus, we can determine the desired probability as the limit of the right-hand side sums when $\varepsilon \to 0$. This limit is exactly the *integral* of the function $\rho(x)$. Thus,

$$P(x \in [a, b)) = \int_a^b \rho(x) dx.$$

How to process random variables: maximum likelihood method. In many practical cases, we do not know the exact probability density function, but we know the *family* of probability density functions to which our (unknown) density function belongs. For example, in the measurement process, we know the measurement result X, and we often know the probability density function $\rho(e)$ of the measurement error $e = X - x$, but we do not the actual value x of the measured quantity (if we knew the exact value of x, then there would no need for a measurement). For every possible values of x, the probability density function for the measurement result $X = x + e$ is $\rho(X - x)$. So, the family of PDF consists of functions $\{\rho(X - x)\}_x$, where $\rho(e)$ is a given function, and x can take any real value. How do we choose x?

In the above example, we assumed that we *know* the probability of different values of error. In some cases, we *do not know* these probabilities, but we know a *family* of possible probability distributions. In general, an m-parametric family of functions can be described as $\{\rho(X, C_1, \ldots, C_m)\}$, where $C_1, \ldots, C_m$ are arbitrary real numbers. In this general case, based on the measurement results, we must choose the values of these parameters $C_1, \ldots, C_m$. How can we do that?

In order to decide how to choose the values of these parameters, let us first consider a *discrete analogue* of this situation, when instead of a PDF, we simply have probabilities. Let us assume that:

- we have only finitely many alternatives $a_1, \ldots, a_k$, and
- we have finitely many hypotheses $h_1, \ldots, h_l$.

Let us denote the probability of i-th alternative a_i according to the hypothesis h_j by p_{ij}. In these terms, if we observe an alternative i, which of the l hypotheses should we choose? It seems natural to choose a hypothesis h_j for which the probability p_{ij} of this alternative a_i is the largest: $p_{ij} \to \max_i$.

For example, if,

- according to hypothesis h_1, a_1 should appear with the probability $p_{11} = 0.1$, and the alternative a_2 with probability $p_{21} = 0.9$; and
- according to hypothesis h_2, a_1 should appear with the probability $p_{12} = 0.8$, and the alternative a_2 with probability $p_{22} = 0.2$,

and we observed a_1, then it is natural to choose hypothesis h_2 that predict this alternative with a high probability 0.8 (i.e, crudely speaking, considers this alternative to be highly probable) and not the competing hypothesis h_1 that predicts a_1 with a low probability $0.1 \ll 0.8$ (i.e., crudely speaking, discounts this alternative as highly improbable).

In the continuous case, it is also reasonable to choose the hypothesis in which the predicted probability of the observed alternative is the largest possible. In this case:

- Hypotheses are combinations of coefficients C_i; each combination $(C_1, \ldots, C_m)$ describes a different hypothesis.
- Alternatives are different real numbers X.

In reality, every measuring device only generates finitely many digits of the measured value, so, we can only get *finitely many* possible values: if the "ideal" measurement result is X, and the actual measuring instrument returns d decimal digits, then this instrument returns X rounded to d digits.

For example, if we measure the value from the interval $[0, 1]$, and the device produces two decimal digits, we can only get 101 different measurement results $0.00, 0.01, \ldots, 0.99, 1.00$. For example, the value 0.278 is rounded to 0.28.

If the instrument produces d digits and results in X, this means that the ideal measuring instrument could produce any value $\tilde{X}$ from the interval

$$[X - (1/2) \cdot 10^{-n}, X + (1/2) \cdot 10^{-n}].$$

The probability of $\tilde{X}$ being in this interval is equal to the integral

$$P = \int_{X-(1/2)\cdot 10^{-n}}^{X+(1/2)\cdot 10^{-n}} \rho(x, C_1, \ldots, C_m) dx.$$

For real-life measuring instruments, the number of digits d is reasonably high, and therefore, the interval $[X - (1/2) \cdot 10^{-n}, X + (1/2) \cdot 10^{-n}]$ (over which we integrate) is very narrow. Thus, we neglect the changes of the functions $\rho(x)$ on this interval and assume that on this interval, $\rho(x) = \rho(X)$. Hence, the integral simply becomes the product of the value $\rho(X)$ and the length 10^{-n} of the integral: $P = \rho(X, C_1, \ldots, C_n) \cdot 10^{-n}$. Thus, choosing the hypothesis for which this probability is the largest means choosing the coefficients $C_1, \ldots, C_m$ for which

$$P = \rho(X, C_1, \ldots, C_m) \cdot 10^{-n} \to \max_{C_1,\ldots,C_m}.$$

Multiplying by a constant does not change which hypothesis is better and which is worse. Thus, the function $P = 10^{-n} \cdot \rho$ attains its maximum at exactly the same values of C_i as the simpler function ρ. Thus, the above idea can be formulated as looking for the value for which the density ρ takes the largest possible value:

$$\rho(X, C_1, \ldots, C_m) \to \max_{C_1,\ldots,C_m}.$$

In this context, the value of a function $\rho(x, C_1, \ldots, C_m)$ for a certain combination $(C_1, \ldots, C_m)$ describes how likely it is that this particular hypothesis is correct: the largest the value of ρ, the more likely the corresponding hypothesis. Because of this interpretation, the function $\rho(x, C_1, \ldots, C_m)$ is called a *likelihood function*, and the formula for determining the values C_i is called a *maximum likelihood method*.

Comment. In this text, we have only provided a "physical" justification of the maximum likelihood method, a justification that is based on a specific physical situation and that uses approximate formulas. In addition to this physical justification, there also exist precise mathematical justifications according to which in many situations, maximum likelihood method is indeed the best method (in some reasonable sense) to determine the parameters of an unknown probability distribution based on the measurement results.

Normal (Gaussian) distribution. In many practical situations, the error distribution is described by a so-called *Gaussian*, or *normal* PDF:

$$\rho(x, a, \sigma) = \frac{1}{\sqrt{\pi} \cdot \sigma} \cdot \exp\left(-\frac{(x-a)^2}{2 \cdot \sigma^2}\right),$$

usually, with $a = 0$. Experimental analysis shows that about half of the measuring instruments have normally distributed errors.

The reason why this particular distribution is so wide-spread is that in many practical situations, the measurement error e is caused by a joint effect of many small independent factors. In other words, $e = e^{(1)} + \ldots + e^{(N)}$, where N is a large integer, and $e^{(1)}, \ldots, e^{(N)}$ are small random variables. There is a theorem (called *central limit theorem*), according to which, under certain reasonable conditions, when $N \to \infty$, the distribution of the sum $e^{(1)}, \ldots, e^{(N)}$ of N small random variables tends to a Gaussian distribution. Thus, if N is large enough, Gaussian distribution is a good approximation to the actual PDF.

Gaussian distribution leads to simple computations. At first glance, the formula that describes the Gaussian distribution may seem somewhat complicated. However, this formula leads to simple data processing methods. Indeed, if we have n independent measurements, then the probability density of the error e_i in i-th measurement is equal to $\text{const} \cdot \exp(-e_i^2/(2 \cdot \sigma^2))$, and the probability density of the errors being $e_1, \ldots, e_n$ is equal to the product of these probabilities: $\rho = \text{const}^n \cdot \exp(-e_1^2/(2 \cdot \sigma^2)) \cdot \ldots \cdot \exp(-e_n^2/(2 \cdot \sigma^2))$ (we take the product because the probability of two independent events occuring together is equal to the product of the probabilities of each of these events). Since $\exp(a) \cdot \exp(b) = \exp(a + b)$, this formula can be re-written as

$$\rho = \text{const}^n \cdot \exp\left(-\frac{e_1^2 + \ldots + e_n^2}{2 \cdot \sigma^2}\right).$$

We are interested in the the the values of the parameters for which this likelihood attains the largest possible value. Since $\exp(-z)$ is a decreasing function, ρ takes the largest possible value if and only if the sum $e_1^2 + \ldots + e_n^2$ takes the smallest possible value.

If this formula looks familiar, it should. In Lesson 3, we have already described the method $e_1^2 + \ldots + e_n^2 \to \min$, under the name of the *least squares method*. The only difference is that in that lesson, we motivated this method by the necessity to have the simplest possible computations. Thus, we arrive at the following two conclusions:

- *For Gaussian distribution, the maximum likelihood method turns into the least squares method.*

- *Gaussian distribution leads to simple computations.*

The problem: methods based on Gaussian distribution are not applicable to positive random variables

General problem. Since Gaussian distribution leads to computationally simple statistical methods, it is desirable to use the corresponding methods whenever possible. However, there are cases when these methods are not applicable. One of these cases is the description of random variables that characterize *positive* physical quantities. For such quantities, Gaussian-based methods are not applicable because for each Gaussian random variable, there is a non-zero probability that this variable takes a *negative* value.

Example: intervals. A natural positive-valued quantity is used to characterize measuring instruments.

- Ideally, to characterize the measuring instrument, we would like to know the *probabilities* of different values of error.

- In principle, *it is possible to get* these *probabilities*: for that, we just test this measuring instrument again and again for different values of the measured quantity.

- In many practical situations (e.g., in manufacturing), this testing would be *too costly* (cost much more than a typical mass-produced instrument itself), so we need to be satisfied with a *simpler description* of this instrument's measuring errors.

The least we can know about the measuring instrument is how different can the measured result X be from the actual value x of the measured quantity. If we do not know any bounds on this difference, then our "measurement" is no good, because this means that for any measurement result X, the actual value x can be arbitrarily different from X. So, we must know the upper bound Δ on the absolute value $|e| = |X - x|$ of the measurement error. If we know this upper bound Δ, and we know the measurement result X, then we can conclude that the actual value x can differ from X by no more than X, i.e., that x is guaranteed to belong to the *interval* $[X - \Delta, X + \Delta]$.

The smaller the upper bound Δ, the narrower the resulting interval and therefore, the more information we get about the actual measured value x. Thus, it is desirable, for each measuring instrument, to find the smallest possible value Δ_{ins} of Δ, i.e., the largest possible absolute value $|e| = |X - x|$ of the measurement error e over all possible situations and over all possible actual values x. Since every measuring instrument has some error, the value Δ_{ins} is always *positive.*

Now, we are talking about mass produced instruments. As the production goes on, conditions change slightly. As a result, the values Δ_{ins} change slightly (and unpredictably) from an instrument to an instrument. Therefore, the value Δ_{ins} is a *random variable.* Since it only takes positive values, we cannot directly apply Gaussian-based methods to process these values.

Remark: interval computations. Special data processing methods have been developed to process the measurement results for which the interval is the only information we get after each measurement. These methods are called *interval computations.* For an introduction into interval computations and into their applications, see, e.g., an edited book R. B. Kearfott and V. Kreinovich (eds.), *Applications of Interval computations* (Kluwer, Dordrecht, 1996).

How to solve the problem, and how continuous mathematics can help

Main idea: re-scaling. We have convenient data processing methods developed for *Gaussian*-distributed random variable, and we have *non-Gaussian* positive-valued random variables. What can we do? To answer this question, let us analyze two similar situations:

- What do we do if we have lengths measured in feet and we have a program that processes lengths in meters? We *re-scale* the values from feet to meters, and apply the existing program. In this case, the rescaling is a linear procedure.

- What do we do if we know the earthquake's magnitude M on a Richter scale, and our program processes magnitudes E in Joule (as required by SI)? We *re-scale* from Richter scale to Joule. In this case, the re-scaling is a non-linear procedure $E = C \cdot \exp(k \cdot M)$ (for appropriate constants C and M).

Similarly, if we have several cases $x^{(1)}, \ldots, x^{(k)}$ of a positive-valued random variable x, it is reasonable to do the following:

- use some re-scaling $y = f(x)$ to *re-scale* these values into the values $y^{(1)} = f(x^{(1)}), \ldots, y^{(k)} = f(x^{(k)})$ that can be arbitrary real numbers; and then

- apply Gaussian-based data processing techniques to the re-scaled values $y^{(1)}, \ldots, y^{(k)}$.

Which re-scaling should we use? In principle, there are many different functions that transform the set of all positive real numbers into the set of all real numbers. Different functions lead to quite different processing results. So, the question is: which of these re-scaling functions $f(x)$ should we choose?

Continuous mathematics helps. We are looking for a function $f(x)$ that transforms a positive-valued random variable x into a Gaussian random variable $y = f(x)$.

It is known that a linear transformation transforms Gaussian variable into a Gaussian one, i.e., if y is a Gaussian random variable, then, for arbitrary C_1 and C_2, the variable $\tilde{y} = C_1 \cdot y + C_2$ is also Gaussian. Therefore, with each function $f(x)$, all functions $C_1 \cdot f(x) + C_2$ also transform x into a Gaussian variable. Therefore, it makes sense to look not for a single function $f(x)$, but for the entire *family* of functions $\{C_1 \cdot f(x) + C_2\}$ that correspond to different values of C_1 and C_2. Out of all possible families of this type, we want to choose a family that is optimal with respect to some reasonable optimality criterion.

If we change a unit with which we measure the quantity x, the relative quality of different families should not change. Changing a unit means changing x to $\lambda \cdot x$ and $f(x)$ to $f(\lambda \cdot x)$. Invariance with respect to these transformations is what we have called, in Lesson 4, *scale invariance*. Thus, it is reasonable to require that the desired family $\{C_1 \cdot f(x) + C_2\}$ should be invariant with respect to some final scale-invariant criterion.

Such optimal families were described in the proof of Theorem 4.2: they correspond either to the function $f(x) = \ln(x)$, or to the function $f(x) = x^\alpha$. For this particular problem, we should exclude the power functions $f(x) = x^\alpha$, because they only maps positive real numbers too positive ones, and we want a transformation that maps positive numbers to all real numbers. So, the only function left is the logarithm $f(x) = \ln(x)$. Thus, we conclude that:

The optimal transformation is the logarithm $f(x) = \ln(x)$.

So, if we have several cases $x^{(1)}, \ldots, x^{(k)}$ of a positive-valued random variable x, we do the following:

- *re-scale* these values into the values $y^{(1)} = \ln(x^{(1)}), \ldots, y^{(k)} = \ln(x^{(k)})$ that can be arbitrary real numbers; and then

- apply Gaussian-based data processing techniques to the re-scaled values $y^{(1)}, \ldots, y^{(k)}$.

Experimental confirmation. In many practical situations in which the random variable only takes positive values, a logarithm of this random variable is indeed normally distributed. In particular, as shown in A. V. Ekimov and M. I. Revyakov, *Reliability of electrical measuring instruments* (Leningrad, Energoatomizdat, 1986, in Russian), this is true for the the worst-case accuracy $\Delta_{\text{ins}} > 0$ of different mass-produced measuring instruments.

Our result provides a *theoretical explanation* for this empirical fact.

Acknowledgments. This appendix is based on a project by A. E. Yurzditsky.

OPTIMAL ROBUST STATISTICAL METHODS

The least squares method (that we have described in Lesson 3) is computationally the simplest method of choosing the best model consistent with the given data. However, its results are sometimes not good enough. If we know the probabilities of different error values, then we can use the maximum likelihood method. In many practical situations, we do not know these probabilities. In this appendix, we show that for such situations, continuous mathematics can help in selecting the best data processing techniques.

Least squares method is not always the best. In Section 3, we have described the *least squares* method, in which the parameters of the model is chosen from the condition that $e_1^2 + \ldots + e_m^2 \to \min$, where e_i is the (absolute value of the) difference between the actual result of i-th measurement and the model's prediction of this result. This method is often very useful, but there are cases when it does not lead to the best possible results. There are two possible reasons for this non-optimal performance:

- One of the reasons why we selected the least squares method from all possible methods is that we were looking for the method that is computationally the simplest.

  However, the fact that the method is computationally simple does not necessarily mean that its results are the best.

- In the previous appendix, we showed that the least squares method is the best data processing method for the case when all the errors are Gaussian.

  However, as we have mentioned in the previous appendix, about half of the measuring instruments have measurement errors whose probability distribution is different from Gaussian. For such distributions, the optimal data processing method is the maximum likelihood method, so, for non-Gaussian distributions, the least square method is not optimal.

Example. Let us give a simple realistic example of such situation. To make this example really simple, we will consider the case when we measure the same physical quantity x several times. Since the measurement error is a random variable, the results $x^{(1)}, \ldots, x^{(m)}$ of these measurements differ from x and from each other. We want to use these measurement results to estimate x.

In this case, i-th measurement error is equal to $e_i = |x^{(i)} - x|$. Therefore, if we use *least squares* method, then we must choose x from the condition that

$$e_1^2 + \ldots + e_m^2 = (x^{(1)} - x)^2 + \ldots + (x^{(m)} - x)^2 \to \min_x.$$

Differentiating the minimized function with respect to x and equating the derivative to 0, we conclude that

$$x = \frac{x^{(1)} + \ldots + x^{(m)}}{m},$$

i.e., that the arithmetic average of the measurement results is the best estimate for x.

There are, however, realistic cases in which this estimate is not the best possible. For example, for measuring instruments, measurement errors are usually small, but with a certain small probability something goes wrong, and we get a measurement result that is drastically different from the measured value (such a result is called an *outlier*).

For example, if the actual value is $x = 1$, the error is usually around 0.01, but in 10% of the cases, we get a 100% error, then after 10 measurements, we may get the measurement results $x^{(1)} = 1.01$, $x^{(2)} = 0.99$, $x^{(3)} = 1.00$, $x^{(4)} = 1.00$, $x^{(5)} = 0.99$, $x^{(6)} = 1.01$, $x^{(7)} = 2.12$, $x^{(8)} = 0.99$, $x^{(9)} = 1.01$, and $x^{(10)} = 1.00$.

■   If we use common sense, we would see that the value $x^{(7)}$ is drastically different from the others. Since we know that outliers are possible, we will simply dismiss this outstanding value as an outlier, and average only the remaining values. This averaging will result in the estimate $x = 1.00$. This estimate is accurate within 2 decimal digits.

- However, if we simply apply the least squares method to all ten measurement results, we get an estimate $x \approx 1.11$. This estimate is much worse than the "commonsense" one: it has only has one correct decimal digit.

Non-Gaussian statistical methods and robust methods.

- If we know the PDF $\rho(e)$ for the error e, then, to determine the parameters of the model, we can use the maximum likelihood method. It is natural to assume that all the measurement errors are independent, and therefore, the probability density for the collection of N measurement results is equal to the product of the N probability densities that describe each of the measurement results: $\rho = \rho(e^{(1)}) \cdot \ldots \cdot \rho(e^{(m)})$. Thus, the parameters of the model can be determined from the condition that this product takes the largest possible value.

    - For Gaussian distribution, to simplify computations, we replaced the maximization of the Gaussian function $\rho(e) = \text{const} \cdot \exp(-e^2/(2 \cdot \sigma^2))$ by an equivalent problem of minimizing the negative logarithm e^2 of this expression.
    - It is, therefore, reasonable to apply the same idea to non-Gaussian distribution as well. If we apply the function $-\ln(z)$ to the above product, then product turns into the sum, and we arrive at the following minimization problem:

$$\psi(e_1) + \ldots + \psi(e_m) \to \min_x, \qquad (E.1)$$

where we denoted $\psi(z) = -\ln(\rho(z))$.

- In many practical situations, however, we do not know the exact PDF. What can we do then?

We need data processing method for the situation when we do not know the exact PDF, methods that will not change much (will be *robust*) if the actual PDF changes. Such methods are called *robust statistical methods*. These methods were first considered by P. J. Huber, in his book *Robust statistics* (Wiley, N.Y., 1981) (for a more recent survey, see, e.g., Chapter 16 of H. M. Wadsworth, Jr. (editor), *Handbook of statistical methods for engineers and scientists* (McGraw-Hill Publishing Co., N.Y., 1990)).

Huber suggested the following natural idea for choosing a robust method: In reality, there is some (unknown) distribution $\rho(e)$, so, the optimal method of determining x coincides with the method (E.1) for $\psi(e) = -\ln(\rho(e))$. So, it is reasonable to use method (E.1) for some $\psi(e)$. Such methods are called *M-methods* (*M* comes from *M*aximum likelihood).

So, we arrive at the following problem:

The problem: what function $\psi(e)$ should we choose?

- In some cases, we have *some* information about the probabilities, and we can try to use this information to select an appropriate function $\psi(e)$.

- In many real-life cases, however, we do not have any information about the probabilities. We will show that in such cases, continuous mathematics can help to choose the best function $\psi(e)$.

Continuous mathematics helps. The only thing we need from the function $\psi(e)$ are the parameters of the model for which the expression (E.1) is the smallest possible; we are not interested in the absolute value of $\psi(e)$.

- If we multiply two real numbers by a positive constant $C > 0$, then their relative order will not change: if we have $a < b$, we will still have $C \cdot a < C \cdot b$, etc. Therefore, if we replace the function $\psi(e)$ by a new function $\widetilde{\psi}(e) = C \cdot \psi(e)$, the resulting expression

$$\widetilde{\psi}(e_1) + \ldots + \widetilde{\psi}(e_m) = C \cdot (\psi(e_1) + \ldots + \psi(e_m))$$

will attain its minimum for exactly the same values of the model's parameters as the original expression $\psi(e_1) + \ldots + \psi(e_m)$. Thus, it does not matter whether we use the original function $\psi(e)$ or the new function $\widetilde{\psi}(e)$: we get exactly the same parameters of the model.

- Similarly, if we *add* one and the same number C to two real numbers, then their relative order will not change: if we have $a < b$, we will still have $C + a < C + b$, etc. Therefore, if we replace the function $\psi(e)$ by a new function $\widetilde{\psi}(e) = C + \psi(e)$, the resulting expression

$$\widetilde{\psi}(e_1) + \ldots + \widetilde{\psi}(e_m) = m \cdot C + (\psi(e_1) + \ldots + \psi(e_m))$$

will attain its minimum for exactly the same values of the model's parameters as the original expression $\psi(e_1) + \ldots + \psi(e_m)$. Thus, it does not matter whether we use the original function $\psi(e)$ or the new function $\widetilde{\psi}(e)$: we get exactly the same parameters of the model.

Thus, what we really want to choose is not a *single* function $\psi(e)$, but the entire *family* of functions $\{C_1 \cdot \psi(e) + C_2\}$ that correspond to different real numbers $C_1 > 0$ and C_2.

Which of the families is optimal? If we replace the measuring unit used to describe the measured value x by a new unit that is C times smaller, then both the measured and the predicted values will become C times larger than before. Thus, the error e, that is defined as the difference between the predicted and the actual values, also gets multiplied by the same constant C: $e \to \tilde{e} = C \cdot e$. It is reasonable to require that the relative quality of the two functions $\psi(e)$ and $\varphi(e)$ should not change if we simply change a measuring unit. Applying the *old* function $\psi(e)$ to the *new* (re-scaled) value $\tilde{C} = C \cdot e$ results in the value $\psi(C \cdot e)$, which is equivalent to applying a *new* function $\tilde{\psi}(e) = \psi(C \cdot e)$ to the *old* values of the model error. Therefore, the above-mentioned requirement can be formulated as follows: if the function $\psi(e)$ is better than $\varphi(e)$, then the re-scaled function $\tilde{\psi}(e) = \psi(C \cdot e)$ should be better than the similarly re-scaled function $\tilde{\varphi}(e) = \varphi(C \cdot e)$. In Lesson 4, we called this property *scale-invariance*.

So, in order to choose the best robust technique, we must choose the family of functions $\{C_1 \cdot \psi(e) + C_2\}$ that is optimal with respect to some scale-invariant final criterion.

In the proof of Theorem 4.2, we have already described such families: they correspond to $\psi(e) = (\pm)\ln(e)$ or to $\psi(e) = e^p$ for some real number p.

The use of $\psi(e) = \ln(e)$ does not lead to any meaningful criterion: Indeed, since $\ln(e_1) + \ldots + \ln(e_m) = \ln(e_1 \cdot \ldots \cdot e_m)$, the condition (E.1) with this function $\psi(e)$ is equivalent to looking either for the minimum, or for the maximum of the product $e_1 \cdot \ldots \cdot e_m$.

- The *minimum* of this product is attained when the product of the errors is 0, i.e., when one of these errors is equal to 0. In other words, if we use this criterion, we should consider *all* the models that predict at least *one* of the observations correctly, no matter how bad this model is for other observations. This is not a reasonable way to choose a model.

- The *maximum* of this product is achieved when all the errors are large (tend to ∞). This is also not a good way to choose a model.

So, we exclude $\psi(e) = \ln(e)$, and we arrive at the following conclusion:

If we have no information about the probabilities of different values of error, then the best method of choosing a model is to choose a model for which

$$e_1^p + \ldots + e_m^p \to \min,$$

where e_i is the absolute value of the difference between the actual value measured in i—th experiment and the model's prediction, and $p > 0$ is a real number.

Such methods are called l^p—*methods.*

Comment. In particular, for $p = 2$, we get the least square method as one of the optimal ones.

Example. If in the above numerical example we take $p = 1$, then one can show that the minimum $|x - x^{(1)}| + \ldots + |x - x^{(m)}|$ is attained when x is the *median* of the values $x^{(1)}, \ldots, x^{(m)}$. In other words, if we sort these values into the non-decreasing sequence $x_{(1)} \leq x_{(2)} \leq \ldots \leq x_{(m)}$, then:

- If m is *odd*, i.e., if $m = 2k-1$ for some integer k, then the optimal value of x, according the l^1-method, is $x = x_{(k)}$ (the mid-element $(1+(2k-1))/2 = k$ of the sorted sequence).

- If m is *even*, i.e., if $m = 2k$ for some integer k, then we cannot take the mid-element (because the middle between 1 and $2k$ is $(1+2k)/2 = k+1/2$, not an integer) but take any value x from the interval $[x_{(k)}, x_{(k+1)}]$.

In particular, for our example, l^1-method leads to $x \in [x_{(3)}, x_{(4)}] = [1.00, 1.00]$, i.e., to the same accurate answer $x = 1.00$ as the commonsense reasoning.

Experimental confirmation of our conclusion.

- l^p-methods have indeed been found to be empirically the best robust M-methods; see, e.g., P. J. Bickel and E. L. Lehmann, "Descriptive Statistics for Nonparametric Models. 1. Introduction" (*Ann. Statist.*, 1975, Vol. 3, pp. 1045–1069). These methods have been actively (and successfully) applied, e.g., to *geophysics*; see A. Tarantola, *Inverse problem theory: methods for data fitting and model parameter estimation* (Elsevier, Amsterdam, 1987).

- A method of type (E.1), with an appropriate function $\psi(e)$, comes from the maximum likelihood method for the probability density $\psi(e) = -\ln(\rho(e))$. Therefore, if we know $\psi(e)$, we can reconstruct the probability density as $\rho(e) = \exp(-\psi(e))$. To be more precise, since we only know the *family* $\{C_1 \cdot \psi(e) + C_2\}$ to which the function $\psi(e)$ belongs, we can conclude that the density is $\rho(e) = \exp(-(C_1 \cdot \psi(e) + C_2))$. In particular, for $\psi(e) = e^p$, we get the probability density $\rho(e) = \exp(-(C_1 \cdot |e|^p + C_2))$. This distribution is similar to the PDF proposed by Weibull and is, therefore, called *Weibull-type distribution*.

It is interesting to know that for the vast majority of the measuring instruments, the PDF of measurement errors is indeed Weibull-type: see, e.g., P. V. Novitskii and I. A. Zograph, *Estimating the measurement errors* (Energoatomizdat, Leningrad, 1991, in Russian), and A. I. Orlov, "How often are the observations normal?" (*Industrial Laboratory*, 1991, Vol. 57, No. 7, pp. 770–772).

Acknowledgments. This appendix is based on several projects; the main contents of these projects is described in the following papers:

- I. S. Kirillova and V. Kreinovich, "Choosing l^p method of calculating the shift parameter on the basis of the applicability criterion" (*Metrological Maintenance of Electrical Measuring Instruments*, Proceedings of the National Institute of Electrical Measuring Instruments (VNIIEP), Leningrad, 1984, pp. 91–97, in Russian).

- I. S. Kirillova, V. Kreinovich, and G. N. Solopchenko, "Distribution-independent estimators of error characteristics of measuring instruments" (*Measurement Techniques*, 1989, Vol. 32, No. 7, pp. 621–627).

- K. Crain, M. Baker, and V. Kreinovich, "A simple uncertainty estimator for tomographic inversions" (*Proc. 2nd Borehole Seismics Conference*, Tohoku University, Sendai, Japan, November 1993).

- M. C. Gerstenberger, *Development of new techniques in crosswell seismic travel time tomography*, M.S. Thesis, Department of Geological Sciences, University of Texas at El Paso, El Paso, TX, July 1994.

- D. I. Doser, K. D. Crain, M. R. Baker, V. Kreinovich, and M. C. Gerstenberger, and J. L. Williams, "Estimating uncertainties for geophysical tomography", *Reliable Computing*, 1995, Supplement (Extended Abstracts of APIC'95: International Workshop on Applications of Interval Computations, El Paso, TX, Febr. 23–25, 1995), pp. 74–75.

- G. L. Shevlyakov and N. O. Vil'chevskiy, "On the choice of an optimization criterion under uncertainty in interval computations - nonstochastic approach", *Reliable Computing*, 1995, Supplement (Extended Abstracts of APIC'95: International Workshop on Applications of Interval Computations, El Paso, TX, Febr. 23–25, 1995), p. 188–189.

- D. I. Doser, K. D. Crain, M. R. Baker, V. Kreinovich, and M. C. Gerstenberger, "Estimating uncertainties for geophysical tomography" (*Reliable Computing*, 1998, to appear).

In particular, these papers describe how we can estimate the value of the parameter p from the experimental data.

F

HOW TO AVOID PARALYSIS OF
NEURAL NETWORKS

Hebb training. One problem with backpropagation is that it is somewhat computationally complicated. Therefore, in some cases, simpler techniques are used, in which each weight w is changed linearly: $w \rightarrow w + c$ for some constant c (this is called *Hebb learning*).

Paralysis: the problem. Sometimes some weights become so large that the output of the corresponding neurons is close to 1; see, e.g., P. Wasserman, *Neural computing: Theory and Practice* (Van Nostrand Reinhold, N.Y., 1989), pp. 90–91. Such neurons are called *saturated*. Saturation extends the training time, bringing the whole training process into a *paralysis*.

How to overcome paralysis. Since *linear* weight transformations cause paralysis, to overcome this paralysis, it is reasonable to use *non-linear* weight transformations $w \rightarrow t(w)$, for some non-linear function $t(w)$.

The main purpose of this transformation is to get rid of large values. Thus, we would like the function $t(w)$ to transform the set of all real numbers $(-\infty, \infty)$ into a bounded interval $[-\Delta, \Delta]$.

What non-linear training function should we choose? There exist many different functions that map the set of real numbers into an interval. Different non-linear function lead to a different quality of learning. So, a natural question arises: *what function $t(w)$ is the best to choose?*

We must choose a family of functions, not a single function. Similarly to Lesson 12, we can argue that we need to choose not a *single* function $t(w)$, but the entire *family* of function $\{g(t(w))\}$, where $g(y)$ are transformations from an appropriate transformation Lie group.

The optimality criterion must be shift-invariant. If a function $t(w)$ is better than a function $\tilde{t}(w)$, then it is reasonable to assume that it will still be better if we first make an additional standard training step $w \to w + c$, and only then apply the non-linear transformation. These two consequent transformation steps are equivalent to a single transformation $w \to t(w + c)$ or $w \to \tilde{t}(w + c)$. Thus, the above property means that if $t(w)$ is better than $\tilde{t}(w)$, then the family generated by the function $t(w + c)$ must be better than the family generated by the function $\tilde{t}(w) = \tilde{t}(w + c)$. In other words, the optimality criterion must be *shift-invariant*.

Result. We are looking for the family $\{g(t(w))\}$ (where $g(y) \in G$) that is optimal with respect to some final shift-invariant criterion. In Theorem 12.1, we have already described all functions from the families that are optimal with respect to such criteria: these are logistic, linear, and exponential functions.

If we add an additional requirement that the function $t(w)$ maps all real numbers into a bounded interval, we thus exclude linear and exponential functions. Hence *the optimal training function must be of the type* $t(w) = C_1 \cdot s_0(c_1 \cdot w + c_0) + C_0$, *where* $s_0(w) = 1/(1 + \exp(-w))$. Experiments have shown that logistic training functions are indeed the best in overcoming paralysis (see, e.g., the above-cited book by Wasserman).

Acknowledgments. This appendix is based on a project by C. Quintana. For details, see V. Kreinovich and C. Quintana, "Neural networks: what non-linearity to choose?" (*Proceedings of the 4th University of New Brunswick Artificial Intelligence Workshop*, Fredericton, N.B., Canada, 1991, pp. 627–637).

G

ESTIMATING COMPUTER PRICES

The problem: choosing a computer. Nowadays, every company that wants to buy new computers has many choices. The faster the processor (the more memory the computer has, etc.), the better its performance, but at the same time, the higher its cost. For different computers, we have different gains (due to improved ability), different losses (cost of the computer), and therefore, different profit expectations.

- For example, a *video store* may be interested in setting up computers that will help potential customers find the videos. There are not many computations involved, and there are no hard limits on the time, so, for this problem, computation *speed* of a computer may *not* be *an issue*, but the quality of a *graphic* interface may be very *important* (as a means to attract customers).

- On the other hand, in *manufacturing* applications, when we need to make *real-time* decisions, *speed is important*, while *graphics* is *not* that *relevant*.

The company would like to select a computer that maximize its profits.

Trying all computers on the market: possible but very time-consuming. The profit depends on the computer price. Therefore, one way to find the best buy is to try all possible types of computers and choose the one that leads to the largest expected profit. The problem with this approach is that there are too many different types of computers, and it is very *time-consuming* to try them all.

A formula for a price would make computer selection easier. Since trying all possible computers is very time-consuming, it is desirable to replace the time-consuming *price table* with a simple *price formula*. Namely, we would like to be able to do the following:

- Find a *formula* that would estimate the price y of the computer based on its characteristics $x_1, \ldots, x_n$ (such as processor speed, memory size, number of pixels on a screen, warranty period, etc.).

- Substitute this price formula into the general formula for profit. As a result, we get a formula $J(x_1, \ldots, x_n)$ that expresses the expected profit J as a function of the computer's characteristics x_i.

- Find the computer characteristics $x_1^{\mathrm{opt}}, \ldots, x_n^{\mathrm{opt}}$ for which the expected profit is the largest possible; and, finally,

- buy a computer with these characteristics.

To utilize this idea, we must find a formula $y = f(x_1, \ldots, x_n)$ that describes the price y as a function of the computer's characteristics x_i.

Linear formulas do not work. Let us first try the simplest possible functions: linear functions $y = c_0 + c_1 \cdot x_1 + \ldots + c_n \cdot x_n$. The linear formula is in good accordance with the fact that different computer characteristics are *independent* from each other; however, the linear price model rarely works.

- Since a linear function is a good *local* approximation to an arbitrary smooth function, such a linear expression works well if we *already know the approximate values* of the computer's characteristics, and we are looking for small variations in x_i.

- However, in most computer purchases, the *range* of possible values of each variable is so *wide* that linear approximation does not work well; for example, the RAM size can differ from 1 RAM to 8 RAM, the computer speed may differ from 30 MHz to 200 MHz, etc.

This non-linearity is actually easy to understand: computer characteristics grow exponentially, and prices do not change that fast:

- A new computer model that has twice as much memory and twice larger speed, usually, does not cost twice as much.

- Definitely, a model that is a hundred times faster does not cost hundred times more.

- And when a new chip appears that is ten times faster than the old one, the cost of the old computer does not immediately drop to one-tenth of the cost of the new one.

An alternative: semi-linear models. Since a *linear* price model does not work, we would like to design a *non-linear* price model. In designing this model, we want to pursue two objectives:

- On one hand, we would like to have a price model, in which, similarly to the linear model $y = c_0 + c_1 \cdot x_1 + \ldots + c_n \cdot x_n$, different computer characteristics x_i are *independent*.

- On the other hand, we want a *non-linear* model.

A natural way to satisfy both objectives is to assume that there *is* a linear relation, but this relation is not between the *original* values x_i and y, but between the *re-scaled* values X_i and Y (that are obtained from x_i and y by some non-linear re-scaling). In other words, we consider a model

$$Y = c_0 + c_1 \cdot X_1 + \ldots + c_n \cdot X_n,$$

where $Y = f(y)$, $X_i = f(x_i)$, and $f(z)$ is some non-linear *re-scaling function*. In terms of $f(x)$, this model has the form

$$f(y) = c_0 + c_1 \cdot f(x_1) + \ldots + c_n \cdot f(x_n).$$

We will call such models *semi-linear*. To use these models, we must choose an appropriate function $f(z)$. How can we choose it?

We must choose not a single function $f(z)$, but a family of functions. Let us assume that a function $f(z)$ is a good fit for the actual computer prices. In other words, if we apply the non-linear re-scaling $z \rightarrow f(z)$ both to the computer characteristics and to the prices, then the re-scaled values X_i and Y are related by a linear formula.

- If, instead of the original re-scaling function $f(z)$, we would use a new re-scaling function $\widetilde{f}(z) = C_1 \cdot f(z)$, then the new re-scaled values are

$\widetilde{Y} = \widetilde{f}(y) = C_1 \cdot f(y) = C_1 \cdot Y$ and $\widetilde{X}_i = C_1 \cdot X_i$. Therefore, multiplying both sides of the original linear dependency by C_1, we get

$$\widetilde{Y} = \widetilde{c}_0 + c_1 \cdot \widetilde{X}_1 + \ldots + c_n \cdot \widetilde{X}_n,$$

where $\widetilde{c}_0 = C_1 \cdot c_0$. In other words, if the re-scaling $f(z)$ leads to a linear dependence, then the re-scaling $\widetilde{f}(z) = C_1 \cdot f(z)$ also leads to a linear dependence.

■ Similarly, if we use a function $\widetilde{f}(z) = f(z) + C_0$, then the new re-scaled values are $\widetilde{Y} = \widetilde{f}(y) = f(y) + C_0 = Y + C_0$ and $\widetilde{X}_i = X_i + C_0$. Therefore, adding C_0 to both sides of the original linear dependency, we get

$$\widetilde{Y} = \widetilde{c}_0 + c_1 \cdot \widetilde{X}_1 + \ldots + c_n \cdot \widetilde{X}_n,$$

where $\widetilde{c}_0 = c_0 - (n-1) \cdot C_0$. In other words, if the re-scaling $f(z)$ leads to a linear dependence, then the re-scaling $\widetilde{f}(z) = f(z) + C_0$ also leads to a linear dependence.

Combining these two cases, we conclude that with every function $f(z)$, the linear relation is achieved by the entire *family* of functions $\{C_1 \cdot f(z) + C_0\}$ that correspond to different values of $C_1 > 0$ and C_0. Thus, our goal is to find the best *family* of this type.

The optimality criterion must be scale-invariant. The relative quality of different families should not depend on the choice of units for measuring characteristics x_i. For example:

■ we can measure memory in KBytes or in MBytes;

■ we can measure speed in cycles per seconds or in flops (floating point operations per second),

■ we can measure prices in dollars or in Deutchmarks, etc.

Changing a measuring unit to a λ greater one changes the numerical values of the measured quantity from z to $\widetilde{z} = z/\lambda$; as a result, in terms of the new units, the transformed value $Y = f(y)$ takes the form $Y = f(\lambda \cdot \widetilde{y})$, i.e., the form $Y = \widetilde{f}(\widetilde{y})$ for a new re-scaling function $\widetilde{f}(z) = f(\lambda \cdot z)$.

The optimality criterion must be invariant with respect to this change, i.e., it must be *scale-invariant*.

Mathematical formulation of the problem. We are looking for a family $\{C_1 \cdot f(z) + C_0\}$ that is optimal relative to some final scale-invariant optimality criterion.

The result: optimal pricing formulas. This mathematical problem has already been solved (in the proof of Theorem 7.1). So, *the optimal price formulas correspond to* $f(z) = \ln(z)$ *or to* $f(z) = z^p$ *for some real number p.* In other words, the optimal price formulas have the form

$$\ln(y) = c_0 + c_1 \cdot \ln(x_1) + \ldots + c_n \cdot \ln(x_n)$$

or

$$y^p = c_0 + c_1 \cdot (x_1)^p + \ldots + c_n \cdot (x_n)^p.$$

This formula indeed describe the computer prices pretty well: see, e.g., H. R. Rao and B. D. Lynch, "Hedonistic price analysis of workstation attributes" (*Communications of the ACM*, 1993, Vol. 36, No. 12, pp. 95–103) and references therein.

ALLOCATING BANDWIDTH ON COMPUTER NETWORKS

Allocating bandwidth: why. The communication abilities of communication networks (in particular, of computer networks) are limited. There are only so many bits per second that a network can transfer. In computer networks, the total amount of bits per second is called its *bandwidth*. (A reader should be warned that the same word *bandwidth* is used in electrical engineering and signal processing to denote a somewhat different notion.)

Several nodes are usually connected to the same connection, and it is quite possible that when one of the nodes wants to send a message, the entire bandwidth is already tied up by messages sent by other nodes. A natural way to avoid such situations and to make sure that some bandwidth is always available for every user is to *allocate* a certain portion of the *bandwidth* to every node.

Allocating bandwidth: the main idea. If a node does not use the allocated bandwidth, then this bandwidth is simply wasted. Therefore, it makes sense to allocate, to each node, the bandwidth proportional to the "average" amount of traffic $\bar{t}$ that this node generates. How to define this "average"?

We usually have records that describe the total sizes $t_1, \ldots, t_n$ of messages send by the node at different moments of time $1, \ldots, n$. So, we must compute $\bar{t}$ from the values t_i.

First try: arithmetic average. At first glance, it may seem that we can simply take the arithmetic average of $t_1, \ldots, t_n$ as $\bar{t}$.

The problem with arithmetic average. It turns out that choosing $\bar{t}$ to be arithmetic average is *not the best* arrangement.

■   This choice of $\bar{t}$ is *reasonable* if the flow of outcoming messages is more or less *steady*, i.e., if all the values of t_i are approximately the same.

■   However, in reality, in addition to the *steady* part, there is a significant *bursty* portion of this flow.

For example, we can have two nodes:

* The first node sends exactly 101 KBytes every day. For this node, the average $\bar{t}$ is equal to 101 KBytes.
* The second node usually sends about 1 KByte a day, but, once a month, it sends 3 MBytes (i.e., 3,000 KBytes). For this node, the arithmetic average is equal to $\bar{t} = 1 + 3,000/30 = 101$ KBytes.

Since the two nodes have the same value of $\bar{t}$, both nodes will be allocated the same bandwidth. As a result:

* 29 days out of 30, the bandwidth allocated to the second node will be un-used, and
* on the 30th day, it will not help much anyway, because on that 30th day, we will be sending 3,000 Kbytes, much more than the allocated average.

So, what do we do?

Towards a better allocation. We want to make an allocation that leads to the *best* results. It is difficult to formalize what "best" means in any precise terms, because a network is a very complicated system, and it is very difficult to predict how exactly different allocations will affect the network's overall behavior. Let $f(t)$ denote the effect of allocating t on the network's behavior.

■   The more information we send, the more if affects the network, and therefore, the function $f(t)$ must be *strictly increasing*.

■   Small changes in t will, in general, lead to small changes in $f(t)$, and thus, it is reasonable to require that the function $f(t)$ be *differentiable*.

If we allocated $\bar{t}$, and we actually send t_i, then the resulting error is equal to $e_i = f(\bar{t}) - f(t_i)$. So, for every choice of $\bar{t}$, we have n different values of error. We want to choose $\bar{e}$ in such a way that all these errors are the smallest possible. We have already encountered this problem in Lesson 3, and we concluded that the best way to handle this situation is to minimize the sum of the squares:

$e_1^2 + \ldots + e_n^2 = (f(\bar{t}) - f(t_1))^2 + \ldots + (f(\bar{t}) - f(t_n))^2$. Minimizing this sum with respect to $\bar{f} = f(\bar{t})$, we conclude that

$$f(\bar{t}) = \frac{f(t_1) + \ldots + f(t_n)}{n}, \qquad (H.1)$$

and therefore, that $\bar{t} = f^{-1}(\bar{f})$.

So, if we know the function $f(t)$, we get the formula for the effective bandwidth $\bar{t}$. The only remaining problem is, therefore, to find the function $f(t)$. This problem is very important because we already know that a wrong choice of $f(t)$ (e.g., the choice of $f(t) = t$) can lead to an inadequate bandwidth allocation. So, which function $f(t)$ should we choose?

We must choose a family of functions $f(t)$, not a single function. Our main objective is to compute the effective bandwidth $\bar{t}$. There are two transformations of $f(t)$ that do not change the effective bandwidth:

- If we *add a constant* C_0 to the function $f(t)$, i.e., if we consider a new function $\tilde{f}(t) = f(t) + C_0$, then, by adding C_0 to both sides of (H.1), we will be able to conclude that

$$\tilde{f}(\bar{t}) = \frac{\tilde{f}(t_1) + \ldots + \tilde{f}(t_n)}{n} \qquad (H.2)$$

  for the exact same effective bandwidth $\bar{t}$. Thus, adding a constant to a function $f(t)$ does not change the resulting effective bandwidth.

- Similarly, if we *multiply* all the values $f(t)$ *by a* positive *constant* C_1, i.e., if we consider a new function $\tilde{f}(t) = C_1 \cdot f(t)$, then, by multiplying both sides of (H.1) by C_1, we will be able to conclude that (H.2) is true for the exact same effective bandwidth $\bar{t}$. Thus, multiplying a function $f(t)$ by a constant does not change the resulting effective bandwidth.

Combining these two transformations, we can conclude that the function $f(t)$ and the function $C_1 \cdot f(t) + C_0$ lad to the same effective bandwidth.

Thus, instead of looking for a *single* function $f(t)$, we will look for the entire *family* of functions $\{C_1 \cdot f(t) + C_0\}$ that correspond to all possible values $C_1 > 0$ and C_0.

The optimality criterion must be shift-invariant. The bursty part of the flow (the part that makes allocation a difficult task) occurs *on top* of the steady flow. The allocation part that is caused by the steady part of the flow is easy. Therefore, we have two options:

- First, we can consider the *entire* message flows $t_1, \ldots, t_n$ at different moment of time.

- Second, we can subtract the steady part t_{st} from each flow value t_i and consider *only the bursty* parts $t'_i = t_i - t_{st}$.

We do not have the exact expression for the optimality criterion, but it is reasonable to assume that the relative quality of different families should not change after this shift $t_i \rightarrow t_i - t_{st}$. Thus, we want an optimality criterion to be *shift-invariant*.

Final mathematical formulation of the problem. We want to find a family $\{C_1 \cdot f(t) + C_0\}$ that is optimal with respect to some final shift-invariant optimality criterion.

Towards the solution.

- Similarly to Proposition 4.1, we can conclude that the optimal family must be shift-invariant.

- Thus, similarly to the proof of Theorem 4.2, we get a functional equation $f(t + t_0) = C_1(t_0) \cdot f(t) + C_0(t_0)$.

- Since we have assumed that the function $f(t)$ is differentiable, then, similarly to the proof of Theorem 4.2, we can conclude that both functions $C_1(t_0)$ and $C_0(t_0)$ are differentiable.

- Differentiating with respect to C_0 and taking $t_0 = 0$, we get a differential equation $df/dt = c_1 \cdot f + c_0$. This equation is of the type that we know how to solve:

    - If $c_1 = 0$, then $f(t)$ is a linear function of t, and so, it leads to the same effective bandwidth as $f(t) = t$.

    - If $c_1 \neq 0$, then $f(t) = k \cdot \exp(c_1 \cdot t) + k'$ and therefore, leads to the same effective bandwidth as the function $f(t) = \exp(c_1 \cdot t)$.

So, if we exclude the case $f(t) = t$, we arrive at the conclusion that $f(t) = \exp(c_1 \cdot t)$.

From the condition $s = f(t) = \exp(c_1 \cdot t)$, we express t in terms of s, i.e., find the inverse function $t = f^{-1}(s) = (1/c_1) \cdot \ln(s)$. So, we arrive at the following solution:

The optimal formula for the effective bandwidth. The optimal formula for the effective bandwidth is:

$$\bar{t} = \frac{1}{c_1} \cdot \ln \left(\frac{\exp(c_1 \cdot t_1) + \ldots + \exp(c_1 \cdot t_n)}{n} \right). \qquad (H.3)$$

How to make this formula computationally easier. The main computational time necessary to implement the formula (H.3) comes from n computations of the values $f(t_i) = \exp(c_1 \cdot t_i)$ for $i = 1, \ldots, n$.

■ To compute each of these values *directly*, we need *two* operations: *multiplication* $c_1 \cdot t_i$ and *exponentiation*.

■ To speed up the computations, we must, therefore, find a *single* computer-supported operation that computes $f(t_i)$. In most programming languages, there is such an operation: a^b. To use this operation, we can represent $\exp(c \cdot t_i)$ as $\exp(c_i \cdot t) = P^{t_i}$, where we denoted $P = \exp(c_1)$.

This formula can be even further sped up if we take into consideration that in the computer, there are two different representations of real numbers:

■ *fixed point* real numbers, that can usually represent only real numbers from the interval $[0, 1]$; and

■ *floating point* real numbers, that can, in principle, represent arbitrary real numbers.

Fixed point real numbers require fewer memory and operations with them are usually faster. Therefore, to speed up computations, it is desirable to use fixed point real numbers (i.e., real numbers from the interval $[0, 1]$) as much as possible.

In the formula $f(t_i) = P^{t_i}$, there are two real numbers: t_i and P.

■ t_i can take any value, so, we must use floating point numbers to represent t_i;

■ since $P = \exp(c_1)$. and $c_1 > 0$, we have $P > 1$; thus, we cannot use fixed point numbers to represent P either.

It is therefore desirable to reformulate this formula by using, instead of P, a number from the interval $[0,1]$. This can be done if we take $p = 1/P = P^{-1}$. Then, $f(t_i) = p^{-t_i}$, and p is a fixed point value. In terms of p, the value c_1 takes the form $c_1 = \ln(P) = \ln(p^{-1})$, and therefore, the formula (H.3) takes the form

$$\bar{t} = \frac{1}{\ln(p^{-1})} \cdot \ln\left(\frac{p^{-t_1} + \ldots + p^{-t_n}}{n}\right). \tag{H.4}$$

Comment. In some cases, instead of the *sample values* t_i, we know the *probabilities* of different values t. In this case, t becomes the *random variable*, the *arithmetic average* of p^{-t_i} can be replaced by a *mean* $E[p^{-t}]$, and the formula (H.4) turns into

$$\bar{t} = \frac{\ln(E[p^{-t}])}{\ln(p^{-1})}. \tag{H.5}$$

The resulting formulas indeed lead to a very successful bandwidth allocation. The formulas (H.3)–(H.5) indeed lead to a very successful bandwidth allocation; see, e.g.:

■ J. Y. Hui, "Resource allocation for broadband networks" (*IEEE Journal on Selected Areas in Communication*, 1988, Vol. 6).

■ F. P. Kelly, "Notes on effective bandwidths", In: F. P. Kelly, S. Zachary, and I. B. Zeidins (eds.), *Stochastic networks: theory and applications* (Oxford University Press, 1996).

■ J. Kleinberg, Y. Rabani, and E. Tardos, "Allocating bandwidth for bursty connection" (*Proceedings of the 29th Annual ACM Symposium on Theory of Computing STOC'97, El Paso, TX, May 4-6, 1997*, ACM, 1997, pp. 664-673).

I

ALGORITHM COMPLEXITY REVISITED

In this appendix, we show that continuous mathematics can help to answer the following question: why is the asymptotic time complexity of algorithms often $\sim n^\alpha$ or $\sim n^\alpha \ln(n)$?

Formulation of the problem. An important characteristic of an algorithm $\mathcal{U}$ is its *time complexity* $t_\mathcal{U}(n)$, defined as the largest time that this algorithm takes for all inputs of length n. An algorithm is usually considered to be *feasible*, if its time complexity is bounded by a polynomial of n, i.e., if $t(n) \leq C \cdot n^\alpha$ for some $C > 0$ and α.

It turns out that for many important feasible algorithms, time complexity is not only *bounded* by an expression Cn^α, but it is also *asymptotically equivalent* either to the expression $C \cdot n^\alpha$, or to a slightly more complicated expression $C \cdot n^\alpha (\log(n))^\beta$. This is true for Fast Fourier Transform, for different algorithms of matrix multiplication and sorting, etc.; see, e.g., J. D. Smith, *Design and Analysis of Algorithms* (PWS-Kent Publishing Co., 1989) and Th. H. Cormen, Ch. L. Leiserson, and R. L. Rivest, *Introduction to algorithms* (MIT Press, Cambridge, MA, 1990). The question is: why?

Solution. Our answer is: one of the main methods of algorithm design is *divide-and-conquer*, in which an application of an algorithm to an input of length n is reduced to an application of this same algorithm to several inputs of half-size. For example, in mergesort, sorting a list of size n is done by sorting two half-lists of this list, and then merging the resulting sorted half-lists. In a more general case, an algorithm consists of several stages, and the application of each stage to the input of length n is equivalent to one or several applications of this same algorithm to inputs of half length.

Let us describe this idea formally. Let m denote the number of types of stages, c_j denote the number of stages of j–th type, and let $t_1(n)$, ..., $t_m(n)$ denote the time complexity of these stages. Then,

$$t(n) = c_1 t_1(n) + ... + c_m t_m(n), \qquad (I.1)$$

and

$$t_i(n) = a_{i1} t_1(n/2) + ... + a_{im} t_m(n/2), \qquad (I.2)$$

where a_{ij} is the number of times that j–th stage has to be repeated on a half-size input as part of the i–th stage for the input of the original size.

For example, for mergesort, we have 2 stages: sorting itself and merging. Here, $t(n) = t_1(n)$ (i.e., $c_1 = 1$ and $c_2 = 0$), $t_1(n) = 2t_1(n/2) + t_2(n)$, and $t_2(n) = 2t_2(n/2)$ (i.e., $a_{11} = a_{22} = 2$, $a_{12} = 1$, and $a_{21} = 0$).

To simplify the system (I.1)–(I.2), let us consider $n = 2^k$. Let us denote $t(2^k)$ by $T(k)$, and $t_i(2^k)$ by $T_i(k)$. Then, the equations (I.1)–(I.2) take the form

$$T(k) = c_1 T_1(k-1) + ... + c_m T_m(k-1), \qquad (I.3)$$

and

$$T_i(k) = a_{i1} T_1(k-1) + ... + a_{im} T_m(k-1). \qquad (I.4)$$

Similarly to the general method of solving systems of linear differential equations with constant coefficients, we can describe a general method of solving systems of linear *difference* (functional) equation with constant coefficients; this methods is described, e.g., in R. Bellman, *Introduction to matrix analysis* (McGraw Hill Co., N.Y., 1970) and in F. Chorlton, *Ordinary differential and difference equations. Theory and applications*, D. van Nostrand, London, Toronto, Princeton, NJ, N.Y., 1965.

As a result, we conclude that each component $T_i(k)$ of the generic solution of the system (I.4) of difference equations is a linear combination of the terms of the following types:

- $\exp(\alpha k)$,

- $k^h \cdot \exp(\alpha k)$,

- $k^h \cdot \exp(\alpha k) \cdot \cos(\beta k)$,

- $k^h \cdot \exp(\alpha k) \cdot \sin(\beta k)$

for real number α and integer h. Therefore, $T(k)$ is also equal to the linear combination of such terms. Substituting $k = \log(n)$ (to be more precise, $\log_2(n)$), and taking into consideration that $\exp(\alpha \log(n)) = [\exp(\log(n))]^\alpha = n^\alpha$, we conclude that $t(n)$ is a linear combination of the terms of the following types:

- n^α,

- $n^\alpha \cdot (\log(n))^h$,

- $n^\alpha \cdot (\log(n))^h \cdot \cos(\beta \cdot \log(n))$,

- $n^\alpha \cdot (\log(n))^h \cdot \sin(\beta \cdot \log(n))$.

So, we have explained why the asymptotics of the time complexity is often of the type n^α or $n^\alpha \cdot (\log(n))^h$.

Comment. A similar idea explains why for parallel algorithms, the required number of processors often depends on the input size as n^α or $n^\alpha \cdot (\log(n))^h$.

Acknowledgments. This appendix is based on a project by O. M. Kosheleva.

J

HOW CAN A ROBOT AVOID OBSTACLES: CASE STUDY OF REAL-TIME OPTIMIZATION

Formulation of the problem. One of the main objectives of a *mobile* robot is to *move* from one place to another. Ideally, the robot should follow the shortest possible trajectory, and thus, arrive at its destination in the shortest possible time.

Heuristic methods are needed. In a realistic shop-like or office-like environment, where the robot has to navigate its way across many static and moving obstacles, finding the optimal trajectory is often very computationally complicated. The fact that some obstacles are moving makes it necessary to re-compute the trajectory in real-time, and precise optimization in real time is, usually, way beyond the abilities of an on-board computer. Since we cannot *precisely* compute the optimal trajectory, it is desirable to use *heuristic*, approximate methods.

Simulating nature as a source of heuristic methods. We have already shown (in lessons about genetic algorithms and neural networks) that one of the important sources of heuristic algorithms is simulating nature: to solve a problem, we find a similar problem that nature "solves", and then try to simulate the way the nature solves it.

For mobile robots, we cannot directly use genetic algorithm. We already know one class of algorithms that solve optimization problems by simulating natural evolution: *genetic algorithms*. In general, genetic algorithms are a good and useful optimization tool; they are successfully applied in many areas including such an important computer-related area as VLSI design (in particular, the optimal design of the computer chips). For mobile robot navigation, however, we cannot directly use genetic algorithms, because these algorithms take a long time to optimize and we need to make decisions in real time. This slowness comes from the very natural process that we are trying to simulate: natural selection is definitely a slow process. Thus, if we want a *faster* optimization algorithm, we must simulate a *faster* natural process.

Simulating physical processes: idea. Which processes are faster than biological ones? Living creatures can be fast, but non-living things can be even faster: planets move around the sun at heartbreaking speeds, elementary particles travel even faster, etc. It is, therefore, reasonable to *simulate the motion of a physical particle.*

Of course, physical particles do not exhibit the ingenious optimizing behavior of biological systems: e.g., an (un-controlled) satellite, if it encounters any obstacle, will simply crash or get destroyed, while a tree seed grows pretty well even if it encounters a stone. Thus, when we turn from genetic algorithms (that simulate biological processes) to algorithms that simulate physical processes, we gain speed, but, as a trade-off, we get not-so-optimal results.

In physics, laws of motion are usually described on two levels:

- On the *theoretical* level, the laws of physics can be formulated in terms of an *optimization principle*. Namely, the actual trajectory of a particle is usually the one that minimizes the value of a certain functional; in physics, this functional is called *action*. For example, the light traveling between the two points follows the shortest path, which means that for the light, this functional is simply the length of the trajectory. For charged particles, the action is more complicated.

  This formulation shows that physical processes are, in effect, solving optimization problems.

- On the *practical* level, the motion of physical particles is described by laws of mechanics that were first discovered by I. Newton. According to Newton's laws, the particle's acceleration $\vec{a}$ is determined by the total *force* acting upon the particle: $\vec{a} = m^{-1} \cdot \vec{f}$ (where m is a constant called *mass*). The force is a sum of forces (attracting and repelling) from different sources: $\vec{f} = \sum \vec{f_i}$.

  Thus, to simulate the particles' motion, we can apply, to the mobile robot, the appropriate forces. Since forces are related to physical fields, the simulation of these forces is called a *field method* in robotics.

Simulating physical processes: towards implementation. The motor(s) of the mobile robot can apply any force within reasonable limits. We want the robot to reach the destination and to avoid all the obstacles. Thus, we need to have an *attracting* force that will lead the robot to the destination and the *repelling* forces that will help the robot avoid all the obstacles.

The *attracting* force is the easiest to get, because it corresponds to the case when we have no obstacles at all. For this case, we can compute the acceleration $\vec{a}$ along the optimal trajectory, and then compute the corresponding force $\vec{f} = m \cdot \vec{a}$.

The main problem is how to generate the forces corresponding to different obstacles. For each obstacle, we want a force that is *repelling* the robot from this obstacle. Thus, this force should be oriented in the direction *opposite* to the direction to the obstacle (to be more precise, in the direction opposite to the direction from the robot to the nearest point on the obstacle). The absolute value f_i of this force should depend on the distance d to the obstacle: $f_i = |\vec{f}_i(d)| = f(d)$. Which function $f(d)$ should we choose?

How to choose a function $f(d)$? The closer the obstacle, the more dangerous it is for the robot, and therefore, the stronger we need to repel it. Thus, the function $f(d)$ must be strictly decreasing.

The numerical value of the function $f(d)$ depends on the choice of the unit for describing force. If we change a unit to a new unit that is C times smaller, then the same dependency that was expressed by a function $f(d)$ in the old units will be described by the new function $C \cdot f(d)$ in the new units. Therefore, if $f(d)$ is a reasonable function to describe the dependency, then $C \cdot f(d)$ is also a reasonable function. Hence, it makes sense to look not for a *single* function $f(d)$, but for the entire *family* of functions $\{C \cdot f(d)\}$, where $f(d)$ is a given function, and C runs over arbitrary positive numbers. Which family is the best?

The comparative quality of two different families should not change if we simply use different units for measuring the distance d. In mathematical terms, replacing an old unit with a new unit that is λ times smaller means replacing each numerical value d by a new value $\lambda \cdot d$. In the main text, we have called this procedure *scaling*. Thus, it is reasonable to require that the family $\{C \cdot f(d)\}$ must be optimal with respect to some criterion that is final and scale-invariant. In Theorem 8.1, we have shown that for such optimality criteria, every function from the optimal family is equal to $f(d) = C \cdot d^\alpha$ for some real numbers C and α.

Since we want the function $f(d)$ to be decreasing, we must take $\alpha < 0$. So, the optimal solution is to apply the force

$$\vec{f} = \sum_i C_i \cdot d_i^{\alpha} \cdot \vec{e}_i,$$

where:

- the sum is taken over all the obstacles i;

- d_i is the distance from the current position of the mobile robot to i-th obstacle;

- C_i is a parameter that characterizes the importance of i-th obstacle (it is large for obstacles that we should not hit at all, and smaller for obstacles hitting which will not be that drastic);

- $\vec{e}_i$ is a unit vector in a direction that is opposite to the direction from the mobile robot to the nearest point on i-th obstacle.

This method, for $\alpha = -2$, indeed leads to an efficient robot control; see, e.g., G. H. Ogasawara, "A distributed, decision-theoretic control system for a mobile robot" (*SIGART Bulletin*, 1991, Vol. 2, No. 4, pp. 140–145).

K

DISCOUNTING IN ROBOT CONTROL: A CASE STUDY OF DYNAMIC OPTIMIZATION

Formulation of the problem. In industrial applications of a robot, we would like to have a control that is *optimal* in some reasonable sense. The corresponding optimality criterion is relatively easy to formulate in the case when we manufacture by a deadline: in this case, our goal is to maximize the *profit P*, i.e., the difference between the cost of what we have produced and the cost of what we have spent on this production. However, most frequently, manufacturing is a *continuous, dynamic* process. In this case, different manufacturing strategies lead to different profits $P(t)$ at different moments of time $t = 1, 2, 3, \ldots$ If one strategy generates more profit now, but another strategy generates more profit next time, which of them is better? To compare these strategies, we must somehow "combine" the profits generated at different moment of time t into a single optimality criterion. Such a combination is called *discounting*. How can we do this?

Let us denote by $f(t)$ the cost that we are willing to pay now for a \$1 profit generated at time $t \geq 0$. Then, the current profit, that is equivalent to the profit $P(t)$ generated at time t, is equal to $f(t) \cdot P(t)$, and the total efficient profit of a manufacturing strategy is equal to $\sum f(t) \cdot P(t)$. Similarly, at each moment of time t, the efficient "cumulative" profit is equal to $\sum f(n) \cdot P(t+n)$. The question is: how to choose the values $f(t)$?

Main idea. There are two ways to view a \$1 profit obtained at a moment of time $t + s$:

- on one hand, by definition of the function $f(t)$, this profit is equivalent to the amount $f(t + s)$ given now;

■   on the other hand, we can view the same profit via the following two-step procedure:

    – since the moment $t + s$ occurs s moments after the moment t, the profit of \$1 at a moment $t + s$ is equivalent to the profit of $f(s)$ dollars at a moment t;

    – the profit of $f(s)$ dollars at moment t is, by itself, equivalent to $f(s) \cdot f(t)$ dollars now.

These two viewpoints must lead to the same equivalent value; thus, we conclude that $f(t+s) = f(t) \cdot f(s)$. In particular, for $s = 1$, we get $f(t+1) = f(1) \cdot f(t)$. This is a linear functional equation with constant coefficients, i.e., an equation that we know how to solve. The natural initial condition comes from the fact that $f(0)$ describe the current cost of \$1 that is earned now; clearly, $f(0) = 1$.

Using the known method of solving functional equations, we conclude that $f(t) = \gamma^t$ for some constant γ ($\gamma = f(1)$). The value γ is called the *discount rate*.

Conclusion. At every given moment of time t, our goal is to maximize the function

$$P = \sum_{n=0}^{\infty} \gamma^n \cdot P(t).$$

This formulas has been successfully used. Such *discounting* is indeed successfully used in robotic applications; see, e.g., S. D. Whitehead, "A framework for integrating perception, action, and trial-and-error learning" (*SIGART Bulletin*, 1991, Vol. 2, No. 4, pp. 174–178).

INDEX